Invertebrate Tissue Culture

Main Speakers and Sess on Chairmen of the Fourth International Conference on Invertebrate Tissue Culture.

Standing, l. to r.: A. Dübendorfer (Switzerland), D. Peters (Netherlands), K.A. Harrap (Great Britain), J.J. Lipa (Poland), P. Faulkner (Canada), J. Mitsuhashi (Japan), S.S. Sohi (Canada), J. Řeháček (Czechoslovakia), J.C. Landureau (France); *seatea l. to r.:* S. Buckley (U.S.A.), K. Maramorosch (U.S.A.), E. Kurstak (Canada), C.M. Williams (U.S.A.). *(C.P.A. Photo, Montreal)*

INVERTEBRATE TISSUE CULTURE

APPLICATIONS IN MEDICINE, BIOLOGY, AND AGRICULTURE

Edited by

Edouard Kurstak

Groupe de Recherche
en Virologie Comparée
Faculté de Médecine
Université de Montréal
Montréal, P.Q., Canada

Karl Maramorosch

Waksman Institute of Microbiology
Rutgers University
New Brunswick, New Jersey, U.S.A.

Academic Press New York San Francisco London 1976

A Subsidiary of Harcourt Brace Jovanovich, Publishers

ACADEMIC PRESS RAPID MANUSCRIPT REPRODUCTION

ACADEMIC PRESS, INC.
111 Fifth Avenue, New York, New York 10003

United Kingdom Edition published by
ACADEMIC PRESS, INC. (LONDON) LTD.
24/28 Oval Road, London NW1

Library of Congress Cataloging in Publication Data

International Conference on Invertebrate Tissue Culture,
4th, Mont Gabriel, Que., 1975
Invertebrate tissue culture.

Sponsored by the Université de Montréal and others.
Bibliography: p.
Includes index.
1. Tissue culture–Congresses. 2. Invertebrates–Cultures and culture media–Congresses. I. Kurstak, Edouard. II. Maramorosch, Karl. III. Université de Montréal. IV. Title.
QH585.I54 1975 592'.08'21 76-14911
ISBN 0-12-429740-4

PRINTED IN THE UNITED STATES OF AMERICA

Contents

Contributors

Aitcken, T.H.G., Yale University School of Medicine, Yale Arbovirus Research Unit, Dept. of Epidemiology & Public Health, New Haven, Connecticut, U.S.A.

Ball, G.H., Department of Biology, University of California, Los Angeles, California, U.S.A.

Bergtrom, G., The Biological Sciences Group, University of Connecticut, Storrs, Connecticut, U.S.A.

Bird, F.T., Canadian Forestry Service, Insect Pathology Research Institute, Sault Ste-Marie, Ontario, Canada.

Brooks, M.A., Department of Entomology, Fisheries & Wildlife, University of Minnesota, St-Paul, Minnesota, U.S.A.

Brown, D.T., Institut für Genetic der Universität zu Köln, Köln, Federal Republic of Germany.

Brown, M., Department of Microbiology and Immunology, Queen's University, Kingston, Ontario, Canada.

Buckley, S.M., Yale University School of Medicine, Yale Arbovirus Research Unit, Dept. of Epidemiology & Public Health, New Haven, Connecticut, U.S.A.

Casals, J., Yale University School of Medicine, Yale Arbovirus Research Unit, Dept. of Epidemiology & Public Health, New Haven, Connecticut, U.S.A.

Cecil, J.T., Osborn Laboratories of Marine Sciences, New York Aquarium, New York Zoological Society, Brooklyn, New York, U.S.A.

Chao, J., Department of Biology, University of California, Los Angeles, California, U.S.A.

Chen, T.T., Department of Biology, Queen's University, Kingston, Ontario, Canada.

Couble, P., Department of Biology, Queen's University, Kingston, Ontario, Canada.

DiCapua, R.A., School of Pharmacy, University of Connecticut, Storrs, Connecticut, U.S.A.

Dübendorfer, A., Zoological Institute of the University of Zürich, Kunstlergasse, Zürich, Switzerland.

Echalier, G., Service de Biologie Animale, Université de Paris VI, Paris, France.

Faulkner, P., Department of Microbiology and Immunology, Queen's University, Kingston, Ontario, Canada.

Filtzer, D., Institut für Genetik der Universität zu Köln, Köln, Federal Republic of Germany.

Fowler, G.L., Department of Biology, University of Oregon, Eugene, Oregon, U.S.A.

Gardiner, G.R., Shell Research Limited, Woodstock Laboratory, Sittingbourne Research Centre, Sittingbourne, Kent, England.

Gilbert, L.I., Department of Biological Sciences, Northwestern University, Evanston, Illinois, U.S.A.

Gliedman, J.B., Institut für Genetik der Universität zu Köln, Köln, Federal Republic of Germany.

Granados, R.R., Boyce Thompson Institute, Yonkers, New York, U.S.A.

Hansen, E.L., Clinical Pharmacology Research Institute, Berkeley, California, U.S.A.

Harrap, K.A., NERC Unit of Invertebrate Virology, South Parks Road, Oxford, England

Hayes, C.G., Yale University School of Medicine, Yale Arbovirus Research Unit, Dept. of Epidemiology & Public Health, New Haven, Connecticut, U.S.A.

Hink, W.F., Department of Entomology, The Ohio State University, Columbus, Ohio, U.S.A.

Hirumi, H., Boyce Thompson Institute, Yonkers, New York, U.S.A.

Hirumi, K., Boyce Thompson Institute, Yonkers, New York, U.S.A.

Igarashi, A., Department of Microbiology, CMDNJ-Rutgers Medical School, Piscataway, New Jersey, U.S.A.

Johannisson, R., Institute for General Biology, University of Düsseldorf, Düsseldorf, Federal Republic of Germany.

Jonathan, J.F., Institut für Genetik der Universität zu Köln, Köln, Federal Republic of Germany.

Kimura, M., Medical Research Institute, Wakayama Medical College, Wakayama, Japan.

Kiss, I., Institute of Genetics, Biological Research Center, Szeged, Hungary.

Knudson, D.L., NERC Unit of Invertebrate Virology, South Parks Road, Oxford, England.

Koeppe, J., Department of Zoology, University of North Carolina, Chapel Hill, North Carolina, U.S.A.

Koo, R., Department of Microbiology, CMDNJ-Rutgers Medical School, Piscataway, New Jersey, U.S.A.

Kurtti, T.J., Department of Entomology, Fisheries and Wildlife, University of Minnesota, St-Paul, Minnesota, U.S.A.

Landureau, J.C., Service de Biologie Animale, Faculté des Sciences, Université de Paris VI, Paris, France.

Laufer, H., The Biological Sciences Group, University of Connecticut, Storrs, Connecticut, U.S.A.

Leloup, A.M., Unité de Morphologie Animale, Université Catholique de Louvain, Louvain, Belgique.

Lichter, W., Department of Microbiology, University of Miami, School of Medicine, Miami, Florida, U.S.A.

Lipman, M., Yale University School of Medicine, Yale Arbovirus Research Unit, Dept. of Epidemiology & Public Health, New Haven, Connecticut, U.S.A.

Lopez, D.M., Department of Microbiology, University of Miami, School of Medicine, Miami, Florida, U.S.A.

Maloney, J.M., Yale University School of Medicine, Yale Arbovirus Research Unit, Dept. of Epidemiology & Public Health, New Haven, Connecticut, U.S.A.

Marks, E.P., Agricultural Research Service, U.S. Department of Agriculture, Metabolism & Radiation Research Laboratory, Fargo, North Dakota, U.S.A.

Mazzone, H.M., Forest Insect and Disease Laboratory, Forest Service, U.S. Department of Agriculture, Hamden, Connecticut, U.S.A.

McIntosh, A.H., Waksman Institute of Microbiology, Rutgers University, New Brunswick, New Jersey, U.S.A.

Mitsuhashi, J., Division of Entomology, National Institute of Agricultural Sciences, Tokyo, Japan.

Miyake, T., Mitsubishi-Kasei Institute of Life Sciences, Minamiooya, Machidashi, Tokyo, Japan.

Nakajima, S., Mitsubishi-Kasei Institute of Life Sciences, Minamiooya, Machidashi, Tokyo, Japan.

Naughton, M., Boyce Thompson Institute, Yonkers, New York, U.S.A.

Nigrelli, R.F., Osborn Laboratories of Marine Sciences, New York Aquarium, New York Zoological Society, Brooklyn, New York, U.S.A.

Norton, P.W., School of Pharmacy, University of Connecticut, Storrs, Connecticut, U.S.A.

Nowock, J., Department of Biological Sciences, Northwestern University, Evanston, Illinois, U.S.A.

Oberlander, H., Agricultural Research Service, USDA, Gainesville, Florida, U.S.A.

Ofengand, J., Roche Institute of Molecular Biology, Nutley, New Jersey, U.S.A.

Padhi, S.B., Waksman Institute of Microbiology, Rutgers University, New Brunswick, New Jersey, U.S.A.

Potter, K.M., Department of Microbiology and Immunology, Queen's University, Kingston, Ontario, Canada.

Ramoska, W.A., Department of Entomology, The Ohio State University, Columbus, Ohio, U.S.A.

Řeháček, J., Institute of Virology, Slovak Academy of Sciences, Bratislava, Czechoslovakia.

Rens, D., Institut für Genetik der Universität zu Köln, Köln, Federal Republic of Germany.

Riddiford, L.M., Department of Zoology, University of Washington, Seatle, Washington, U.S.A.

Riedel, B., Institut für Genetik der Universität zu Köln, Köln, Federal Republic of Germany.

Rogers, R., The Biological Sciences Group, University of Connecticut, Storrs, Connecticut, U.S.A.

Rosales-Sharp, M.C., Department of Biology, University of New Mexico, Albuquerque, New Mexico, U.S.A.

Ruggieri, G.D., Osborn Laboratories of Marine Sciences, New York Aquarium, New York Zoological Society, Brooklyn, New York, U.S.A.

Schlesinger, R.W., Department of Microbiology, CMDNJ-Rutgers Medical School, Piscataway, New Jersey, U.S.A.

Shenk, T.E., Department of Biochemistry, Stanford University, School of Medicine, Stanford, California, U.S.A.

Sigel, M.M., Department of Microbiology, University of Miami, School of Medicine, Miami, Florida, U.S.A.

Silverman, P.H., Department of Biology, University of New Mexico, Albuquerque, New Mexico, U.S.A.

Smith, J.B., Institut für Genetik der Universität zu Köln, Köln, Federal Republic of Germany.

Sohi, S.S., Canadian Forestry Service, Insect Pathology Research Institute, Sault Ste-Marie, Ontario, Canada.

Speyer, G., Boyce Thompson Institute, Yonkers, New York, U.S.A.

Stollar, V., Department of Microbiology, CMDNJ-Rutgers Medical School, Piscataway, New Jersey, U.S.A.

Strauss, E., Department of Entomology, Ohio State University, Columbus, Ohio, U.S.A.

Stockdale, H., Shell Research Limited, Woodstock Laboratory, Sittingbourne Research Centre, Sittingbourne, Kent, England.

Summers, M.D., Cell Research Institute and Department of Botany, University of Texas, Austin, Texas, U.S.A.

Volkman, L.E., Cell Research Institute and Department of Botany, University of Texas, Austin, Texas, U.S.A.

Wellham, L.L., Department of Microbiology, University of Miami, School of Medicine, Miami, Florida, U.S.A.

Williams, C.M., Biological Laboratories, Harvard University, Cambridge, Massachusetts, U.S.A.

Wyatt, G.R., Department of Biology, Queen's University, Kingston, Ontario, Canada.

Wyatt, S.S., Department of Biology, Queen's University, Kingston, Ontario, Canada.

Preface

The present volume comprises the Proceedings of the IV International Conference on Invertebrate Tissue Culture, held June 5-8, 1975 at Mont Gabriel, Quebec, Canada. The scheduling of this conference, as well as the choice of location, were coordinated so as to follow the Annual Meeting of the Tissue Culture Association held at the Université de Montréal. Since the Proceedings of the earlier conferences were not widely distributed, it seems appropriate to present here a brief history of this series.

The First International Conference on Invertebrate Tissue Culture was held October 22-24, 1962 at Montpellier, France. It was organized under the auspieces of the Centre National de la Recherche Scientifique, the Institut National de la Recherche Agronomique of France, the Institut Pasteur, and the Université de Montpellier. The papers presented at this meeting appeared as special volume of the Annales des Epiphyties, Paris, 1963.

The Second Conference was held at Villa Carlotta, Tremezzo, Como, Italy, September 9-10, 1967, under the sponsorship of the Italian Academy of Sciences and the Instituto Lombardo (Baselli Foundation). The Proceedings of this conference were published in a separate, soft-cover volume in Milano in 1968.

The Third Conference was held in Smolenice near Bratislava, Czechoslovakia, June 22-25, 1971, under the auspieces of the Slovak Academy of Sciences and the Institute of Virology at Bratislava. The Proceedings appeared in 1973 as a separate volume, produced by the Publishing House of the Slovak Academy of Sciences.

The scope of the Fourth International Conference was to discuss invertebrate organ, tissue, and cell culture, its limitations, pitfalls, present and potential applications in medicine, biology, and agriculture, in studies of morphogenesis, differentiation, viruses, symbionts, parasites, and neurophysiology. This time, the sponsorship included the Université de Montréal, the National Institutes of Health in Bethesda, Maryland, U.S.A., the National Research Council of Canada, Ottawa, the Medical Research Council of Canada, Ottawa, and the Faculty of Medicine of the Université de Montréal.

The program was arranged by Prof. E. Kurstak of the Université de Montréal and Prof. K. Maramorosch of Rutgers – the State University of New Jersey, who acted as conference co-chairmen. They were assisted by an International Advisory Committee, consisting of Prof. C. Barigozzi (Italy), Dr. A. Dübendorfer (Switzerland), Prof. G. Echalier (France), Dr. P. Faulkner (Canada), Dr. T.D.C. Grace (Australia), Prof. E. Hadorn (Switzerland), Dr. J. Mitsuhashi (Japan), Dr. J. Peleg (Israel), Dr. D. Peters (Netherlands), Dr. J. Řeháček (Czechoslovakia), Dr. K.A. Harrap (U.K.), and Dr. I.V. Tarasevich (U.S.S.R.). Papers were presented by 62 scientists from Belgium, Canada, Czechoslovakia, Federal Republic of Germany, France, Hungary, India, Japan, Netherlands, Poland, Switzerland, United Kingdom, U.S.A., and U.S.S.R.

New applications of invertebrate tissue culture in medicine and in biology, the latest developments in mollusc tissue culture, cell cloning, and their applications to the study of viruses and microspiridians were among the highlights of the conference. The meeting brought together the outstanding and most active research leaders in invertebrate cell, tissue and organ culture and provided a direct contact of these experts with younger workers and attending graduate students. New avenues of research were explored and suggestions made for novel approaches to advance the field. Throughout the conference discussions were intense and illuminating, lasting until late at night, and raising new questions and challenges. In order to make this information available to the widest possible scientific community, it was decided to publish the invited and contributed papers, in an expanded, edited form, as a hard-cover book, rather than as conference proceedings. A volume devoted to the basic aspects of invertebrate tissue culture and complementing the present one, is being published at the same time by Academic Press under the title: *Invertebrate Tissue Culture in Basic Research.*

The editors felt that it was of utmost importance to publish the original papers, often in a much longer version, as well as the comprehensive and incisive reviews of invertebrate tissue culture applications with no delay, and at a reasonable price. These goals were achieved, thanks to the generous support of the original sponsors of the Conference, and the efficiency of the Publisher.

The editors hope that this book will be useful and stimulating, and will provide in a single volume the latest results obtained in the diverse areas of research pursued by the leading exponents of invertebrate tissue culture from America, Europe, Asia, and Australia. The presentation of the latest techniques used in laboratories around the world will also be of immediate value in furthering studies in infectious diseases and possibly lead to the development of new methods of disease control.

The volume provides the most recent information on sophisticated laboratory methods and on numerous utilizations of invertebrate cell culture techniques. Applications to the study of arboviruses, malarial parasite biology, the use of snail cell lines in Schistosoma work, applications of these techniques to embryology, genetics, endocrinology and physiology are only a few of the fascinating areas included. This book will be of interest to many, including researchers and students in medical and biomedical sciences, such as virology, immunology, pathology, parasitology, endocrinology, developmental biology, microbiology, entomology, plant pathology, and biological control of vectors.

The presentations of currents and of the most recent results of the leading invertebrate tissue culture experts, their own interpretations and original conclusions, as well as the inclusion of numerous illustrations make this book a timely source of information and bring into sharp focus the rapidly moving frontier and new directions of invertebrate tissue culture.

The chairmen of the Fourth International Conference on Invertebrate Tissue Culture and editors of this book wish to express their gratitude to all contributors for the effort and care, as well as promptness with which they have prepared their chapters; to the Faculty of Medicine of the Université de Montréal, to the Université de Montréal, to the Institute of Allergy and Infectious Diseases, National Institutes of Health, Bethesda, Maryland, to the National Research Council of Canada and to the Medical Research Council of Canada, for their financial support; to Dr. Maurice L'Abbé, Vice-Recteur for Research of the Université de Montréal, to

Dr. Pierre Bois, Dean of the Faculty of Medicine, to Dr. Gaston de Lamirande, Vice Dean for Research of the Faculty of Medicine of the Université de Montréal, and to Dr. Sorin Sonea, Director of the Department of Microbiology and Immunology of the Université de Montréal, for their support in the organization of the conference and editing matters. Our thanks are extended to Miss Ghislaine Montagne for her part in editing this volume and last, but not least, to the staff of Academic Press for their part in the production of the volume.

Prof. Edouard Kurstak
Prof. Karl Maramorosch

I

APPLICATION IN MEDICINE

Chapter 1

ARBOVIRUS STUDIES IN INVERTEBRATE CELL LINES

S.M. Buckley, C.G. Hayes, J.M. Maloney,
M. Lipman, T.H.G. Aitcken, and J. Casals

I. Introduction

By definition, "arboviruses are viruses which are maintained in nature principally or to an important extent, through biological transmission between susceptible vertebrate hosts by hematophagous arthropods" (WHO Scientific Group, 1967). At the end of 1974, 350 arboviruses were recognized. These heterogeneous viruses, incorporated into a general system of virus classification (Casals, 1971), have been arranged for the most part into groups of related, but distinct agents. In as much as characterized by serology, morphology, morphogenesis, biochemistry and biophysics, arboviruses belong to six taxons: alphavirus, bunyavirus, flavivirus, iridovirus, orbivirus, and rhabdovirus. The range of natural vectors extends from mosquitoes to ticks, phlebotomines and *Culicoides* (Casals, 1971). Since the double break-through with regard to haemolymph-free medium (Mitsuhashi-Maramorosch, 1964) and establishment of stable cell lines from *Aedes* mosquitoes (Grace, 1966; Singh, 1967; Peleg, 1968), *in vitro* studies of viruses in invertebrate tissue culture systems have increased in parallel with the production of new invertebrate cell lines. According to Hink (1972), Diptera cell lines are represented by 14 species with primary explants derived from embryos, larvae, imaginal discs, adult ovaries, or adult species. Two invertebrate species, *Aedes malayensis* and *Aedes pseudoscutellaris*, both belonging to the *scutellaris* subgroup of the subgenus *Stegomyia*, have been used recently for the establishment of two new *Aedes* cell lines by Varma *et al.* (1974). A real break-through, however, is the establishment of three cell lines from the tick *Rhipicephalus appendiculatus* by Varma *et al.* (1974). Thus, arbovirus studies can be carried out in parallel in cell lines derived from two important vectors, i.e., mosquito and tick.

Recent and comprehensive reviews have covered arthropod cell cultures and their value in arbovirus *in vitro* studies (Singh, 1971, 1972; Yunker, 1971; Řeháček, 1972; Dalgarno and Davey, 1973; Buckley, 1976). In this presentation, a detailed survey of the literature is, therefore, not attempted. Examples indicative of the prevalent direction of research are presented.

II. Kinetics of arbovirus multiplication

Studies of quantitative aspects of arbovirus multiplication in invertebrate cell lines logically depend on preliminary investigations with regard to growth or non-growth of an agent in a particular invertebrate cell system. The susceptibility of Diptera cell lines to members of the different taxons of arboviruses has been reviewed recently (Buckley, 1976). Briefly, the following determinants were found to be important: 1) innate differences between invertebrate cell lines, and 2) heterogeneity of viruses with regard to a) vector and b) presence or absence of a virion envelope. Enveloped arboviruses are sensitive to the action of sodium deoxycholate or relatively resistant (Theiler, 1957; Casals, 1971; Borden *et al.*, 1971); unenveloped arboviruses (orbiviruses) are relatively resistant (Borden *et al.*, 1971). Thus, in as much as reported, unenveloped orbiviruses multiply in Singh's *Aedes albopictus* cells (Singh, 1967), regardless of vector; mosquito- or *Culicoides*-borne enveloped alpha, bunya-, flavi-and rhabdoviruses reproduce also. However, enveloped tick-borne flaviviruses or enveloped tick-or phlebotomine-borne bunyaviruses fail to multiply. Table 1 summarizes results obtained in the *Aedes albopictus* cell line. In infection experiments with the tick cell line TTC-243, Varma *et al.* (1974) reported replication of the flaviviruses West Nile, Louping ill, Langat as well as of the still unclassified tick-borne virus Quaranfil. While West Nile virus has been isolated from different mosquitoes, Taylor *et al.*, (1956) succeeded in infecting *Ornithodorus savigni* with West Nile virus and on one occasion were able to obtain transmission of the virus following experimental feeding of the infected ticks on susceptible vertebrates. It would be of great interest to investigate growth or nongrowth of African swine fever virus, a tick-borne DNA virus of the iridovirus taxon, in a tick cell line.

TABLE 1

Growth of arboviruses in Singh's AEDES ALBOPICTUS cell cultures

Taxon	No. of arboviruses tested	Proven or suspected vector*	Sensitive to SDC or relatively resistant	Growth
alphavirus	8	mosquito	sensitive	yes (8/8)
bunyavirus	3	mosquito	sensitive	yes (3/3)
flavivirus	11	mosquito	sensitive	yes (11/11)
orbivirus	3	mosquito	relatively resistant	yes (3/ 3)
rhabdovirus	4	mosquito	sensitive	yes (4/4)
bunyavirus	3	tick	sensitive	no (0/3)
flavivirus	3	tick	sensitive	no (0/3)
orbivirus	5	tick	relatively resistant	yes (5/5)
not classified	5	tick	sensitive	no (0/5)
bunyavirus	3	phlebotomine	sensitive	no (0/3)
orbivirus	9	phlebotomine	relatively resistant	yes (9/9)
rhabdovirus	1	*Culicoides*	sensitive	yes (1/1)
orbivirus	1	*Culicoides*	relatively resistant	yes (1/1)

* In addition to having been isolated from mosquitoes, bunyavirus Ganjam and some rhabdoviruses (VSV subgroup) have also been isolated from ticks and phlebotomine flies respectively.

A review pertaining to kinetics, biochemistry, and ultrastructural studies in invertebrate and invertebrate cell cultures with special reference to alphaviruses such as Semliki Forest, Sindbis, eastern equine encephalitis, western equine encephalitis, Venezuelan equine encephalitis and to a flavivirus, Kunjin, has been published by Dalgarno and Davey (1973). Optimal togavirus replication in invertebrate and vertebrate cell systems was, apparently, related to optimum temperature of cell growth. However, invertebrate cell lines maintained by individual investigators as well as heterogeneity of arboviruses, viral strain, viral passage history, multiplicity of infection are factors determining the specifics of kinetics of viral growth *in vitro* in addition to temperature of incubation. Examples with regard to arboviruses belonging to five different taxons have been discussed in detail by Buckley (1975).

III. Persistence

Speculating on origin and evolution of arboviruses, Andrews (1973) states: "it seems almost certain that one is dealing either with insect-parasites which have become secondarily adapted to living in vertebrates of the other way round. The first alternative seems preferable". In analogy to the *in vivo* situation, inapparent persistent infection of mosquito cell cultures may be induced with any arbovirus capable of replication in a given cell system (Buckley, 1976) such as *Aedes albopictus* and *Aedes aegypti* (monolayer; hollow vesicles) (Singh, 1967), *Aedes aegypti* (Peleg, 1968), and *Aedes w-albus* (Singh and Bhat, 1971).

1. Alphavirus. Production of carrier cultures of Singh's *Aedes albopictus* cell line infected with different arboviruses has been described first by Banerjee and Singh (1968). The same investigators, subsequently, reported loss of mouse virulence in chikungunya virus, an alphavirus from the carrier culture of *Aedes albopictus* cell line (Banerjee and Singh, 1969). We have been able to confirm their results, using the same chikungunya virus (strain I 634029; 4th mouse passage) and an *Aedes albopictus* subline brought personnaly to the Yale Arbovirus Research Unit by Singh in Spring 1970. Briefly, primary infection of *Aedes albopictus* monolayer cultures was initiated with approximately 20 plaque forming units (PFU) of the Oriental strain. Cultures were incubated at 30°C for 4 days and thereafter at room temperature (20-25°C). Infected carrier cultures as well as uninoculated control cultures were transferred at weekly intervals by a 1:8 split ratio. Cells in persistently infected cultures were indistinguishable from the uninoculated cultures with regard to morphology and growth potential. At each transfer level, fluid phases of two cultures were pooled, centrifuged (2000 RPM for 10 minutes) and assayed for the presence of virus in 2-day-old mice, in BHK-21 cell cultures (Karabatsos and Buckley, 1967) and in Vero cell cultures (Karabatsos, 1969). The infectivity titers expressed as "dex" (decimal exponent) (Haldane, 1960) per ml are listed in Table 2. Titers in mice started to decrease as of the 5th transfer; fluids of the 9th and 10th transfers induced an occasional death in mice which were observed for 14 days post-inoculation. No significant plaque forming unit (PFU) or tissue cytopathic dose $(TCD)_{50}$ titer differences were noticed in the vertebrate cell lines with one exception (3rd transfer; BHK-21) in which poor quality cultures were used. In the Vero plaque assays, plaque sizes were uniform up to the 3rd transfer in which a small plaque (SP) variant was observed, measuring 3 x 3 mm in diameter. The numbers of SP increased during subsequent transfers with 6/10 plaques measuring 3 x 3 mm and 4/10 plaques representing large plaques (LP), measuring 5 x 5 mm and 6 x 6 mm, at the 8th transfer level (10^{-5} dilution). In a previous study (Buckley, 1973a) with the high mouse brain passage chikungunya virus (Ross strain; African

TABLE 2

Infectivity titers of chikungunya virus, strain I 634029, mouse brain tissue stock, and of I 634029-infected fluid phases of transfer levels No. 1-10 AEDES ALBOPICTUS carrier cultures.

Material	LD_{50}*	dex per ml PFU	TCD_{50}
chik. (I 634029) (stock)	8.5	9.0	9.3
1st transfer	6.2	7.5	6.5
2nd transfer	4.9	7.4	6.2
3rd transfer	5.7	6.4	4.5
4th transfer	5.0	5.2	5.3
5th transfer	3.4	6.5	5.5
6th transfer	2.6	7.2	6.3
7th transfer	2.5	5.5	5.7
8th transfer	2.2	6.2	6.5
9th transfer	trace	6.9	5.7
10th transfer	trace	6.9	6.3

* determined in 2-day-old mice

origin), it was shown that continued subculturing of *Aedes albopictus* carrier culture virus for 800 days *in vitro* resulted in a change of virulence for 3-day-old mice. With the African strain in contrast to the Indian strain, the extent to which virulence was lost depended upon the transfer level and specifically upon plaque size as analyzed in Vero cells. Two *in vitro* markers, SP and LP variants of chikungunya virus were characterized. Progenies of the LP variant were comparable in mouse pathogenicity to the parent strain, whereas progenies of the SP variant failed to induce illness in newborn mice. In a more recent study (Buckley *et al*, 1976), the LP variant tented to modify toward the SP variant. On a comparative basis, the Indian strain of chikungunya (I 634029) has modified much more easily with regard to pathogenicity for newborn mice than the African strain (Ross). Serologically, it has been shown by Casals (1961) that Oriental strains of chikungunya differ slightly from African strains.

Attonuation during persistence in mosquito cell cultures has been reported also for Semliki Forest virus by Peleg (1971) and by Davey and Dalgarno (1974).

2. Rhabdovirus. Artsob and Spence (1974a) established a persistent infection with vesicular stomatitis virus (VSV), rhabdovirus taxon, in Singh's *Aedes aegypti* and *Aedes albopictus* cell lines. In the rabies serogroup (Shope *et al.*, 1970; Shope, 1974; Shope *et al.*, 1975), a subgroup within the rhabdovirus taxon, Obodhiang virus (isolated from *Mansonia uniformis)* and kotonkan virus (isolated from *Culicoides*) (Kemp *et al.*, 1973) multiply in Singh's *Aedes albopictus* cell line, but not in *Aedes aegypti* cells (Buckley, 1973b). A third member, Mokola virus (isolated from *Crocidura* sp.) (Kemp *et al.*, 1972) behaves similarly (Buckley, 1976); at present, there is no other evidence indicative of an arthropod cycle in the maintenance of Mokola virus. According to Murphy (1974), the *in vivo* site of budding for the two presumptive arboviruses is the plasma membrane, whereas Mokola virus buds

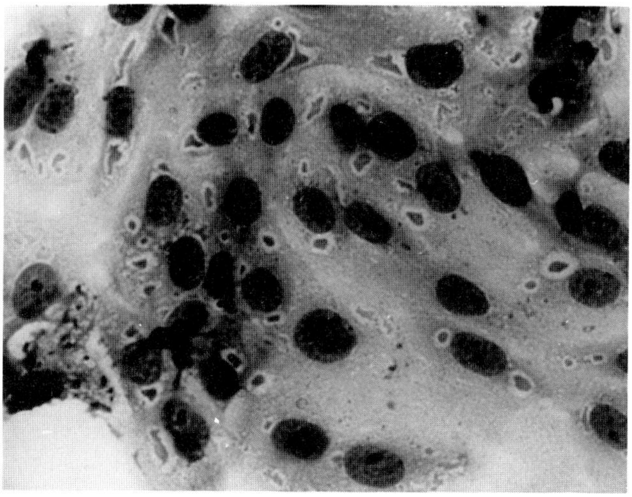

Fig. 1. Infection of Vero cells with cloned Obodhiang virus. Giemsa stain. Typical Negri-like bodies localized in the cytoplasm.

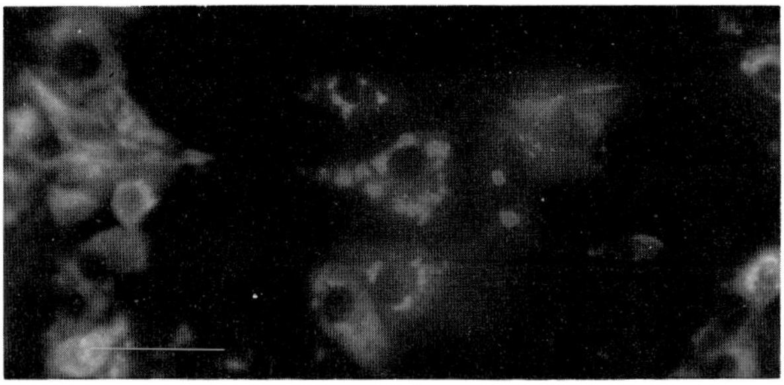

Fig. 2. Infection of Vero cells with cloned Obodhiang virus. Viral antigen localized specifically in the cytoplasm by immunofluorescence.

through and acquires its outer coat both from plasma membrane as well as from the endoplasmic reticulum. Negri-like bodies were found with Mokola virus in infected monkey brain (Percy *et al.*, 1971). In infected Vero cell cultures, cytoplasmic inclusion bodies were found by immunofluorescence as well as by a tinctorial method (Giemsa) with both kotonkan and Obodhiang viruses. Figures 1 and 2 illustrate results obtained with Obodhiang virus. The methods used here were the same as described in detail elsewhere (Buckley and Clarke, 1970; Buckley and Casals, 1970). Persistent infection of *Aedes albopictus* cells with cloned kotonkan, Obodhiang and Mokola viruses (Buckley and Tignor, 1975) was established as follows: mosquito cells were infected with cloned kotonkan and Obodhiang viruses at a multiplicity of infection (MOI) of one PFU per 1,000 cells. Primary infection was successful as assayed both by plaque assay in Vero cells and mouse inoculation with centrifuged fluid phase of infected cultures. These cultures were kept at room temperature (20-25°C) for 24 months prior to initiation of transfers.

Primary infection of mosquito cells with cloned Mokola virus was established at a MOI of one PFU per 20 cells. Transfers of the infected cells was carried out approximately three months following primary infection. Subsequently, all three carrier cultures were transferred regularly at 2-week-intervals by a low split of 1:2. Uninfected control cells were transferred in parallel. Two temperatures of incubation were routinely combined: 30°C for 4 days followed by room temperature (20-25°C) for ten days; carrier cultures were fed once between transfers. As determined by infectious center assay in Vero cells by methods described previously (peleg, 1969; Likikova and Buckley, 1971) only a small percentage of cells was infected as documented in Table 3. Figures 3, 4, and 5 show *Aedes albopictus* cells persistently infected with cloned kotonkan, Obodhiang and Mokola virus and stained by the indirect method (Weller and Coons, 1954) of the fluorescent antibody technique. While specific staining of kotonkan- and Obodhiang-infected cells was diffuse, bright cytoplasmic bodies as well as diffuse immunofluorescence was observed in Mokola-infected cells. Mokola carrier cultures remained pathogenic for 2-day-old mice, whereas kotonkan-and Obodhiang carrier cultures lost pathogenicity for infant mice in as much as tested; moreover, Obodhiang-and kotonkan-persistently infected *Aedes albopictus* cells failed to produce antibodies in mice following one intracerebral inoculation. All carrier cultures infected *Aedes aegypti* mosquitoes in one experiment in which kotonkan virus was recovered both by plaque assay in Vero cells as well as by mouse inoculation 12 days post-inoculation, Obodhiang 9 days and Mokola 11 days after the inoculation of mosquitoes, respectively. Aitken (unpublished observation) has shown that both kotonkan and Obodhiang viruses infect *Aedes aegypti* by intrathoracic inoculation and multiplied in salivary glands. At the time of the writing of this communication, Mokola virus

TABLE 3

Percentage of AEDES ALBOPICTUS cells containing infective kotonkan, Obodhiang, and Mokola virus

Virus	Strain	Transfer level of infected cells	Cells per ml of tested cell suspension (X 10^6)	PFU* per ml of tested cell suspension (X 10^3)	Ratio of infective to non infective cells	Percentage of infective cells
kotonkan	Ib An 23380	6	7.50	2.5	1/3000	0.03
kotonkan	Ib An 23380	10	1.84	10.0	1/184	0.55
kotonkan	Ib An 23380	11	5.80	39.0	1/149	0.77
Obodhiang	Sud Ar 1154-61	6	7.60	27.5	1/276	0.36
Obodhiang	Sud Ar 1154-61	8	9.40	10.0	1/940	0.11
Obodhiang	Sud Ar 1154-61	10	1.14	37.5	1/30	3.33
Mokola	Ib An 27377	2	8.60	190.0	1/45	2.22
Mokola	Ib An 27377	5	9.00	425.0	1/21	4.76
Mokola	Ib An 27377	8	19.60	940.0	1/20	5.00

* PFU = plaque forming units (plaque assay in Vero cells)

has propagated persistently in the *Aedes albopictus* for 15 months; although an arthropod vector has not been found, *in vitro* evidence presented here points to the fact that Mokola virus in all probability is an arbovirus, related to rabies (Shope *et al.*, 1970) and known to be pathogenic for homo sapiens (Familusi and Moore, 1972).

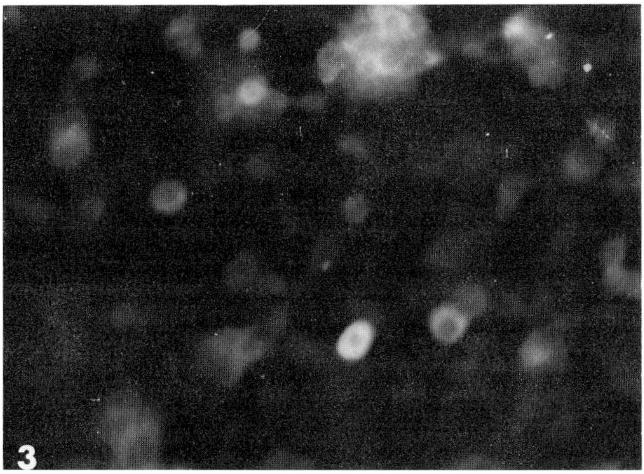

Fig. 3. Immunofluorescence pattern of *Aedes albopictus* cells persistently infected with cloned kotonkan virus; transfer # 18.

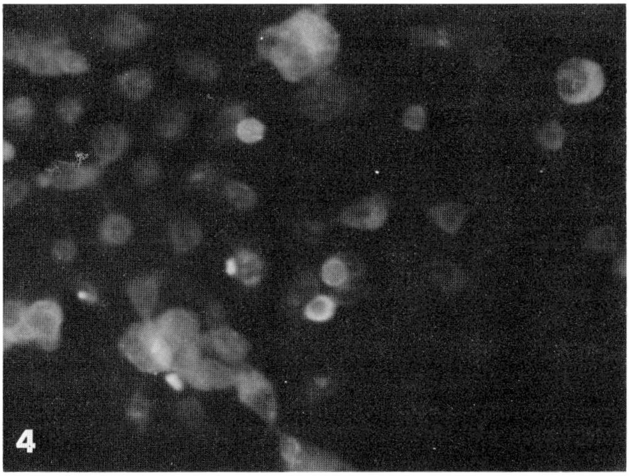

Fig. 4. Immunofluorescence pattern of *Aedes albopictus* cells persistently infected with cloned Obodhiang virus; transfer # 18.

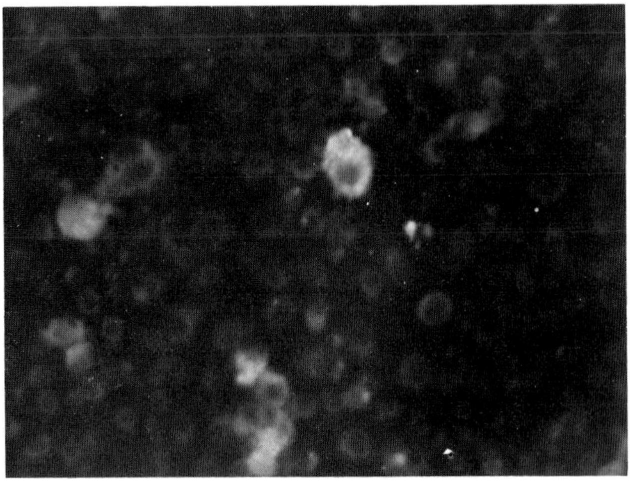

Fig. 5. Immunofluorescence pattern of *Aedes albopictus* cells persistently infected with cloned Mokola virus; transfer #12.

IV. Diagnostic tools

Mosquito cells have been found refractory to infection with viruses not belonging to arbovirus taxons (Singh, 1972); thus, invertebrate cell systems can be used advantageously in assisting in the identification of unknown virus strains isolated in the field or generally in the characterization of viruses. During Lassa virus studies (Buckley and Casals, 1970), the fact that the agent failed to multiply in *Aedes* cells made it likely that the agent was not an arbovirus. Likewise, the fact that Mokola virus, rabies serogroup (Shope *et al.*, 1970), propagated persistently in mosquito cells for 15 months favors the hypothesis of an arbovirus although an arthropod vector may never be found.

Singh's *Aedes* cell lines have been an aid to us recently in the characterization of an unknown virus strain (U. V. Str.) which was kindly sent to the Yale Arbovirus Research Unit (YARU) by the Microbiological Research Establishment, Porton Down, Salisbury, England. The U.V. Str. had been isolated from "a pool of 100 *Mansonia uniformis* collected in Kampong Tijirak, Sarawak on 15 July, 1969" (Simpson, personal communication). Apparently, adaptation to mice was a little troublesome. At YARU, a tentative diagnosis of a rhabdovirus was quickly surmised based on the fact that the U.V. Str. multiplied best in *Aedes aegypti* cells. It has been shown by Buckley (1969), Singh (1971) and Artsob and Spence (1974b) that in the rhabdovirus taxon, viruses belonging to the vesicular stomatitis subgroup multiply in both *Aedes* cell lines with the *Aedes aegypti* cell line apparently being more sensitive. Briefly, *Aedes* cells as well as vertebrate BHK-21 and Vero cells grown in large Leighton tube cultures were inoculated with a 1:1000 dilution of the infected mouse brain suspension. Two days post-inoculation, the infected fluid phases were assayed for virus in 2-day-old mice. The infected cells were fixed with acetone and stained by the indirect method (Weller and Coons, 1954) of immunofluorescence. The procedures used in this laboratory have been described in detail elsewhere (Buckley and Clarke, 1971). Specific staining was most brilliant in the cytoplasm of infected *Aedes aegypti* cells as demonstrated in Figure 6 and also shown in Table 4. Moreover, undiluted infective fluid phase of such *Aedes aegypti* cultures killed all mice with an average survival time of 1.9 days. Examination of sectioned infected BHK-21 and *Aedes aegypti* cells in the electron microscope revealed rhabdovirus-like structures. Multiplication of the U.V. Str. in *Aedes aegypti* cells was not inhibited by the incorporation of 5-bromodeoxyuridine in the medium (Webb *et al.*, 1967); hence the virus probably contains ribonucleic acid. A persistent infection was established with the U.V. Str. in *Aedes albopictus* cells in addition to *Aedes aegypti* cells; thus, it is unlikely that the agent belongs to the rabies serogroup, subgroup of the rhabdovirus taxon, in as much as it has been shown that kotonkan, Obodhiang and Mokola virus infect *Aedes albopictus* cells only (Buckley, 1976), but don't multiply in *Aedes aegypti* cells. The fact that the U.V. Str. multiplied in the mosquito cells strongly suggests that it is a true arbovirus. The agent may be a new rhabdovirus, as it did not react by complement-fixation with any of the immune reagents available in the laboratory. Presentation here of some of the characteristics of the U.V. Str. is not in any way intended to supercede the definitive description and identification of the virus which will be done by the Microbiological Research Establishment, Porton Down, Salisbury, England.

TABLE 4

Mouse pathogenicity and immunofluorescence observed in invertebrate and vertebrate cell systems with U.V.Str.

Cells inoculated with infected mouse brain suspension	Mouse Pathogenicity		Immunofluorescence* (cytoplasmic)
	Mortality 2-day-old-mice	Average Survival time (days)	
Aedes aegypti	16/16	1.9	+++
Aedes albopictus	12/16	1.9	++
BHK-21	5/16	3.4	+
Vero	1/16	3.0	+

* Immunofluorescence graded on a scale of +++, ++, and +

Pavri and Ghose (1969) showed that untreated fluid phases obtained from *Aedes albopictus* cell cultures inoculated with arboviruses can be used satisfactorily as antigens for viral identification in the complement-fixation (CF) test. Subsequently, Singh and Paul (1969) and Casals and Buckley (1973) used this technique successfully to identify dengue viruses, types I to IV, isolated directly in mosquito cells from field material. Recently, Ajello *et al.* (1975) have extended this technique to West Nile virus. Briefly, CF antigen with titers up to 1:128 was demonstrated in *Aedes albopictus* cell cultures inoculated with the Egypt 101 strain, 11th mouse brain passage. The CF antigen was specific. There were no cross reactions with tick-borne encephalitis mouse serum. The sensitivity of the system was documented by demonstrating that West Nile virus diluted to 8 dex (Haldane, 1960) multiplied in the mosquito cells as ascertained both by Vero plaque assay as well as by CF reaction. A satisfactory CF antigen has recently been obtained in this laboratory also with fluid phases of *Aedes albopictus* cell cultures inoculated with an Oriental strain of chikungunya virus (I 634029). In our hands, such antigens have not been anti-complementary at any time.

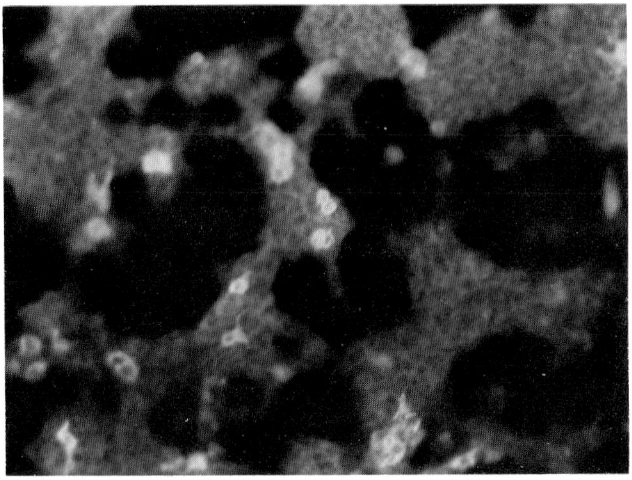

Fig. 6. Immunofluorescence pattern with U.V.Str. in *Aedes aegypti* cells; first passage.

V. Contamination

In vertebrate cell cultures, a cytopathic effect (CPE) has been described with arboviruses of the six taxons. Under fluid medium, CPE consists of moderate to marked cell destruction (Scherer and Syverton, 1954; Buckley, 1964; Karabatsos and Buckley, 1967), whereas under nutrient agar overlay plaque formation is induced (Dulbecco, 1952; Stim, 1969). Addition of specific antibody to cell cultures along with the viral inoculum prevents the development of CPE or plaques. With invertebrate cell cultures, CPE (Paul *et al.,* 1969; Suitor and Paul, 1969; Varma *et al.,* 1974) or plaques (Suitor, 1969; Cory and Yunker, 1972; Yunker and Cory, 1975) have also been described in Singh's *Aedes albopictus* cell line (Singh, 1967) and recently in two new cell lines established from *Aedes malayensis* and *Aedes pseudoscutellaris* (Varma *et al.,* 1974). What is puzzling with regard to cellular damage is the fact that the extent of the CPE seems to depend on the individual sublines used and also on the type of container (i.e. glass-or plastic vessel) in which the *Aedes albopictus, A. malayensis* or *A. pseudoscutellaris* cells are grown (Suitor and Paul, 1969; Varma *et al.,* 1974). According to Dalgarno and Davey (1973), arbovirus-infected *Aedes albopictus* cells only show CPE "when stressed in a particular way". Thus, dengue virus, type 2, failed to produce CPE in the hands of Sinarachatanant and Olsen (1973), whereas CPE with the same virus was reported by Paul *et al.* (1969) and confirmed by Suitor and Paul (1969) as well as by Sweet and Unthank, 1971). That the CPE caused by a stress situation might be dependent on latent viral contamination came to light when two institutions, The Boyce Thompson Institute, Yonkers (Hirumi, personal communication) and Purdue University, West Lafayette (Webb, personal communication) reported spontaneous syncytia formation in two sublines of uninoculated *Aedes albopictus* cells, here designated "Hirumi" and "Webb" sublines. In the meantime, isolation of chikungunya virus contaminating the "Webb" subline has been described (Cunningham *et al.,* 1975).

We report here on some further studies with regard to the "Hirumi" subline and its contaminating agent, isolated and identified by CF test and plaque reduction neutralization test as chikungunya virus and characterized as apathogenic for newborn mice. As determined by infectious center assay in Vero cells by methods described (Peleg, 1969; Libikova and Buckley, 1971), approximately 1.6% of all the cells were infected with chikungunya virus which was localized specifically in the cytoplasm as determined by the indirect method (Weller and Coons, 1954) of immunofluorescence by procedures described in detail (Buckley and Clarke, 1970) (See Figure 7). The fluid phase of the "Hirumi" cell line, transfer #3, induced hazy plaques (Figure 8) in Vero cells. In attempting to restore mouse pathogenicity, the chikungunya isolate was plaque-purified in Vero cells and passed for 17 passages in BHK-21 cells. At this time, large and clear plaques were obtained in Vero cells (Figure 9); however, the virus strain failed to induce illness in 2-day-old mice by intracerebral inoculation. Subsequently, blind virus passages in newborn mice were carried out at 3-4 day intervals. Mouse brain suspensions were plaque-assayed in Vero cells for presence of virus. Some of the inoculated mice were observed for 24 days after inoculation, then sacrificed under deep ether anesthesia and their respective sera examined for development of neutralizing antibodies by plaque reduction neutralization test in Vero cells. The results are summarized in Table 5. Virus was present through seven passages and disappeared thereafter. No deaths were observed. The plaque forming unit (PFU) titers varied from 2.2 to 4.9 dex per ml. Mice of the first to the sixth passage level developed neutralizing antibodies. Inactivated, undiluted sera reduced the plaque counts from 60 to 100%; however, as can be seen from the reciprocal of serum titers obtained by the 50% plaque reduction method, the virus was a poor immunogen.

Ten serial *in vivo* passages of the avirulent chikungunya virus were made thereafter in *Aedes aegypti*. Briefly, female mosquitoes were inoculated intrathoracically with approximately 25,000 PFU of the cloned virus. Seven days after inoculation, ten mosquitoes were removed, pooled and triturated in 2.0 ml of diluent (phosphate buffered saline, pH 7.2, containing 0.75% bovine albumin, fraction V). The undiluted supernatant of this suspension was inoculated intrathoracically into a new batch of mosquitoes and also plaque-assayed in Vero cells. Mosquitoes from each passage levels were allowed to feed on newborn mice; in addition, mosquito suspensions were inoculated intracerebrally into infant mice. The latter were observed for development of illness, then sacrificed 21 days after inoculation as described above for assays of neutralizing antibodies. The results are summarized in Table 6. Briefly, the virus replicated well in *Aedes aegypti* at all passage levels with no appreciable change in the maximum PFU titer obtained in the mosquitoes throughout the experiment. Plaques in Vero cells were also monitored at each passage level for change in size, since the cloned preparation used initially had been selected for large plaque size in an attempt to restore mouse pathogenicity. However, during the ten serial *in vivo* passages both large and small plaques were produced. This characteristic was constant without undergoing any selection whatsoever during these experiments. The virus remained avirulent for 2-day-old mice through all ten passage levels in mosquitoes. Specific neutralizing antibodies were present in the sera of mice inoculated intracerebrally with low dilutions of infected mosquito suspensions and sacrificed three weeks post-inoculation. Positive transmission of virus also occurred as evidence by the development of neutralizing antibodies in mice exposed to the bites of infected mosquitoes at various passage levels.

Based on these results it can be stated that an apparently stable mutant of chikungunya virus has been isolated from the uninoculated *Aedes albopictus* "Hirumi"

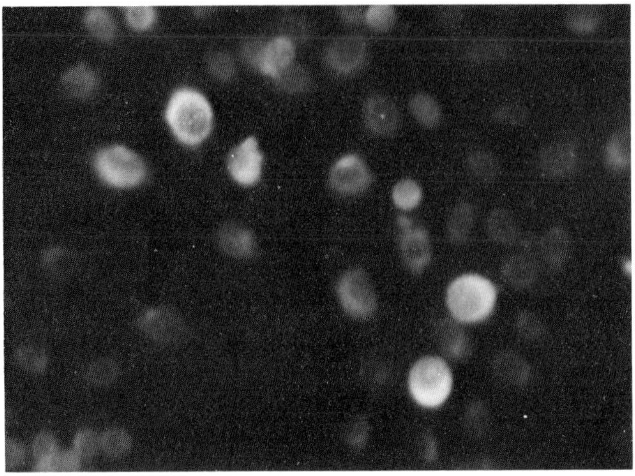

Fig. 7. Immunofluorescence pattern of *Aedes albopictus* cells, "Hirumi" subline, Transfer # 3, persistently contaminated with an avirulent chikungunya virus strain.

subline. The main characteristics are apathogenicity for newborn mice as well as poor immunogenicity. In the latter respect, this virus strain differs from the attenuated chikungunya virus (Ross strain) (Buckley, 1973) and the attenuated Semliki Forest virus (Peleg, 1971), both characterized as good immunogens.

On a national and international level, this laboratory has distributed Singh's *Aedes* cell lines since 1968. It is remarkable that the majority of investigators have requested additional shipments of cells due to the fact that their cell line "had become refractory" to the particular virus they had been studying. This state of refractoriness may be explained, perhaps, by homologous interference. As pointed out by Peleg (1975), cells in Sindbis (alphavirus) persistently infected cultures, subcultured or not, are indistinguishable from cells in uninoculated control cultures of the same age in respect to growth potential and morphology. Yet these cells are resistant to superinfection by the homologous virus. These two investigators have found further that the transition to a state of cellular resistance in Sindbis virus infected *Aedes aegypti* cultures coincided with the appearance in the culture of a small plaque-forming mutant designated SV-S. The cloned SV-S was shown to interfere with the growth of the wild strain of Sindbis virus (SV-W) in a number of invertebrate and vertebrate cell cultures. Thus, the accidental, unintentional contamination of an invertebrate cell line with any arbovirus studied in any laboratory at any time appears to be the gravest hazard encountered in invertebrate cell culture work.

VI. Conclusions

Two aspects appear to warrant special consideration. Firstly, of the large number of registered arboviruses (350 as of 1974), conclusive evidence that they are *de facto* arthropod-borne viruses is available for relatively few. Classification of viruses, generally, is based essentially on the properties of the virion and the interaction of virus and host at the cellular level. Usually, serologic evidence is the

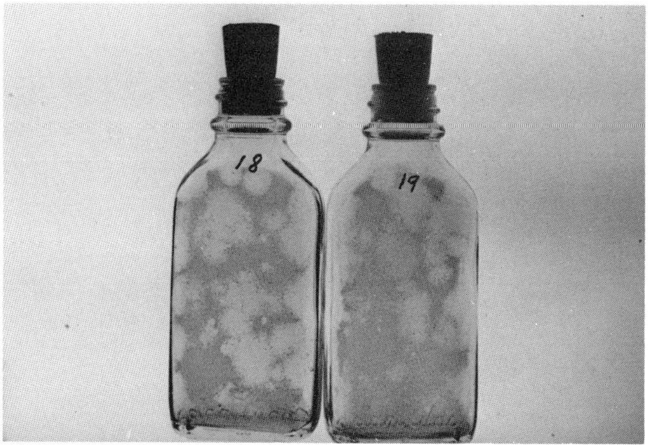

Fig. 8. Plaque morphology observed in Vero cells with the fluid phase of *Aedes albopictus* cells, "Hirumi" subline, Transfer # 3.

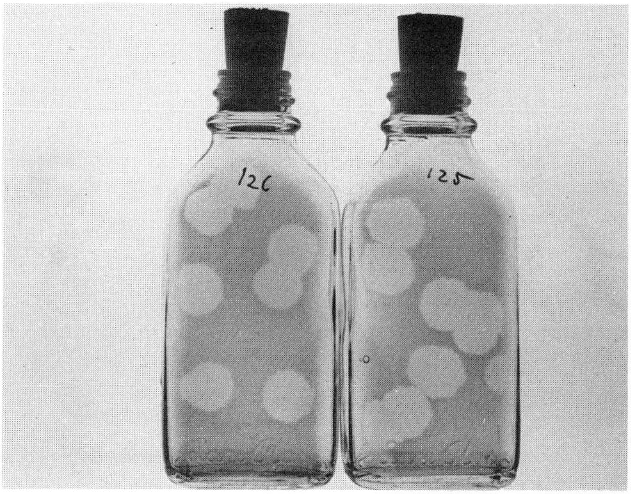

Fig. 9. Plaque morphology observed in Vero cells with the avirulent chikungunya virus strain following cloning in Vero cells and 17 additional high-dilution passages in BHK-21 cells; apathogenic for newborn mice.

most complete line of characterization next to morphology and morphogenesis. We have demonstrated here the practicality and usefulness of mosquito cells as a host system in the rapid characterization of an unknown virus strain isolated from *Mansonia uniformis.* With the establishment of mosquito- and more recently of three tick-cell lines, the availability of invertebrate tissue culture as a diagnostic tool in addition to mosquitoes should be mandatory in any laboratory engaged in arbovirus studies. At the present time, these established invertebrate cell lines are as

TABLE 5

Infectivity titers obtained during blind passages of mouse-avirulent chikungunya as well as production of complement-fixing (CF) or neutralizing antibodies (NT)

Number of blind passage	Infectivity titers (dex per ml) LD_{50} (mouse)	Infectivity titers (dex per ml) PFU (Vero)	Antibody production CF (serum dil. 1:2)	Antibody production NT % plaque reduction (serum undiluted)	Antibody production NT 50% reduction titer**
1	0	4.7	0	90	8
2	0	4.3	0	100	< 2
3	0	4.8	0	75	< 2
4	0	2.2	0	90	8
5	0	4.9	0	90	8
6	0	3.4	0	60	< 2
7	0	3.1	nd	nd	nd
8	0	0	nd	nd	nd
9	0	0	nd	nd	nd
10	0	0	nd	nd	nd

* Mice bled 20 to 24 days after inoculation

** Serum titers expressed as the reciprocal of the highest serum dilution giving a 50% reduction in the plaque count

TABLE 6

Infectivity titers obtained during passages of mouse-avirulent chikungunya in AEDES AEGYPTI, as well as transmission status and plaque size at individual passage levels obtained in Vero cells

Number of mosquito intrathoracic passage number	Infectivity titers (dex per ml) LD_{50} (mouse)	Infectivity titers (dex per ml) PFU (Vero)	Plaque size range (mm)	Plaque size mean (mm)	Transmission* status
1	0	6.4	1-7	4.3	+
2	0	5.9	3-8	5.5	+
3	0	5.3	3-6	4.5	+
4	0	5.8	1-7	3.3	+
5	0	5.7	1-6	3.9	+
6	0	5.5	2-6	3.7	+
7	0	5.8	1-6	3.0	nd**
8	0	5.9	1-5	2.3	nd
9	0	5.6	2-5	3.5	nd
10	0	5.3	1-5	2.6	nd

* Mice bled 21 days after being fed on by infected mosquitoes

** nd = not done

easy to handle as vertebrate cell lines. Secondly, one of the really serious hazards linked with the use of mosquito- or maybe also tick-cell lines is the fact that any arbovirus capable of replication in one or the other invertebrate cell system is also able to induce a persistent infection in analogy to the *in vivo* situation where arthropods, once infected, remain infected for life. Thus, extraneous contamination of invertebrate cell lines should be monitored on a routine basis by electron microscopy, immunofluorescence and infectious center assay in a vertebrate cell system.

Acknowledgments

We are gratefully acknowledging the skillful technical assistance of Mrs. M. Garrison, Mrs. V. Parcells, Mr. C. Mullen, Mrs. E. Gilson, Mrs. M. Malhoit, and Mrs. K. Gilson. Microphotographs were kindly supplied by G. Martine. We are indebted to Mrs. Carmel Bierwirth for her competent assistance with the typescript.

Previously unpublished studies of the authors were supported by the United States Army Medical Research and Development Command (DADA-17-12-C-2170), National Institute of Allergy and Infectious Diseases (PHS-RO-1-AI 10984), The World Health Organization, and The Rockefeller Foundation.

VII. References

Ajello, C., Gresikova, M., Buckley, S.M., Casals, J. (1975). *Acta Virol.*, in press.

Andrews, C. (1973). In: Viruses and Invertebrates (A.J. Gibbs, ed.), 1-13. American Elsevier Publishing Co., New York.

Artsob, H., and Spence, L. (1974a). *Acta Virol. 18,* 331.

Artsob, H., and Spence, L. (1974b). *Canad. J. Microbiol. 20,* 329.

Banerjee, K. and Singh, K.R.P. (1968). *Indian J. Med. Res. 56,* 812.

Banerjee, K., and Singh, K.R.P. (1969). *Indian J. Med. Res. 57,* 1003.

Borden, E.C., Shope, R.E., and Murphy, F.A. (1971). *J. Gen. Virol. 13,* 261.

Buckley, S.M. (1964). *Proc. Soc. Exp. Biol. & Med. 116,* 354.

Buckley, S.M. (1969). *Proc. Soc. Exp. Biol. & Med. 131,* 625.

Buckley, S.M. (1973a). In: Proc. Third International Colloquium on Invertebrate Tissue Culture (J. Řeháček, D. Balskovic, and W.F. Hink, eds.), 307-324. Slovak Academy of Sciences, Bratislava.

Buckley, S.M. (1973b). *Appl. Microbiol. 25,* 695.

Buckley, S.M. (1976). In: Invertebrate Tissue Culture (K. Maramorosch, ed.). Academic Press, Inc., New York.

Buckley, S.M., and Casals, J. (1970). *Am. J. Trop. Med. & Hyg. 19,* 680.

Buckley, S.M., and Clarke, D.H. (1970). *Proc. Soc. Exp. Biol. & Med. 135,* 533.

Buckley, S.M., and Tignor, G.H. (1975). *J. Clin. Microbiol. 1,* 241.

Buckley, S.M., Singh, K.R.P., and Bhat, U.K.M. (1975). *Acta Virol. 19,* 10.

Casals, J. (1971). In: Comparative Virology (K. Maramorosch and E. Kurstak, eds.), 307-333. Academic Press, Inc., New York.

Casals, J. (1961). Tenth Pacific Science Congress, Honolulu, Hawaii. Abstracts of Symposium papers, page 458.

Casals, J., and Buckley, S.M. (1973). *Dengue Newsletter for the Americas, PAHO, 2,* 6.

Corey, J., and Yunker, C.E. (1972). *Acta Virol. 16,* 90.

Cunningham, A., Webb, S.R., Buckley, S.M., Casals, J. (1975). *J. Gen. Virol. 27,* 97.

Dalgarno, L., and Davey, M.W. (1973). In: Viruses and Invertebrates (A.J. Gibbs, ed.), 245-270. American Elsevier Publishing Company, New York.

Davey, M.W., and Dalgarno, L. (1974). *J. Gen. Virol. 24,* 1.

Dulbecco, R. (1952). *Proc. Nat. Acad. Sci.* (U.S.A.). *38,* 747.

Familusi, J.B., and Moore, D.J. (1972). *Afr. J. Med. Sc. 3,* 93.

Grace, T.D.C. (1966). *Nature 211,* 366.

Haldane, J.B.S. (1960). *Nature 187,* 879.

Hink, W.F. (1972). In: Invertebrate Tissue Culture (C. Vago, ed.), 363-387. Academic Press, Inc., New York.

Karabatsos, N. (1969). *J. Trop. Med. & Hyg. 18,* 803.

Karabatsos, N., and Buckley, S.M. (1967). *Am. J. Trop. Med. & Hyg. 16,* 99.

Kemp. G.E., Lee, V.H., Moore, D.L., Shope, R.E., Causey, O.R., and Murphy, F.A. (1973). *Am. J. Epidemiology 98,* 43.

Libikova, H., and Buckley, S.M. (1971). *Acta Virol. 15,* 393.

Mitsuhashi, J., and Maramorosch, K. (1964). *Contr. Boyce Thompson Inst. 22,* 435.

Murphy, F.A. (1974). In: Viruses, Evolution and Cancer (E. Kurstak and K. Maramorosch, eds.), 699-722. Academic Press, Inc., New York

Paul, S.D., Singh, K.R.P., and Bhat, U.K.M. (1969). *Indian J. Med. Res. 57,* 339.

Pavri. K.M., and Ghose, S.N. (1969). *Bull. WHO 40,* 984.

Peleg, J. (1968). *Virology 35,* 617.

Peleg, J. (1969). *Nature 221,* 193.

Peleg, J. (1971). In: Current Topics in Microbiology and Immunology (E. Weiss, ed.), 155-161. Springer-Verlag, New York.

Peleg, J. (1975). Conference on Pathobiology on Invertebrate vectors of Disease, March 17-18, The New York Academy of Sciences, abstract # 17. New York City, N.Y.

Percy, D.H., Bhatt, P.N., Tignor, G.H., and Shope, R.E. (1971). *Vet. Path. 10,* 534.

Řeháček, J. (1972). In: Invertebrate Tissue Culture (C. Vago, ed.), 279-320. Academic Press, Inc., New York.

Scherer, W.F., and Syverton J.T. (1954). *Am. J. Pathol. 30,* 1075.

Shope, R.E. (1975). In: The Natural History of Rabies (G.M. Baer, ed.), Academic Press, Inc., New York, in press.

Shope, R.E., Murphy, F.A., Harrison, A.K., Causey, O.R., Kemp, G.E., Simpson, D.I.H. and Moore, D.L. (1970). *J. Virol. 6,* 690.

Shope, R.E., Buckley, S.M., Aitken, T.H.G., and Tignor, G.H. (1975). Proc. Third International Congress of Virology, September, Madrid, Spain.

Sinarachatanant, P., and Olsen, L.C. (1973). *J. Virol. 12,* 275.

Singh, K.R.P. (1967). *Curr. Sci. 36,* 506.

Singh, K.R.P. (1971). In: Current Topics in Microbiology and Immunology (E. Weiss, ed.), 127-133, Springer-Verlag, New York.

Singh, K.R.P. (1972). In: Advances in Virus Research (K.M. Smith, M.A. Lauffer, and F.B. Bang, eds.), 187-206, Academic Press, Inc. New York.

Singh, K.R.P., and Bhat, U.K.M. (1971). *Experientia 27,* 142.

Singh, K.R.P., and Paul, S.D., (1968). *Curr. Sci. 37,* 65.

Stim, T.B. (1969). *J. Gen. Virol. 5,* 329.

Stollar, V. (1975). Conference on Pathobiology on Invertebrate Vectors of Disease, March 17-19, The New York Academy of Sciences, abstract # 18, New York City, N.Y.

Suitor, E.C. (1969). *J. Gen. Virol. 5,* 545.

Suitor, E.C., and Paul, F.J. (1969). *Virology 38,* 482.

Sweet, B.H., and Unthank, H.D. (1971). In: Current Topics in Microbiology and Immunology (E. Weiss, ed.), 150-154, Springer-Verlag, New York.

Taylor, R.M., Work, T.H., Hurlbut, H.S., and Rizk, F. (1956). *Am. J. Trop. Med. & Hyg. 5,* 579.

Theiler, M. (1957). *Proc. Soc. Exp. Biol. & Med. 96,* 380.

Varma, M.G.R., Pudney, M., and Leake, C.J. (1974a). *Trans. R. Soc. Trop. Med. & Hyg. 68,* 374.

Varma, M.G.R., Pudney, M., and Leake, C.J. (1974b). *J. Med. Ent. 11,* 698.

Weller, T.H., and Coons, A.H. (1954). *Proc. Soc. Exp. Biol. (N.Y.) 86,* 789.

Webb, P.A., Johnson, K.M., Mackenzie, R.B., and Kuns, M.L. (1967). *Am. J. Trop. Med. & Hyg. 16,* 531.

WHO Scientific Group (1967). W.H.O. Tech. Rept. Ser. No. 369.

Yunker, C.E. (1971). In: Current Topics in Microbiology and Immunology (E. Weiss, ed.), 113-126, Springer-Verlag, New York.

Yunker, C.E., and Cory, J. (1975). *Appl. Microbiol. 29,* 81.

Chapter 2

TICK TISSUE CULTURE AND ARBOVIRUSES

J. Řeháček

I. Introduction

Ticks are vectors of many microorganisms (viruses, rickettsiae and protozoan parasites) pathogenic to man and animals. There is no doubt, that the culture of tick tissues and cells *in vitro* can be effective in investigations of a number of interesting and important problems concerning the relationships between these pathogens and their man and animal hosts.

The purpose of this lecture is to familiarise you with the present status of tick cell and tissue culture in arbovirus investigations.

II. Organ culture and its application in virology

1) General considerations.

Very little attention has been previously paid to the use of organ cultures from ticks for cultivation of viruses. It is proposed that such cultures offer a reasonable substrate for studies of many interesting problems in cell - virus relationships.

2) Preparation of organ culture.

Attempts at the cultivation of tick organs *in vitro* succeeded only in their temporary survival. The first experiments on maintaining tick tissues *in vitro* were reported by Weyer, 1952. He cultivated explants of connective tissues and probably those of other organs of *Rhipicephalus bursa* (for cultivation of rickettsiae) by the hanging drop method in a medium consisting of human or rabbit plasma with the addition of rabbit spleen or testes extracts. Cultures were kept at 31 - 32°C. The organs were viable only for a very short time.

Organ cultures from ticks were also developed as a medium for cultivation of Eastern Equine Encephalomyelitis (EEE) virus (Řeháček, 1958a,b, Řeháček and Pešek, 1960). The various organs of half - engorged females of *Dermacentor pictus, Dermacentor marginatus* and *Ixodes ricinus* cultivated in TC 199 medium at 25°C retained their viability for about 30 days.

The organ cultures were developed further from half-engorged females of D. pictus and *I. ricinus* for the study of *Coxiella burneti* development from its filterable particles (Kordová and Řeháček, 1959). The media used in these experiments were Parker 199, and Parker 199 containing 25% heated horse serum with or without hemolymph from cockroaches. Contraction of ovaries and Malpighian tubuli were observed for a period of 10 days and a slight proliferation of fibroblast-like cells was observed in a few ovary cultures.

A series of experiments with organ cultures of adult *Rhipicephalus appendiculatus* ticks were performed by Martin and Vidler, 1962. The explants of tick organs were maintained in a medium consisting of Hanks'balanced salt solution, amino acids and vitamins of Eagle's basal medium with 20% ox serum. Some of the explants survived without active growth up to 170 days.

The organs of several hard tick species - *Hyalomma anatolicum excavatum, Hyalomma dromedarii, Rhipicephalus sanguineus, Rhipicephalus evertsi, Boophilus microplus, Boophilus decoloratus* and *Boophilus annulatus* were prepared for studies on the maintenance of piroplasms *in vitro* (Hoffmann and Köhler, 1968, Hoffmann *et al.*, 1970). The explants of ovaries survived in the medium composed of Hanks' solution and a modified mixture of vitamins and amino acids (after Eagle) at 28°C for 28 - 82 days; salivary glands for 26 - 56 days; intestinal tract for 12 - 46 days and Malpighian tubuli for 13 - 17 days.

The explants prepared from male and female genital glands of *Dermacentor andersoni* and *B. microplus* survived in HLH medium or Eagle's medium, both with 10% fetal calf serum for as long as two months and both served as the sources of fibroblast and epithelial cells (Řeháček, 1971).

3) Cultivation of arboviruses in organ cultures.

Only the EEE and Newcastle Disease (NDV) viruses were cultivated in surviving explants of various organs of three species of ticks (Řeháček, 1958a, Řeháček and Pešek, 1960, Blaskovic and Řeháček, 1962). No multiplication of the EEE virus occurred in organ cultures of *D. pictus, D. marginatus* and *I. ricinus* ticks inoculated with 10^4 $TCID_{50}$ of virus, but virus survived in an almost undiminished titer for six days. When the cultures were inoculated with $10^{2.5}$ of $10^{0.6}$. $TCID_{50}$ of virus, its multiplication was noted in the connective tissue and hypodermis of *D. pictus* and *I. ricinus,* whereas the other tissues only slowed down the virus inactivation. The same amounts of NDV inoculated into tick organ cultures did not multiply. The fact that it was possible to demonstrate EEE in the connective tissues and hypodermis of *D. pictus* in such a small dose of the virus which was not detectable in highly susceptible substrates as icer inoculated mice or chick embryo cells, might be useful in virus isolation experiments.

4) Prospects of organ culture in virological research.

Most virologists look for the most convenient substrate for their work i.e. sufficient amount of available cells, high sensitivity and economy. In our opinion, the tick organ cultures despite their laborious preparation are such a substrate, suitable namely for special studies as observations of how the virus particles pass through the gut cells and how they enter the cells in which they multiply. Although tick organ cultures have been recently essentially abandoned, I propose that for experiments on the pathogenity of various viruses for ticks, the isolated organs maintained *in vitro* might be recommended.

III. Hemocyte culture and its application in virology

1) General considerations.

Hemocytes of ticks *in vitro* have been rarely used as a substrate for the cultivation of pathogens. The reason for this was probably the small amount of available cells.

2) Preparation of hemocyte culture.

Surviving hemocytes of several argasid ticks *(Ornithodoros lahorensis, O. papillipes* and *Argas persicus)* for *in vitro* studies of phagocytosis of bacteria were used by Sidorov, 1960. In these experiments hemocytes survived only for several days in hemolymph obtained from the same tick species.

The first hemocyte culture in greater volume was prepared from *Hyalomma asiaticum* developing adults (Řeháček, 1963). The cells were cultivated in a medium composed of equal parts of Eagle's and Vago and Chastang's media plus 3% dextrane or 10% calf serum. The cells did not multiply but survived in good condition for about one week.

D. andersoni females, half - engorged on rabbits, served as a good source of hemocytes in the experiments of Cory and Yunker, 1971. The hemocytes set up in HLH medium with 10% fetal calf serum survived for 72 - 74 days. They became fibroblast - like, spindle shaped or rounded, being well dispersed or in clumps. The same results were achieved in our laboratory.

3) Cultivation of arboviruses in tick hemocytes in vitro.

The cultures prepared from hemocytes of *H. asiaticum* showed multiplication of TBE virus, but to a lower extent than the cultures from tissues of developing adults containing epithelial and fibroblast - like cells (Řeháček, 1963)

Dermacentor andersoni hemocyte cultures were used successfully for the cultivation of Colorado tick fever (CTF) virus. Extracellular virus was recovered from cultures as long as 62 and 71 days after inoculation. Two peaks of virus proliferation were demonstrated. First, the higher peak occurred between the 1st - 3rd weeks and the second, lower peak occurred 6 weeks after inoculation. The virus did not cause any cytopathic changes in hemocytes. Virus growth curves resembled those seen in CTF virus infected primary cultures of developing adults (Yunker and Cory, 1967), but they were lower than in the latter cultures caused probably by a lesser amount of cells in culture or by absence of cells able to multiply the virus.

4) Prospects of hemocyte culture in virology.

Because of hemocytes long viability and good sensitivity to viruses, we recommend their use for also growing other pathogens such as protozoan parasites and rickettsiae. However, the limited number of cells available will always interfere with a broader use of hemocyte cultures in microbiology.

IV. Primary tissue culture and its application in virology

1) General considerations.

This type of culture includes the cultivation of cells and tissues from donors when significant multiplication occurs for an undertermined time period. It requires more specifically defined conditions than those mentioned above.

2) Source of tissues and cells.

Most of the experiments concerned with primary tick cell culture employ developing tissues of adults within nymphs undergoing metamorphosis. This material consists of tissues, cells and hemocytes of the whole preimaginal tick organism with the exception of cells and tissues of Malpighian tubuli and digestive tract which are discarded during dissection. This material is either used as an explant of the total body content, or the cells are separated by gentle pipetting, the use of trypsin, or by a combination of both methods. The best source of the highest number of growing cells is the engorged nymph in which the developing adult is clearly visible, e.g. the developing frontal part of the imaginal body and the legs.

Because of the long time, necessary for completion of the tick life cycle, nymphs are not always available when needed. Experiments with *D. andersoni* ticks indicated that metamorphosing nymphs held at -11°C for many weeks provide no less growing tissues than unrefrigerated tissues (Yunker and Cory, 1965).

3) Preparation of tissues and cells for cultures.

Engorged nymphs are in running water, disinfected by immersion in 70% ethanol for a few minutes and repeatedly washed in steril water. Then the developing adults are separated from the metamorphosing nymphs, washed in sterile saline and cut into small pieces. After treatment with 0.25% trypsin the tissues are magnetically stirred in 0.25% trypsin prewarmed to 27°C for 10 minutes (Varma and Wallers, 1965 and Varma and Pudney, 1969a) at room temperature. The tissue fragments are agitated gently until the fluid shows even cloudiness, with only nerve ganglia remaining conspicuous. The suspension is sedimented by centrifugation at 800 r.p.

m. for 7 minutes, the trypsin removed with the supernatant and after washing in saline the cells are ready for seeding.

4) Effects of media composition on tissue culture growth.

The first description of proliferation of fibroblast - like cells from tick explants was observed in *D. marginatus* (Řeháček, 1958, a,b.). The hanging drop method and a temperature of 25 - 28°C was used in this experiment. The explants were mostly cultivated in Trager's medium and Hanks'medium plus 0.5% lactalbuminhydrolysate with 0.1% yeast hydrolysate.

Attempts were done to improve the composition of the nutrient medium by adding sera or embryonal extracts from several invertebrates and to prepare synthetic media composed of the same constituents found in the tick hemolymph. Because of high glycoprotein content of tick serum, various sugars and polysacharides were added to the medium. Marked improvement in the growth of tick tissues was found in synthetic Eagle plus Vago and Chastang's medium used in a ratio of 1:1, and in the same medium to which 5% dextran (M.W. 60.000) or tick egg extract was added (Řeháček and Hâna, 1961). Further improvement was achieved by enrichment of the media with 10% calf serum (Řeháček, 1962, Varma and Pudney, 1967). Yunker and Cory, 1967 successfully examined the HLH medium (Grand Island Biol. Co.) to which 10% heat - inactivated normal rabbit serum, 10% whole chicken - egg ultrafiltrate, and 10 mg/ml bovine plasma albumin were added. Cellular outgrowth was observed in this medium up to 246 days and survival of tissues, as evidenced by contractions, for as long as 263 days.

We have prepared a new medium for the cultivation of tick cells based on the results of chemical analyzes of the aminoacids, sugars and salts in the hemolymph of *B.microplus, Argas lagenoplastis* and *H. dromedarii* containing vitamins of the B complex, organic acids, 1% Antheraea pernyi hemolymph and 1% bovine plasma. The cells of *R. sanguineus* tick were viable in this medium for nine months (Řeháček and Brzostowski, 1969a).

Varma and Pudney, 1973 cultivated successfully tissues and cells from developing adults of *R. appendiculatus* in a medium VP 12 of their own formulation. The cells remained healthy up to 4 months and were subcultured four times.

When tested medium in experiments with *D. marginatus* and *H. dromedarii* ticks, only negative results were obtained (Řeháček, unpubl. results). Also VP 12 medium in which primary cell cultures of *H. dromedarii* were grown successfully, was proved unsatisfactory for the growth of *R. appendiculatus* and *B. microplus* cells. On the contrary L - 15 medium, in which cells of *R. appendiculatus* and *B. microplus* grew well proved less satisfactory for cells from *H. dromedarii* (Varma *et al.*, 1975). It is evident that the successful growth of tick cells depends not only on the media but predominantly on the tick species used.

5) Cell composition and growth in culture.

Observation by Yunker and Cory, 1967 indicate that the production of outgrowth is biphasic. The initial cellular response is the migration of hemocytes and their adherence to the vessel wall. These cells survive for a few weeks and are gradually overgrown by cells arising from organs and tissues, which produce sheet - like complexes.

Martin and Vidler, 1962 recognized four cell types in the culture of *R. appendiculatus.* The fibroblast - like cells were seen in cultures from 24 hours up to 167 days after seeding. The slender elongate type continued to appear in varying numbers for about 60 - 70 days. The third type of cells observed were small polymorphic epithelial - like cells that began to appear in cultures from the 10th to the 35th - 40th day and disappeared by the 50th - 85th day. The fourth type of cells was of epithelial character, appearing in cultures after 11 - 27 days and continuing up to 175 days.

In cell culture from developing adults of *D. andersoni* ticks different types of vesicles were seen from the first day after seeding the explants (Řeháček, 1971).

6) Electron microscope study of tick cells in vitro.

Cell cultures prepared from developing adults of *R. sanguineus* ticks were investigated with the electron microscope. The purpose of this study was to determine what ultrastructural differences, if any, exist between tick cells and other invertebrate cells and also to establish whether tick cells of this species act as carriers of various tick specific viruses or other pathogens. It was shown in one week old cultures that the cells tested do not possess special arrangements of organelles and in general do not differ in ultrastructure from *in vitro* cultures of other arthropod cells. No inclusions, viral particles or other pathogens were found in any of the cells examined (Filshie and Řeháček, unpubl. results).

7) Utilization of the aminoacids and sugars by tick cells in vitro.

The study of utilization of the aminoacids and sugars in the medium by the cells of R. sanguineus has shown that aminoacids can be divided into those which were utilized to a significant extent and those in which no change, or an increase in concentration occurred. The first group consisted of leucine, methionine, threonine, phenylalanine, proline, glutamic and aspartic acids. The second one consisted of isoleucine, valine, cystine, tyrosine, arginine, histidine, lysine, glycine, serine and alanine. Sugars in the medium were utilized by the cells in culture, glucose disappeared to a greater extent than inositol - 43,5% of glucose was metabolized over 10 day period compared to 30,1% of the inositol (Řeháček and Brzostowski, 1969b, Řeháček, 1969).

Varma and Pudney (1969a) in measuring the uptake of glucose as a possible indicator of growth and metabolism of the tick cells *in vitro* found that the glucose uptake was proportional to the number of cells. When comparing the uptake of glucose by cells growing in tubes with or without coverslips, they found that the cells grown under a coverglass utilized significantly larger amounts of glucose; by six days, 50% of the glucose was used up and by the 12 days this had increased only to 70%.

These results indicate that although the growth rate was low, the cells *in vitro* were actively metabolizing the amino acids and sugars.

8) Use of primary cell culture for the cultivation of viruses.

a) Cultivation of different viruses.

The marked multiplication of TBE virus in cultivated *H. dromedarii* cells (Řeháček, 1962) stimulated further studies with this and other viruses in tissue cultures from ticks. It was found that mosquito - borne viruses, namely WEE, EEE, Sindbis and Semliki Forest virus multiplied in cultures of *H. dromedarii* very well at approximately similar rate of 0,5 - 1 log unit per day. Using small amounts of virus as inoculum (1 - 10 mouse LD_{50} of CPD_{50}), by the 8th day of cultivation the amounts were 10^4 - 10^5 times greater. Langat (TP 21), Japanese encephalitis, St.Louis encephalitis and yellow fever viruses multiplied in these cultures to a lesser extent.

Kyasanur Forest Disease, Powassan, Omsk hemorrhagic fever, and West Nile viruses multiplied relatively well and both subtypes of TBE virus and louping - ill virus multiplied very well. The highest titres of the viruses were again obtained with small inocula (1-10 mouse LD_{50} per 0,03 ml), the virus increment being approximately of the order of 0,5 - 1 log unit per day. Viruses other than arboviruses (EMC, polio, vaccinia, NDV and pseudorabies), did not multiply in tick tissue cultures. An exception was LCM virus which multiplied in tick cells very well, similar to viruses transmitted in nature by ticks (Rehacek, 1965b).

The results of cultivation of CTF virus in cultures of *D. andersoni* have shown about a 4½ log units increase of extracellular virus from 6 - 10 days up to 4 - 5 weeks of cultivation. The virus was recovered in diminishing quantities for as long as 159 days in the medium and 166 days in triturated tissues (Yunker and Cory, 1967). Titers in whole nymphs fed on viremic hamsters remained about the same level from drop - off to molting, but titers in cultures prepared from these ticks increased about five log units 2 weeks after seeding. It was shown that a tissue culture from *D. andersoni* is a very sensitive system for the detection of small amounts of CTF virus, because less than 0,1 of suckling mouse icer LD50 can be propagated to high titers. The 98 and 124 days propagation of the CTF virus (strain Florio 2) in tick tissue cultures did not alter the virulence of the virus.

H. dromedarii tissue cultures were also successfully used for cultivation of the Tribec virus (Kemerovo group of arboviruses). Starting from the second day post - infection the virus titer had increased about 4 log units. These values were detectable until the 14th days(the conclusion of the experiment). A clear - cut specific fluorescence was seen in the cytoplasm of both epithelial - like and fibroblast - like cells from the second to seventh day after infection. At the beginning only a few fluorescing granules with perinuclear localization were found, but by the fourth day they increased in number and blended together into larger, bright fluorescing masses, filling the whole cytoplasm (Rehacek *et al.*, 1969) (Fig. 1).

The Lanjan virus propagated in cultures of *H. dromedarii* increased from 1,4 log units on the 1st day after inoculation to 4,8 log units on the 10th day after inoculation. Twenty days after infection when the experiment was discontinued, the virus titre was 3,5 log units. Quaranfil virus increased in the same culture from a titre of 1,2 log units at 2 days after inoculation to 3,5 log units on 10th day after inoculation. At 20 days after infection the virus titre was still 2,4 log units (Varma and Pudney, 1969b).

In all experiments with the viruses mentioned above, the appearance of tick tissue cultures was not affected by viral infections as shown by the absence of CPE and inclusions. In comparison with vertebrate tissue cultures all tested viruses multiplied slowly in tick tissue cultures perhaps due to the low temperature of cultivation.

b) Cultivation of the TBE virus, western type.

Since the TBE virus, western type, is spread in several parts of Europe and sometimes causes serious disease, much attention was paid to this virus in experiments.

No differences were noted between the rate of multiplication of virus strains either adapted or not adapted to HeLa cells in *H. dromedarii* and *D. marginatus* tissue cultures, from which only the latter tick species is known as a vector of this virus (Řehaček, 1963).

In many experiments the dose of virus used successfully for the inoculation of tick tissue cultures was so small that it was not detectable in icer inoculated mice or HeLa cells. Therefore the sensitivity of chick embryo cells, which appeared to be the highest for the propagation of the TBE virus, was compared with that of primary tick tissue cultures. It was found that with inocula of 30 and 3 IFD_{50}, the percentage, of infected chick embryo cells and tick cell cultures was about equal. With inocula containing 0,3 and 0,03 IFD_{50} of the virus, the respective percentages of infected tick cell cultures were 76 and 18, whereas that of chick embryo cell cultures were 34 and 3 (Řeháček and Kožuch, 1964). The virus propagation occurred only in the cytoplasm of cells and was concentrated around the nucleus. The positive fluorescence in tick cells as well as the virus yields from these cells were in direct relation to the time following infection and the dose of inoculated virus (Fig. 2 and Table 1.) (Řeháček, 1965c).

The results indicate that tick cell cultures are actually one of the most susceptible systems for detecting small amounts of this virus.

c) Isolation of the TBE virus in tick cells *in vitro* from various materials collected in nature.

The test materials from nature (blood and suspensions of brains of various animals and ticks) were separately inoculated into 2 - 5 days old tick cell cultures (without washing or change of medium followed by 7 - 9 days incubation) with chick embryo cell cultures, the materials were left to adsorb for 2 hours, then washed, and supplied with fresh medium and incubated for 5 days. Following the selected intervals, the culture fluids from either tick or chick embryo cell cultures were intracerebrally inoculated into suckling mice, which were observed for developing symptoms of infection. Of a total 187 samples, five strains of TBE virus were isolated: one from the blood of *Talpa europaea*, one from the blood of *Apodemus flavicollis* and three from *I. ricinus* ticks. All the strains were isolated by both methods. Toxic effects of mammalian blood or brain and tick suspensions were not observed in tick or in chick embryo cell cultures. The results obtained indicate that tick tissue cultures are as susceptible to TBE virus and as suitable for isolation experiments as chick embryo cell cultures (Řeháček and Kožuch, 1969).

TABLE 1.

Dectection of TBE virus in H. dromedarii tissue cultures

Inoculum	Virus yields and immunofluorescence at days following infection		
	2	5	9
$6{,}5^{++}$	$3{,}5^{+}$	$5{,}5^{+}$	$5{,}5^{+}$
2,5	$1{,}5^{+}$	$4{,}0^{+}$	$6{,}5^{+}$
1,5	0,5	$3{,}0^{+}$	$4{,}5^{+}$
0,5	neg	$2{,}0^{+}$	$4{,}5^{+}$
0,05	neg	neg	$5{,}0^{+}$
0,005	neg	neg	neg

++ log LD_{50}/0,03 ml (mouse icer)

\+ positive immunofluorescence

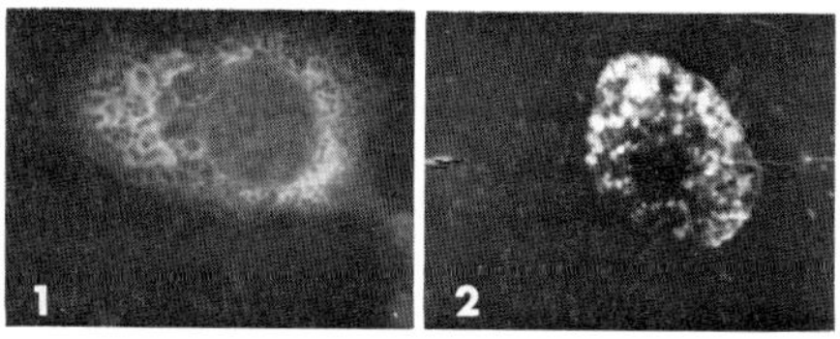

Fig. 1. Cell from *H. dromedarii* primary culture, 6 days after the infection with Tribeč virus (IMF method).

Fig. 2. Cell from *H. dromedarii* primary culture, 10 days after the infection with TBE virus (IMF method)

d) Maintenance of the TBE virus in tick cell cultures.

Tick tissue cultures of *H. dromedarii* were used as a substrate for the laboratory maintenance of the western type of TBE virus. Thirty four passages at weekly intervals were performed during a one year period. The virus titres varied between 3,0 and 6,5 log units during the passages. The virus titers at the 6,10 and 15th passages were at the same level, at the 20th passage the titer in subcutaneously injected mice was 0,5 log unit higher, but it decreased about one log unit at the 25th and 30th passage. It can be concluded from these results that TBE virus can maintained in tick tissue cultures as a virus stock for laboratory experiments without detectable change in its virulence (Řeháček, 1973).

e) Mixed infection with TBE and Kemerovo viruses in tick tissue cultures.

The problem of dual infection with arboviruses of the same biological vector arose with the discovery of mixed foci of arbovirus infections. The western type of TBE virus and the virus of Kemerovo complex both occur in *I. ricinus* ticks in Slovakia. This finding prompted us to examine whether these viruses can cause infection of tick cells and tissues *in vitro*. Virus amounts used as inoculum were 10^3 of TBE virus and 10^4 of Kemerovo virus. In cultures of *H. dromedarii* where the viruses were inoculated at intervals of 3 - 7 days no interference was demonstrated. However, a marked mutual interference in multiplication of both viruses as documented by growth curves and decreased percentage of infected cells was found in cultures inoculated simultaneously. This effect seems to be related to the period of virus adsorption on cell surfaces or to the pinocytosis phase and is certainly not related to the production of interferon (Libíková and Řeháček, 1974).

9. Prospects of tick primary cultures in virology

Almost all problems concerning interactions between cells and viruses can be investigated in primary tick cultures, however, the same problems can be studied in in tick all lines and the preferential use of the latter is expected in the future.

V. Passage of tick cell cultures.

1) General considerations.

Attempts to establish continuous tick cell lines had been unsuccessful for a long time. One of the most promising types of culture for establishing cell lines was proposed to be embryonic tissues and cells, or cells from male and female genital glands; both offering an abundance of cells in an active mitotic state.

2) Culture of embryonic cells.

Only two contributions are concerned with the establishment of embryonic cell cultures. The first culture was prepared from tissues and cells taken from *H. asiaticum* eggs (Medvedeeva *et al.*, 1972). The best source of cells was thought to be the eggs kept at 18 - 20°C for 10 - 14 days after oviposition. The cultures primarily of cells in suspension in the medium of Mitsuhashi and Maramorosch were subcultured 10 times before dying off (Medvedeeva, pers. com.).

The second embryonic cell culture was reported from *B. microplus* ticks (Pudney *et al.*, 1973). The cells were obtained from the eggs laid by females the surface of which were sterilized before oviposition with a 1: 10 solution of Roccal containing 1% benzalkonium chloride. The cells were cultivated in HLH medium with 10% fetal calf serum. About a week after seeding, almost all tissue explants had attached to vessel surface and large granular fibroblast - type cells appeared between the explants. Vesicles and tube like forms were common in the cultures. Four weeks later the surface of the vessels were covered with closely packed, distinct, round or flattened epithelial - type cells. Subcultures were done successfully in Leibovitz L - 15 medium supplemented with 10% tryptose phosphate broth and 15% fetal calf serum. Most cells had diploid chromosome numbers, 21 for male and 22 for female. The cells died off after 20 subcultures.

3) Cell culture from ovaries of female ticks.

Among the various kinds of tissues and cells of adult ticks tested *in vitro*, most attention has been paid to ovarial tissues, which produce well - formed multiplying epithelial cells.

Investigations on the cultivation of ovaries from *D. andersoni* (Řeháček, 1971), *D. pictus* and *D. marginatus* were done recently in our laboratory. The medium used was HLH with 10% fetal calf serum previously heated at 56°C for 60 minutes. The small pieces of ovaries taken from partially engorged females were cultivated at 28 - 30°C in T flasks. After migration of a few fibrobalst - type cells during the first days, discrete small colonies of epithelial - type cells appeared on days 7 - 10, which further slowly increased in the number. Two - four weeks later these colonies became very dense covering almost completely the surface of the culture vessels (Figs 3-4). Most of these cells were subcultured 5 - 9 times, but, after the 7th - 9th passage the cells ceased to multiply and became very large and granular and died off.

4) Cultivation of viruses in tick embryonal cells.

Of the cultures mentioned above, only embryonal cells of *B. microplus* employed in virus replication studies were found to support propagation of Dugbe virus. The chronic infection with this virus was established without visible inclusions in the cells and without visible cytopathic effect (David - Vest, 1974).

5) Prospects of tick passage cultures in virology.

Obtaining cells for repeated subcultures is relatively easy and holds promise of an emergent cell line. The establishment of perfect monolayers of these cells in culture vessels offers a suitable substrate for the cultivation of various pathogens as well as viruses. However, most virologists will prefer to use the simplest type of cultures, i.e. established cell lines.

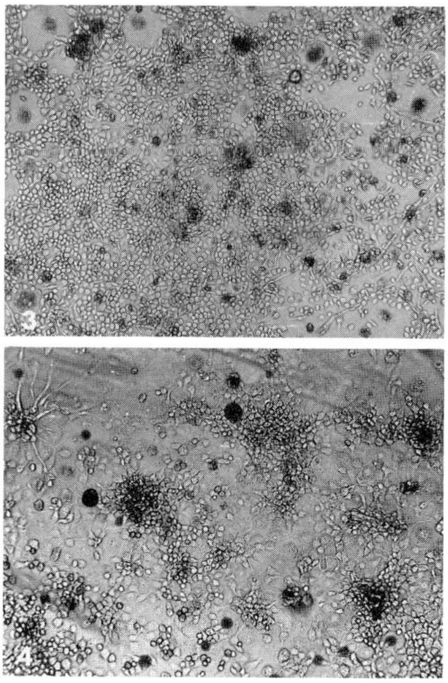

Fig. 3-4. D. pictus cell culture from ovaries, 7th passage.

VI. Cell lines from tick tissues.

1) General considerations.

Most attempts to cultivate tick tissues and cells *in vitro* is directed to the establishment of cell lines.

2) Cell lines from the tick R. appendiculatus.

The first tick cell lines were established from *R. appendiculatus* ticks (Varma *et al.*, 1975). The method used for the preparation of tissues and cells for the establishment of cell lines was the same as used for primary cultures.

Of a total 24 primary cultures seeded, 3 became established as cell lines, i.e. TTC - 219, TTC - 243 and TTC - 257. They are maintained in Leibovitz L-15 medium supplemented with 10% tryptose phosphate broth and 10% fetal calf serum. In March 1974 (the time when the paper concerning the establishment of cell lines was submitted for publication) the cells were at the 54th, 41st and 34th subcultures.

In the early subcultures the epithelial cells were predominant. With progressive subcultures the cells tended to aggregate in dense clumps which provided most of the cells for further passages. The cultures consist of round or epithelial - type cells with irregular outlines and spindle - shaped cells. Chromosome preparations of cells showed mixed ploidy. Most of cells namely in TTC - 243 were diploid with the male chromosome complement of 2n = 21 and the female complement of 2n = 22 chromosomes.

3) Cultivation of arboviruses in R. appendiculatus cell lines.

Preliminary results of cultivation showed that in cell line TTC - 243 West Nile virus increased in titer from $10^{1,4}$ on the 1st day to $10^{3,7}$ on the 10th day after inoculation, the Langat virus from $10^{1,4}$ on the 1st day to $10^{5,4}$ on the 4 - 6th day, the louping - ill virus from $10^{1,6}$ on the 1st day to $10^{5,6}$ on the 4th day, and the Quaranfil virus from $10^{2,8}$ on the 1st day to $10^{4,6}$ on the 4th day. All these viruses multiplied in cells without producing any detectable cytopathic effect.

4) Prospects of tick cell lines in virology.

The establishment of three cell lines by Varma *et al.*, 1975, provides evidence that is promising for the development of tick cell lines for use in investigating cell interactions with a variety of microorganisms.

VII. Conclusions

Tissue cultures prepared from blood-sucking arthropods have progressed as a tool in arbovirology during the last few years. This has been accomplished mostly by the establishment of several mosquito cell lines and during this year by the establishment of tick cell lines. The preferential use of arthropod cell cultures in arbovirology is supported by their ease of preparation and high sensitivity to pathogens.

At present most specialists in invertebrate cell culture research have abandoned the use of organ and primary cultures because of the laborious methods or preparation. More attention is directed toward the development with cell lines. Although they are ideal from the standpoint of easier cultivation methods, the genuine character of arthropod cells has often changed during the serial passage of cells. Sometimes such changes may not have an effect on the cultivation of viruses, but it may be proposed that to explain the relationship of vectors to pathogens, it would be more plausible to use explanted organs rather than dispersed and possibly transformed cells. However, at present there is no scientific basis for this suggestion.

Although good primary cultures have often been prepared from various tick species, they may not be widely used in microbiological research because of preference for established cell line.

Significant progress has been noted in the area of tick tissue culture with regard to virus studies. Perhaps the most promising from the standpoint of practical virology are the findings of *in vitro* tick cultures that could support the reproduction of viruses and their higher sensitivity to viruses in comparison to that of vertebrates tissues and cells *in vivo* and *in vitro*.

The use of tick tissue culture in virology appears promising for the isolation of viruses from natural sources, for eventual differentiation among arbo and other viruses, for the investigation of eventual changes in the properties of viruses following their cultivation in tick tissues and for the establishment of attenuated strains of viruses for possible vaccine production.

In multiplication of agents to be used for the preparation of antigens and for other practical application could include cultivation of tick or other arthropod pathogens which might be of value in arthropod control programs.

It is most likely that the tick tissue cultures, (primarily tick cell lines), will assume a position in modern virology and there is optimism that many investigations utilising these techniques will yield important results in the various areas of microbiology.

VIII. References

Blašković, D., and Řeháček, J. (1962). *Biological Transmission of Disease Agents, Academic Press, Inc.* , New York, p. 135.

Cory, J., and Yunker, C.E. (1971). *Ann. Ent. Soc. Amer. 64,* 1249.

David · West, T.S. (1974). *Arch. Ges. Virusforsch. 44,* 330.

Hoffmann, G., and Köhler, G. (1968). *Zeitschr. Parasitenkunde, 31,* 8.

Hoffman, G., Schein, E., and Jagow, M. (1970). *Z. Tropenmed. Parasitol. 21,* 46.

Kordová, N., and Řeháček, J. (1969). *Acta virol. 3,* 201.

Libíková, H., Řeháček, J., and Rajčáni, J. (1974). *Čs. Epidem. 23,* 332.

Martin, H.M., and Vidler, B.O. (1962). *Exp. Parasitol. 12,* 192.

Medvedeeva, G.I., Beskina, S.R., and Grokhovskaya, I.M. (1972). *Med. Parasitol., Moscow, 41,* 39.

Pudney, M., Varma, M.G.R., and Leake, C.J. (1973). *J. Med. Ent. 10,* 493.

Řeháček, J. (1958a). *Ph. D. Thesis.*

Řeháček, J. (1958b). *Acta Virol. 2,* 253.

Řeháček, J. (1962). *Acta Virol. 6,* 188.

Řeháček, J. (1963). *Ann. Epiphyties, 14,* 199.

Řeháček, J. (1965a). *J. Med. Ent. 2,* 161.

Řeháček, J. (1965b). *Acta Virol. 9,* 332.

Řeháček, J. (1965c). Actual problems of viral infections. XII. Scient. meeting of the Institute of poliomyelitis and viral encephalitides, Moscow, p. 459.

Řeháček, J. (1969). *Proc. 2nd Int. Congr. Acarology, 1967,* p. 455.

Řeháček, J. (1971). In: Arthropod cell cultures and their application to the study of viruses. *Curr. Topics in Microbiology and Immunology, 55.* (E. Weiss, ed.). Springer-Verlag, New York, p. 32.

Řeháček, J. (1973). *Proc. 3rd Intern. Coll. Inv. Tissue Culture, Smolenice,* (1971), p. 439.

Řeháček, J., and Brzostowski, H.W. (1969a). *J. Insect Physiol. 15,* 1431.

Řeháček, J., and Brzostowski, H.W. (1969b). *J. Insect Physiol. 15,* 1683.

Řeháček, J., and Hána, L. (1961). *Acta Virol. 5,* 57.

Řeháček, J., and Kožuch, O. (1964). *Acta Virol. 8,* 470.

Řeháček, J., and Kožuch, O. (1969). *Acta Virol. 13,* 253.

Řeháček, J., and Pešek, J. (1960). *Acta Virol. 4,* 241.

Řeháček, J., Rajčáni, J., and Gresikova, M. (1969). *Acta Virol. 13,* 439.

Sidorov, V.E. (1960). *Zh. Mikr. Epid. Immunol., Moscow, Nr. 6,* 91.

Varma, M.G.R., and Pudney, M. (1969a). *Proc. 2nd Intern. Congr. Acarology, 1967,* p. 637.

Varma, M.G.R., and Pudney, M. (1969b). *Int. Symp. Tick-borne arboviruses (excluding group B),* Smolenice 1969.

Varma, M.G.R., and Pudney, M. (1973). *Proc. 3rd Intern. Colloq. Invert. Tissue Culture,* Smolenice, (1971), p. 135.

Varma, M.G.R., Pudney, M., and Leake, C.J. (1975). *J. Med. Ent. 11,* 698.

Varma, M.G.R., and Wallers, W. (1965). *Nature (London), 208,* 602.

Weyer, F. (1952). *Zbl. Bakt. Orig. 1, 159,* 13.

Yunker, C.E., and Cory, J. (1965). *J. Parasitol. 51,* 686.

Yunker, C.E., and Cory, J. (1967). *Exp. Parasitol. 20,* 267.

Chapter 3

MORPHOGENESIS OF SINDBIS VIRUS IN CULTURED MOSQUITO CELLS

D.T. Brown, J.F. Smith, J.B. Gliedman,

B. Riedel, D. Filtzer, and D. Renz

I. Introduction

We have undertaken an extensive comparative study of the development of the group A arbovirus Sindbis in cultured vertebrate and invertebrate cells. This study which began nearly four years ago is presently examining the genetics, biochemistry, molecular biology, and morphology of virus infected tissue cultured cells of the vertebrate and invertebrate hosts. Sindbis virus is ideally suited for a comparative study of this type as this particular group A virion has been extensively studied in the vertebrate cells (Pfefferkorn and Shapiro, 1974). Our experiments have been carried out with cultured chick embryo fibroblast (CEF) and baby hamster kidney (BHK-21) cells as representatives of the vertebrate system, and continuous cultured larvae cells of *Aedes albopictus* originally prepared by Singh (1967) and provided by Drs. I. Snyder (Walter Reed) and S. Buckley (Yale University) and The American Type Culture Collection. In each case the observations for the two vertebrate cell systems were similar as were generally the observations made in the mosquito cells obtained from the three sources. Some differences were detected in the invertebrate cell lines primarily with respect to growth rate and the amount of virus produced. The reasons for these differences are not clear. In an attempt to place the following discussion in an appropriate frame of reference it will be necessary to first briefly summarize results obtained in our laboratory and by others with vertebrate host cells.

II. Results and discussion

Sindbis is a structurally and chemically simple virus which has single stranded RNA as its genetic material. The RNA genome is contained within a capsid which has icosahedral symmetry and is composed of 92 identical subunits (Acheson and Tamm, 1967; Brown and Gliedman, 1973) which contain multiple copies of a single protein having a molecular weight of 30,000 (Scheele and Pfefferkorn, 1970). The capsid is itself enclosed within a membrane, the lipid moiety of which is derived from the host membranes. The viral membrane in turn contains two "envelope proteins" which are coded for by the virus and are glycosylated in a pattern which suggests that the sugars are added by host enzymes (Sefton and Burge, 1974). The envelope proteins reside on the outer surface of the viral envelope and are anchored into the envelope by a small hydrophobic region which protrudes into the outer member of the envelope bilayer (Utermann and Simons, 1974). The morphogenesis of Sindbis virions in vertebrate host cells takes place in three distinct steps (see Fig. 1). The viral polypeptides are synthesized from a species of RNA which has the same polarity as the RNA contained in the infecting virion but is only one third the size (26 S or interjacent RNA) (Simmons *et al.*, 1972). The structural proteins are produced as a single polypeptide from which the capsid protein is cleaved before the nascent polypeptide is released from the polyribosome complex (Scheele and Pfefferkorn, 1970; Schlesinger and Schlesinger, 1973). The capsid protein is rapidly incorporated into viral cores which are seen free in the cell cytoplasm in the electron microscope (Acheson and Tamm, 1967; Brown and Smith, 1975). The remainder of the polypeptide is destined to be sequentially cleaved and glycosylated, and the products are inserted into the plasma membrane of the infected cell (Schlesinger and Schlesinger, 1973; Schlesinger *et al.*, 1972; Sefton and Burge, 1974; Sefton *et al.*, 1973). The insertion of virus specific proteins into the vertebrate plasma membrane occurs rapidly and can be detected by antigenic probes and heamadsorbtion hours before infectious virus is released. The membrane of the infected cell contains one of the proteins recovered in the viral envelope (E_1) and a precursor to the other (PE_2) (Jones *et al.*, 1974; Sefton *et al.*, 1973). After the insertion of the partially processed virus protein into the plasma membrane the viral nucleocapsids migrate to the inner surface of this structure and attach to some unidentified virus specific component associated with it (Brown and Smith, 1975). The association of the nucleocapsid with the inner surface of the host plasma membrane seems to be strong as capsid-membrane complexes can be isolated from cells in which the attached nucleocapsids remain bound to the membrane after purification by density gradient centrifugation (Brown and Smith, 1975). The final stage in virus morphogenesis occurs as the virus nucleocapsid is wrapped in a fragment of the modified host membrane as it is released into the surrounding media (Acheson and Tamm, 1967; Brown *et al.*, 1972; Brown and Smith, 1975).

Little is known regarding the mechanism of this "budding" process. A final cleavage of one of the virus precursor polypeptides ($PE_2 \longrightarrow E_2$) (Jones *et al.*, 1974; Schlesinger and Schlesinger, 1972; Sefton *et al.*, 1973) and the displacement of host-membrane proteins in the portion of the viral membrane destined to become the viral envelope seen to accompany budding (Figs. 1,2) (Brown *et al.*, 1972). The attachment of the capsid to the plasma membrane does not in itself cause the final processing cleavage of the precursor polypeptide (PE_2) or the displacement of host-membrane proteins (Brown *et al.*, 1972; Brown and Smith, 1975; Jones *et al.*, 1974). The reactions following the attachment of the nucleocapsid to the plasma membrane appear to move sequentially around the nucleocapsid moving it from a position in the cell cytoplasm to one outside of the cell. This process is accompanied by the cleavage of the precursor polypeptide, the displacement of the host membrane proteins and the intimate association of the capsid protein with the developing viral envelope. The growth of Sindbis virus in cultured vertebrate cells produces gross cytopathic effects and the replicative cycle is terminated by death and lysis of the host cell 14-20 hours after infection.

A number of investigations of Arbovirus growth in cultured mosquito cells, have revealed the rate and amount of virus produced in this system to be similar to that observed in cultured vertebrate cells. The cultured mosquito cells, unlike their vertebrate counterparts, are generally able to survive infection by Arboviruses and

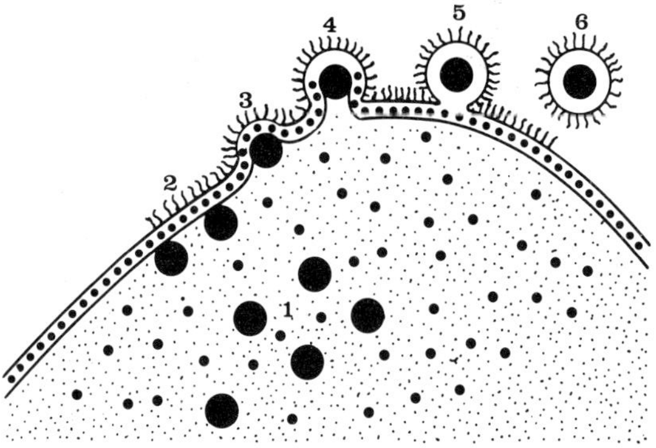

Fig. 1. Schematic representation of Sindbis virus morphogenesis in BHK or CEF (vertebrate cells). 1. Nucleocapsids are assembled in the cytoplasm of the infected cell. 2. Completed nucleocapsids attach to inner surface of host plasma membrane which has been altered by the addition of virus proteins (spikes), the membrane contains interior membrane host-glycoprotein beads. 3. Capsid begins to bud through plasma membrane which still contains glycoprotein beads. 4. A more advanced stage of budding, interior membrane beads are missing from some areas of the developing envelope. 5. Partial release of virion by fusion of inner leaflet of envelope. Virion is still attached to host by continuity of outer leaflet. 6. Free virion interior of envelope membrane is free of glycoprotein beads. Compare to Figs. 2 and 12.

infection ultimately results in the establishment of a persistently infected cell population which, in terms of its growth characteristics, is similar to non-infected cells (Buckley, 1969; Peleg, 1969; Stevens, 1970; Raghow *et al.*, 1973).

In our laboratory maximum yields of virus (ca. 4,000 PFU/cell) were obtained from the mosquito cells at 21-36 hours post infection. After this initial burst of virus production the titer of the culture media falls by about one log and remains at this lower level for several days. An investigation of the morphogenesis of Sindbis virus in the cultured mosquito cells by electron microscopy was undertaken in the hope that the lack of cytocidal effects of virus infection in these cells could be in part explained by differences in the morphology of the infected cells at the ultrastructural level.

The first noticeable morphological event occurring in the infected mosquito cells was the appearance of membrane-rich vesicular structures in which, as time progressed, viral nucleocapsids could be found (Figs. 3,4). The large numbers of cytoplasmic nucleocapsids readily found in infected vertebrate host cells (Acheson and Tamm, 1967; Brown and Smith, 1965) were not found in the infected mosquito cells. Freeze-etching of the vesicular structures in the infected mosquito cells revealed that the internal membranes contained the classical distribution of interior membrane glycoprotein beads (Marchesi *et al.*, 1972; Tillack *et al.*, 1972) (Fig. 5). As time progressed the vesicles were found to contain many mature virions as well as partially mature forms and free nucleocapsids (Fig. 6). Envelopment of the nucleocapsids seemed to occur through interaction of the capsids with the membranes in the vesicles and not by budding of cytoplasmic nucleocapsids into the vacuoles. At times later than 20 hours after infection some cells contained many electron dense vesicles (Fig. 7) which when examined at high magnification proved to contain large numbers of mature virions (Fig. 8). The composition or origin of the

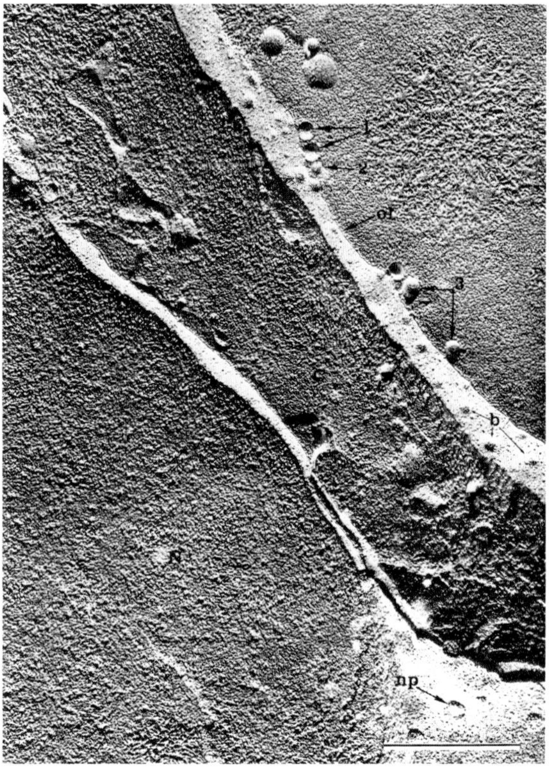

Fig. 2. An electron micrograph of a chick embryo fibroblast cell after infection with Sindbis virus. The cleaved plasma membrane (oL) has many nearly mature virions (1,2,3) which are still attached to the cell by the continuity of their envelopes with the cell membrane. A number of developing viral buds are apparent (b). At this time in development the interior of the membrane of the bud still contains the glycoprotein beads which are characteristic of the rest of the plasma membrane. The beads are not present in the interiors of the viral envelopes (1,2,3). C. cytoplasm N, nucleus, Np nucleopore. Magnification bar is 0.5 μm, from Brown et al. 1972 by permission of the American Society of Microbiology.

electron dense material which completely surrounds and partially obscures the mature virions in these vacuoles is not understood.

The electron dense, virus-containing vacuoles described here are similar to those described by Raghow and coworkers in mosquito cells infected with Semliki forest virus and Ross River virus (both group A virions) (Raghow *et al.*, 1973). Raghow and coworkers suggested that these virions do not contribute to extracellular titres as they are destroyed when the vesicle fuses with lysozomal structures. We found no evidence for destruction of these vesicles but rather found that the virus produced in these structures is released from the host by fusion of the vesicular structure with the plasma membrane. As the vesicles fuse with the cell surface the uniform interior appeared to break up releasing virions with a coating of the electron dense

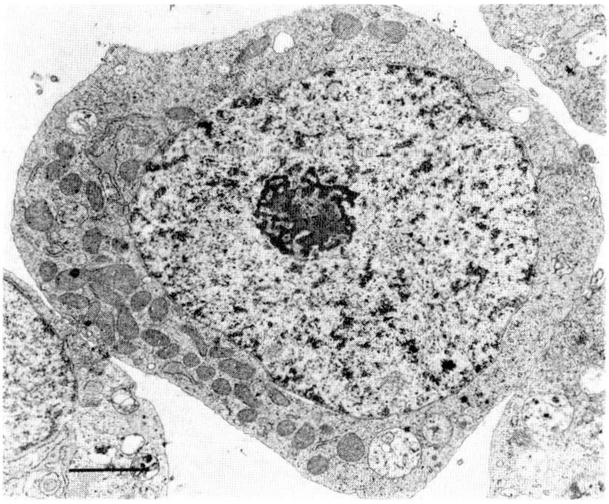

Fig. 3. An uninfected *Aedes albopictus* cell after ultrathin sectioning. Magnification bar is 1 μm.

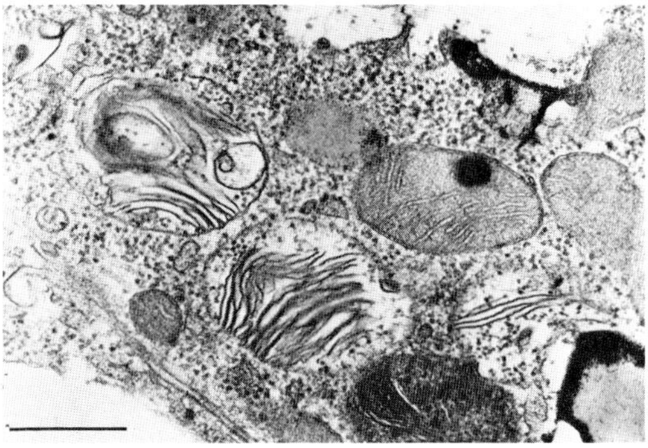

Fig. 4. Ultrathin section of vesicles found in the cytoplasm of an *Aedes albopictus* at 4 hours post infection with Sindbis virus. The vesicles contain many transverse membranes and ribosomes. The membranes to the right of the photograph are not yet contained in a limiting membrane. Magnification bar = 0.5 μm. This and all subsequent experiments were made with a multiplicity of 50 PFU virus/cell and were incubated at 28°C.

material (Fig. 9). This electron dense material was not strongly bound to the new virions as examination of the particles before and after density gradient purification revealed them to be free of surface contamination (Fig. 10). The virions produced in the first forty hours after infection were morphologically indistinguishable from virus produced from BHK or CEF cells. At later times the infected cells produced, in addition to the normal size virion, two smaller particles 80% and 59% the size of the normal virion (Brown and Gliedman, 1973). These particles were found to be noninfectuous for both BHK and mosquito cells when purified away from the normal particles. These particles could not be found in uninfected mosquito cells in spite of a number of attempts to induce them chemically and by irradiation. Infection of

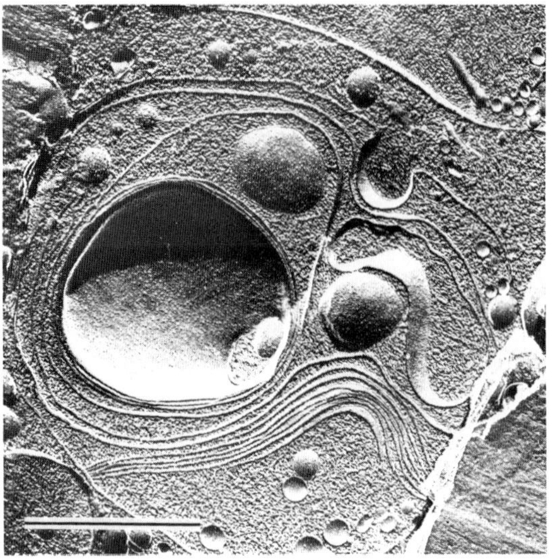

Fig. 5. A cytoplasmic membrane-containing vesicle similar to those in Fig. 4 after freeze-etching. The membranes within the vesicle contain a typical distribution of interior glycoprotein beads (compare with Fig. 2). Magnification bar = 1 μm.

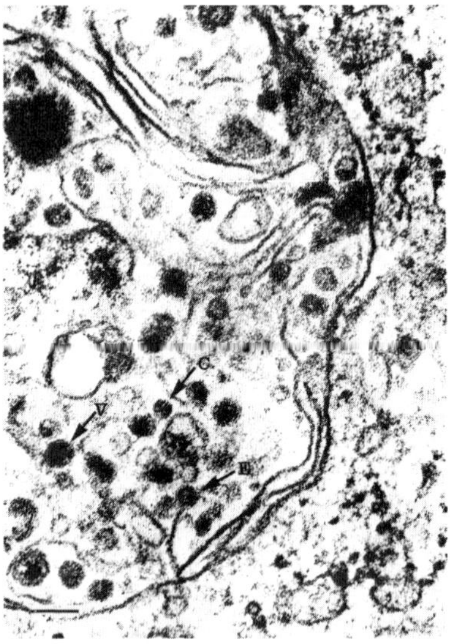

Fig. 6. A vesicle found in infected *Aedes albopictus* cells at 12 hours after infection with Sindbis virus. The vacuole contains virus nucleocapsids (C), partially enveloped virions (E) and mature virions (V). Magnification bar = 100 nm.

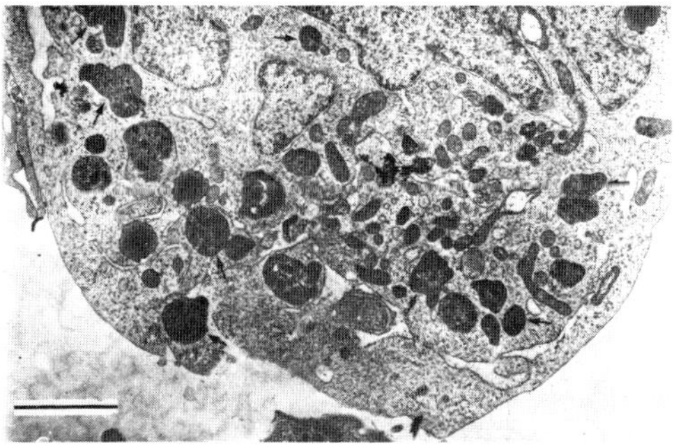

Fig. 7. *Aedes albopictus* cells typical of those found at 14-21 hours after infection. The cell contains a large number of electron dense bodies which examined at higher magnification contained large numbers of virions (arrows). Magnification bar = 1 μm.

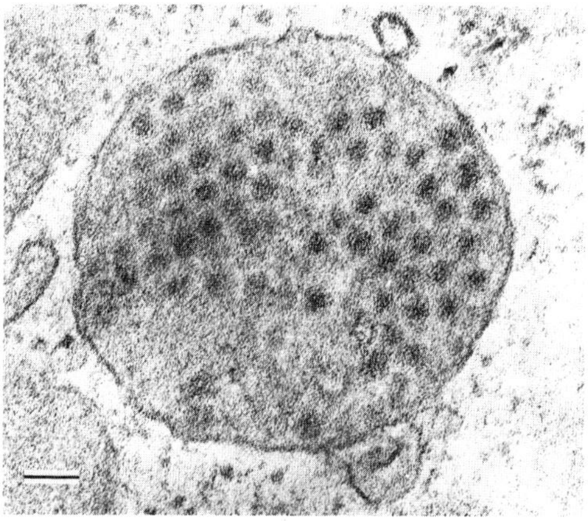

Fig. 8. A high magnification of a vesicle of the type seen in Fig. 7. The vesicle is filled with electron dense material in which mature virions are seen packed in a paracrystaline array. Magnification bar = 100 nm.

the *Aedes albopictus* cells with a number of Sindbis temperature sensitive mutants at nonpermissive temperature produced no particles of this type suggesting that Sindbis is not providing a helper function for some latent virus-like particle. Attempts to establish the presence of RNA in the smaller virions by labeling with uridine have been negative. We have not been able to determine if these particles play any role in the establishment of the persistent state of infection or if they are in any way related to the interfering agent produced by this infected cell system (Stollar and Shenk, 1973).

In less than one percent of the cells examined in this study a few cytoplasmic nucleocapsids were detected and occasionally capsids could be found near the surface of the cell in a configuration which was suggestive of budding (Fig. 11).

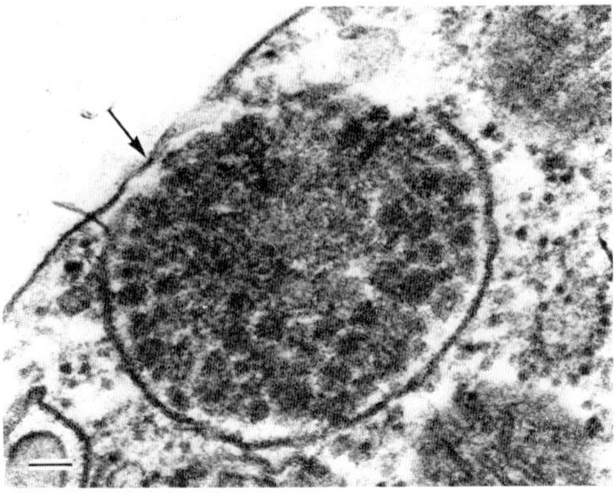

Fig. 9. A virus containing vesicle fusing with the cell surface (arrow). The uniform appearance of the vesicle is broken up during the process of virus release. The virions appear coated with the electron dense contents of the vesicle. Magnification bar = 100 nm.

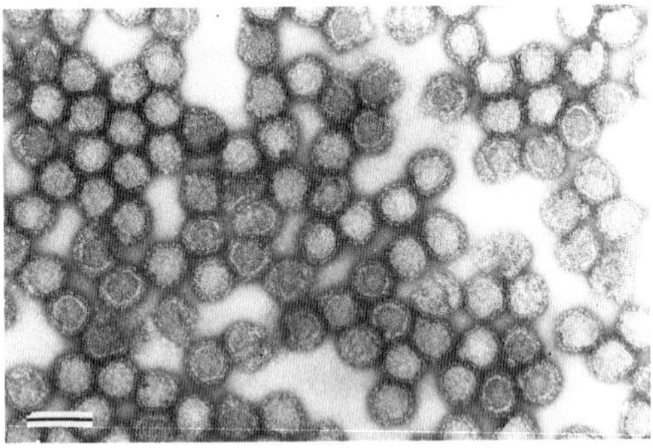

Fig. 10. Mature Sindbis virions from *Aedes albopictus* cells at 24 hours after infection. The virions were purified by isopycnic density gradient centrifugation in potassium tartrate, negatively stained with phosphotungstic acid. Magnification bar = 100 nm.

An exhaustive search for true budding figures utilizing freeze-etching techniques was negative. Considering the ease with which this process is detected in vertebrate cells utilizing either freeze-etching (Fig. 12) (Brown *et al.*, 1972), or thin sectioning (Acheson and Tamm, 1967; Brown and Smith, 1975) we conclude that under conditions employed in our laboratory little virus is produced by this route.

The release of mature virions from the infected cell by fusion of the vesicular structures with the cell surface and the observation that envelopment of the virions takes place in these vesicles presented us with the possibility that during the process of virus release partially enveloped structures might also be released from the cells. Examination of virus structures discharged into the growth medium in the electron microscope revealed a variety of intermediates in the process of envelopment (Fig. 13). These structures were found in rather large numbers in the

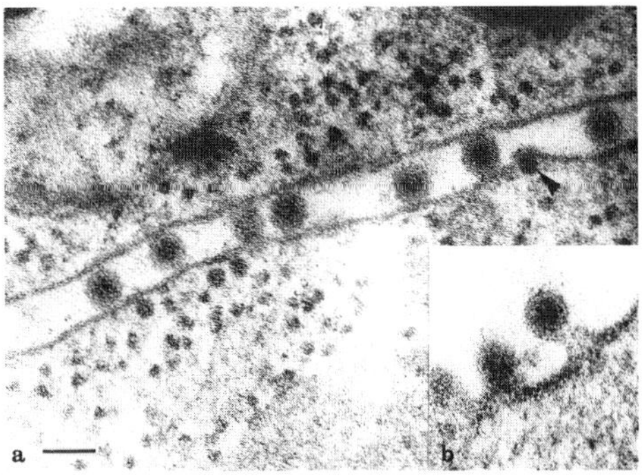

Fig. 11. Thin section of *Aedes albopictus* cells with surface-associated virions. a) Two infected cells. Mature virions are seen in the space between them. The lower cell has nucleocapsids in the cytoplasm. One of which is associated with the plasma membrane as though budding. b) An apparent "late stage" in budding of virions from the surface of an infected *Aedes albopictus* cell. Magnification bar = 100 nm.

growth medium and could be concentrated in potassium tartrate density gradients on the "light" side of the major virus band. Large fragments of membranes were seen with attached nucleocapsids. In some instances three sizes of capsids could be found attached to one side of the membrane (Fig. 13a). These structures probably correspond to the capsids of the three morphological types of virus produced by this cell system (Brown and Gliedman, 1973). The membrane surface away from the side of capsid attachment was free of spike-like structures suggesting that either the morphological change in the membrane resulting in the appearance of spikes has not taken place or that they had been lost from the membrane during purification. Membrane-nucleocapsid complexes could also be found which seemed to be in a somewhat more advanced state of maturation (Fig. 13b). These structures differed from that described above in that the nucleocapsids were partially enclosed in the membrane fragment as though envelopment had been interrupted. In addition, the outer surface of the membrane was coated with spike structures over its entire surface including those areas where no envelopment of virus was taking place. Large numbers of closed membraneous structures containing more than one nucleocapsid were also found (Fig. 13c) suggesting that the process of envelopment in the mosquito vesicles is not as precise as the budding process occurring on the surface of vertebrate host cells which do not produce multicapsid structures under normal conditions.

The total alteration of the surface morphology of the vesicular membranes after the attachment of the nucleocapsids suggested that in this system, unlike the vertebrate system, topological alteration of the membranes could take place in areas not actively involved in viral envelopment. That this is the case was also suggested by freeze-etching of the membranes in the virus-induced vesicles at times late after infection (Fig. 14). Some of the stacked membranes could be seen possessing larger than normal numbers of the interior membrane beads while other interior membrane surfaces were completely free of the particles. Examination of virus purified from the mosquito system revealed that like the vertebrate system the interior of the viral envelope is free of these interior membrane particles (not shown). These observations suggest that unlike the vertebrate cell system des-

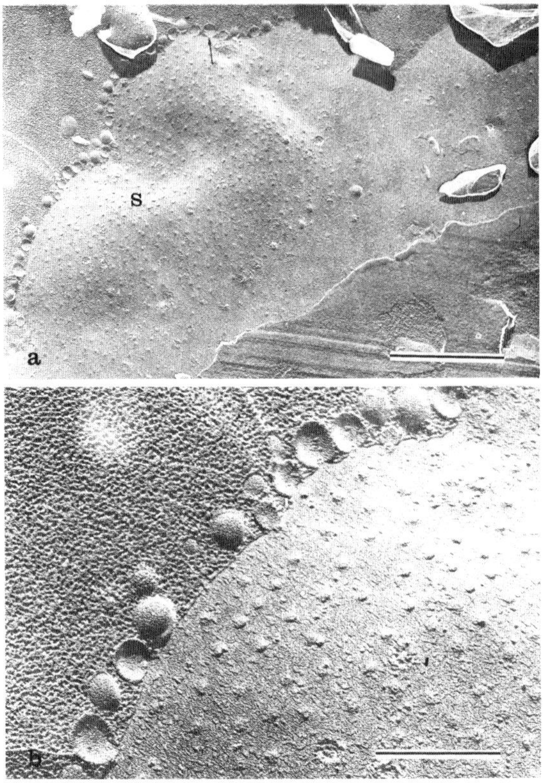

Fig. 12. A freeze-etch replica of the plasma membrane of a chick embryo fibroblast cell after infection with Sindbis virus. At the edge of the cell a large number of nearly mature virions can be seen (a). On the interior surface of the host plasma membrane many large stalks (s) (Figs. a + b) can be detected. These represent points of attachment of the viral envelope to the plasma membrane. They are left on the surface of the inner leaflet of the membrane bilayer after the outer leaflet is removed by the cleaving process. These stalks can also be seen at the base of some of the cleaved peripheral virions. Magnification bar a = 1.0 μm, b = 200 nm; from Brown *et al.* 1972 by permission of the American Society of Microbiology.

cribed above an entire membrane surface can be reorganized by the appearance of spikes on its surface and the complete loss of preexisting membrane proteins from the membrane interior.

As suggested at the outset of this discussion, one would like to be able to correlate these described morphological observations (summarized in Fig. 15) to the ability of the mosquito cells to survive infection by Arboviruses and to the establishment of the persistently infected cell population. In this respect the production of progeny virions in internally located vesicles is appealing as it may isolate some of the biochemical processes which are essential to virus production but

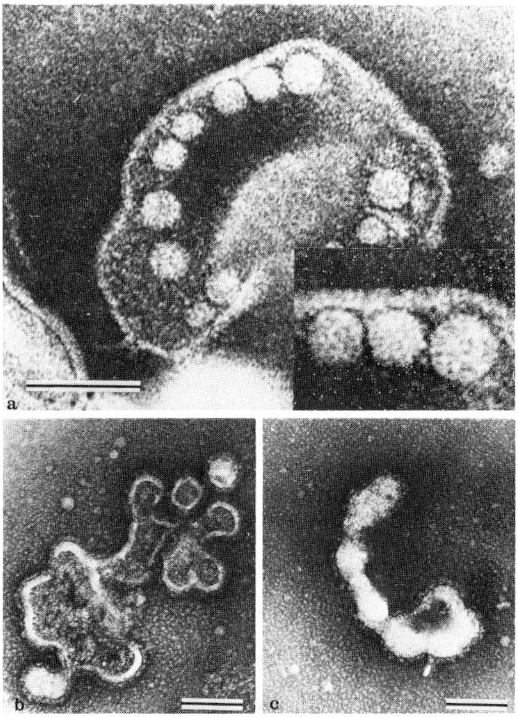

Fig. 13. Immature membrane capsid complexes released from Sindbis infected *Aedes albopictus* cells. a) Large membrane fragment with attached nucleocapsids. The capsids have 3 distinct sizes and are made up of distinct subunits (insect). b) Membrane fragment with partially enveloped capsids and spikes. c) A multicapsid containing viral envelope. Magnification bars = 100 nm.

toxic to the host-cell. The presence of ribosomes in the vesicular structures suggests that viral protein synthesis could occur in these structures. It has also been suggested that the presence of double-stranded RNA in cells may be responsible for the cytopathic effects produced by RNA viruses (Cordell-Stewart and Taylor, 1971; Garwes *et al.*, 1975). Thus, the sequestering of replicative intermediates to enclosed factories in the infected cell cytoplasm could minimize this toxic effect. The restriction of the process of envelopment of viral nucleocapsids to the vesicular structures could also eliminate the necessity of the modification of the host plasma membranes by insertion of virus proteins, a process which occurs early in infected vertebrate cells and might contribute to the overall cytopathic process. We have found that mosquito cells actively producing Sindbis virus do not heamadsorb goose red cells under conditions in which similarly infected BHK or CEK cells readily do so. Electron microscopy confirmed that nearly all of the cells in these experiments were infected, suggesting that the surface of the infected mosquito cells does not contain the viral heamadsorbing protein in detectable quantities.

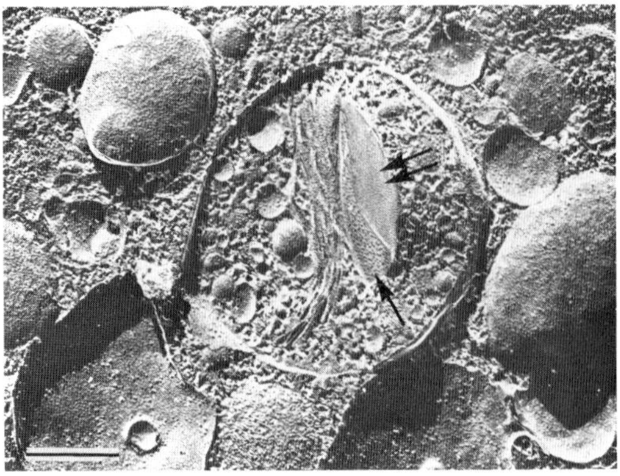

Fig. 14. A freeze-etch replica of a virus induced cytoplasmic vacuole 20 hours after infection with Sindbis virus. One of the vesicular membranes has a large number of interior membrane beads (single arrow) while another is smooth (double arrow) (compare to Fig. 5). Magnification bar = 1 μm.

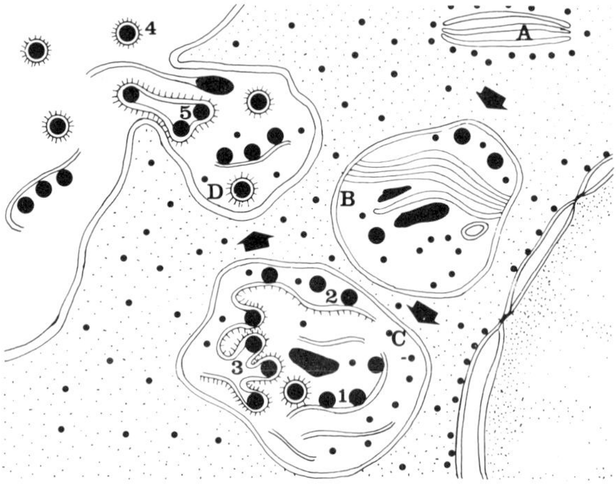

Fig. 15. A schematic diagram of the possible sequence of events occurring after infection of *Aedes albopictus.* Membranes are found within the cytoplasm (A) and are subsequently enclosed in a limiting membrane (B) forming a vesicle in which ribosomes and a few nucleocapsids attach to the membranes in the vesicles (1) and the membranes undergo a topological rearrangement resulting in the appearance of viral spikes (2). The attached nucleocapsids are enveloped by the modified membrane (3) producing mature virions. The virus-induced vesicle fuses with the cell plasma membrane (D) releasing mature virions (4), membranes with attached nucleocapsids (5), and intermediates in envelopment.

A proposal that sequestering of virus production to internal vesicles accounts for the ability of mosquito cells to survive infection is confused by the occasional observation of cytoplasmic nucleocapsids and what appear to be virions budding from the cell surface. It is difficult to assess at the level of electron microscopy the relative importance of these cells in the overall process of virus production.

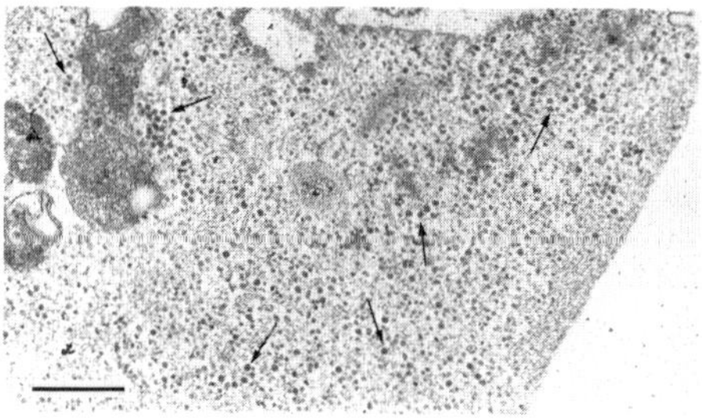

Fig. 16. Area of an *Aedes albopictus* cell after infection for 20 hours with Sindbis virus in the presence of colcemid. Numerous nucleocapsids are seen in the cell cytoplasma (a few are marked with arrows). Magnification bar = 0.5 µm.

It is possible, as Raghow and coworkers have suggested (Raghow *et al.*, 1973), that most of the cells synthesize cytoplasmic nucleocapsids and that they are immediately and efficiently transported out of the cytoplasm by budding through the plasma membrane. This entire process would have to occur much more rapidly than in the vertebrate host, in order to reduce appreciably the probability of observing this process in the electron microscope. Following this logic the few cells in which this pathway is morphologically suggested might represent a situation in which this process is slowed or arrested for physiological reasons. On the other hand, to suggest that the considerable amount of virus that is assembled in vesicles, which are easily detected in the electron microscope, does not contribute to the extracellular virus, is to imply that a system which is so efficient on the one hand in maturing cytoplasmic nucleocapsids by budding through the plasma membrane is completely wasteful on a major portion of the total virus it produces. One must, of course, consider models which propose that both processes of virus production do occur and that whether one pathway of virus production or the other prevails in a particular cell is decided by some unknown factors. In this regard we have found that treatment of Sindbis virus infected mosquito cells with colcemid results in the production of many free cytoplasmic nucleocapsids in most of the cells examined (Fig. 16). This treatment, however, did not increase the frequency of virus budding from the cell surface, but did result in some reduction in the total virus yields.

III. Conclusion

An electron microscope study of infected cells by pulse chase autoradiography with protein and RNA precursors is in progress. Hopefully, such a program will clarify the location of protein and RNA synthesis in the infected cells and further establish the intracellular fate of these products after synthesis.

Acknowledgments

This research was supported by the Deutsche Forschungsgemeinschaft SFB 74.

IV. References

Acheson, N.J. and Tamm, I. (1967). *Virology 32,* 128.

Brown, D.T. and Gliedman, J.B. (1973). *J. Virol. 12,* 1534.

Brown, D.T., Waite, M.R.F. and Pfefferkorn, E.R., (1972). *J. Virol. 10,* 524.

Brown, D.T., and Smith, J.F. (1975). *J. Virol. 15,* 1262.

Buckley, S.M. (1969). *Proc. Soc. Exptl. Biol. Med. 131,* 25.

Cordell-Stewart, B., and Taylor, M.W. (1971). *Proc. Nat. Acad. Sci. U.S.A. 68,* 1326.

Garwes, D.J., Wright, P.J. and Cooper, P.D. (1975). *J. Gen. Virol. 27,* 45.

Jones, K.M., Waite, M.R.F. and Bose, H.R. (1974). *J. Virol. 13,* 809.

Marchesi, V.T., Jackson, R.L., Segrest, J.P. and Kaltane, I. (1972). *Federation Proceeding 32,* 1833.

Peleg, J. (1969). *J. Gen. Virol. 5,* 463.

Pfefferkorn, E.R., and Clifford, R.L. (1964). *Virology 23,* 217.

Pfefferkorn, E.R., and Shapiro, D. (1974). In: Comprehensive Virology (H. Fraenkel-Conrat and R. Wanger, ed.), Vol. 2, 171-230. Plenum Publishing Corporation, New York, London.

Raghow, R.S., Davey, M.W., and Dalgarno, L. (1973). *Archiv. Ges. Virusforschung 43,* 165.

Raghow, R.S., Grace, T.D.C., Filshie, B.K., Bartely, W. and Dalgarno, L. (1973). *J. Gen. Virol. 21,* 109.

Scheele, C.M., and Pfefferkorn, E.R. (1970). *J. Virol. 5,* 329.

Schlesinger, M.J., and Schlesinger, S. (1973). *J. Virol. 11,* 1013.

Schlesinger, M.J., Schlesinger, S. and Burge, B.W. (1972). *Virology 47,* 539.

Schlesinger, S., and Schlesinger, M.J. (1972). *J. Virol. 10,* 925.

Sefton, B.M., and Burge, B.W. (1974). *J. Virol. 12,* 1366.

Sefton, B.M., Wickus, G.C. and Burge, B.W. (1973). *J. Virol. 11,* 730.

Simmons, D.T., and Strauss, J.H., Jr. (1972). *J. Mol. Biol. 71,* 599.

Singh, K.R.P. (1967). *Current Science 36,* 506.

Stollar, V., and Shenk, T.E. (1973). *J. Virol. 11,* 592.

Stevens, T.M. (1970). *Proc. Soc. Exper. Biol. Med. 134,* 356.

Tillack, T.W., Scott, R.E., and Marchesi, V.E. (1972). *J. Exptl. Med. 135,* 1209.

Utermann, G., and Simons, K. (1974). *J. Mol. Biol. 85,* 569.

Chapter 4

ESTABLISHED MOSQUITO CELL LINES AND THE STUDY OF TOGAVIRUSES

V. Stollar, T. E. Shenk, R. Koo, A. Igarashi, and R. W. Schlesinger

I. Introduction

Those members of the togavirus group which are transmitted by insects have the remarkable and uncommon ability of being able to multiply in species as far apart in the evolutionary scale as mosquitoes and man. The bite of an infected mosquito and the simultaneous transmission of virus can result in serious illness and sometimes death, in man as well as in other mammalian species. Examples of diseases caused by togaviruses and transmitted by mosquitoes are Eastern and Western equine encephalitis, among the alphavirus subgroup and dengue, yellow fever, and Japanese encephalitis among the flavivirus subgroup.

During the past years much has been learned about the structure, the replication, the molecular biology, and the morphogenesis of the togaviruses, especially about certain of the alphaviruses. In the laboratory, Sindbis and Semliki Forest viruses have proven very useful model systems and have the advantage of being relatively avirulent for man. In a recent review, Pfefferkorn and Shapiro (1974) have collected and summarized what is known of the structure and replication of the togaviruses. Most of the experimental systems, however, have employed vertebrate cell hosts, either whole animals such as mice, or cultured cells derived from one of several different species. Relatively speaking, much less is known about the interaction between togaviruses and the mosquito host. The establishement by Singh, and by Peleg of cell lines derived from *Aedes albopictus* and *Aedes aegypti* represented a great step forward, and one that made it possible to use insect tissue culture systems for the study of the replication of togaviruses.

The availability of these mosquito cell lines will also undoubtedly prove useful for the study and identification of new viral agents, some of which will likely prove pathogenic for man.

Results and discussion

The system we have chosen to study most intensively is that of Sindbis virus replication in the *Aedes albopictus* cell line of Singh (1967). Some reference will also be made to experiments with the *Aedes aegypti* cell line of Peleg (1968).

Our Sindbis virus stocks, originally derived from the HR strain of Burge and Pfefferkorn (1966), have been plaque-purified and grown in chick cells at a low initial multiplicity in order to exclude the production of defective viral particles.

The *Aedes albopictus* cell line (Singh) derived from mosquito larvae was maintained originally in the medium devised by Mitsuhashi and Maramorosch (1964) (MM medium) but more recently has been adapted to a medium composed of 9 parts Eagle's minimal medium (Eagle, 1959) (including non-essential amino acids) and 1 part MM medium. In each case fetal calf serum was added to a final concentration of 5 to 10%.

II. The acute infection of A. albopictus cells with Sindbis virus.

1. Growth curve of Sindbis virus in Aedes albopictus cells.

It has generally been observed that in the face of high virus yields, and viral antigen in the majority of the cells, infected mosquito cell cultures infected with alphaviruses continue to grow, and do not show any obvious cytopathic effect (CPE), (Stevens, 1970).

Figure 1 demonstrates a growth curve of Sindbis virus in *A. albopictus* cells maintained at 28°C. The latent period, the kinetics of replication, and the final yield do not differ substantially from what we observe when virus is grown in chick embryo fibroblasts (CEF) or in hamster cells. No cytopathic effect was visible, in keeping with previous observations made on mosquito cells infected with Sindbis virus or other alphaviruses (Davey *et al.*, 1973). Other experiments in our laboratory have shown 1) that the infected mosquito cells continue to grow at a rate comparable to that of the uninfected cells, and 2) that by 24 hours after infection at least 75-80% of the cells contain viral antigen as demonstrated by the fluorescent antibody method.

Since the infected mosquito cells are not killed but continue to grow, a chronically infected culture results, with continuous virus production. Such chronically infected cultures have been maintained in our laboratory for up to 2 years. Some properties of these cultures and of the virus produced by them will be described in a later section.

2. Lack of sialic acid in Sindbis virus grown in mosquito cells.

Togaviruses are composed of RNA, protein, carbohydrate in the form of sugar residues on the envelope glycoproteins, and lipid. The viral RNA and protein are both specified by the viral genome. In contrast, the evidence is good that both the lipid and carbohydrate components of the viral membrane are determined largely by the host cell. To expand, precisely which sugar residues are added to the viral glycoproteins is probably a function both of the nature and the activity of the cellular glycosyl transferase enzymes. It might be expected that comparative studies of Sindbis virus grown in mosquito and vertebrate cells would show the effect of such host modifications much more dramatically than if one compared Sindbis virus grown in 2 different vertebrate species.

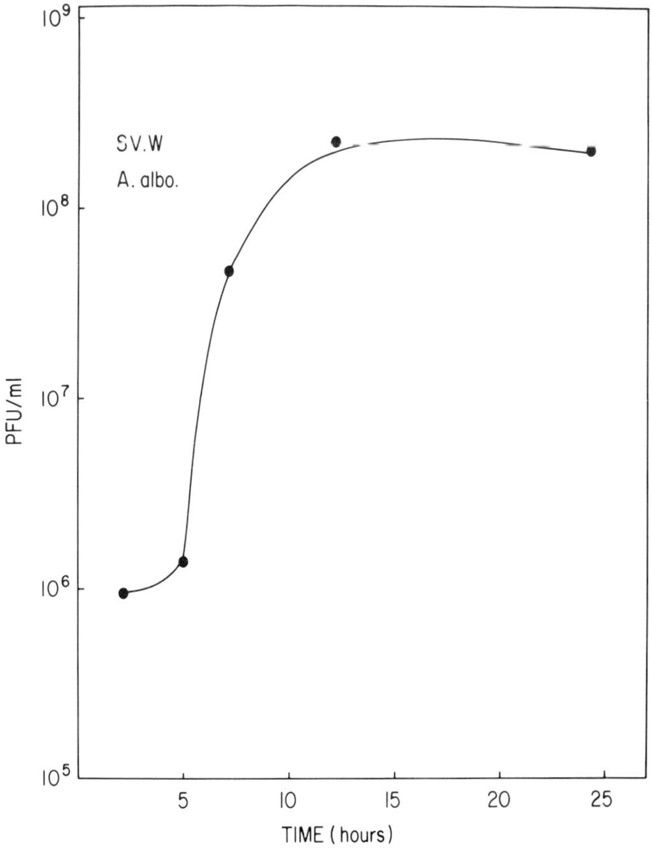

Fig. 1. Growth of Sindbis virus in *Aedes albopictus* cells at 28°. The input multiplicity was approximately 30 pfu/cell.

Warren (1963) and Schlesinger *et al.* (1961) have shown that in contrast to vertebrate cells, nearly all invertebrate cells and tissues examined lack sialic acid. If the addition of sialic acid to viral glycoprotein is a "host function" it seemed likely that Sindbis virus grown in mosquito cells would also lack sialic acid and thus differ remarkably from virus grown in BHK21 or chick cells.

Before examining virus, we wished to be certain that the cultured mosquito cells indeed did lack sialic acid. In the first experiment, *A. albopictus* cells, and hamster cells (BHK-21) were grown in the presence of radioactive glucosamine (a good precursor of sialic acid in vertebrate cells) (Kraemer, 1967). The cells were harvested, washed, and then incubated with neuraminidase. The release of TCA-soluble counts (above the control value) was taken as evidence for the presence of sialic acid. In the case of the hamster cells, substantially more TCA soluble counts were released upon incubation with neuraminidase than were released in buffer only; when mosquito cells were treated under similar conditions, there was no difference in the TCA soluble radioactivity released from control cells and from cells incubated

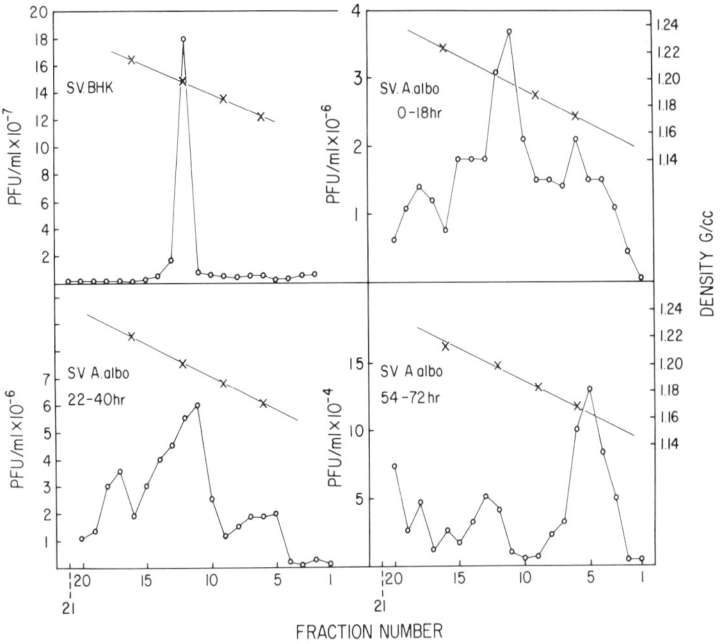

Fig. 2. Sucrose-D_2O equilibrium gradients of Sindbis virus from BHK cells and *A. albopictus* cells:

BHK cells were infected at an input multiplicity of approximately 50 pfu/cell and maintained at 37°. Culture medium was harvested after 22 hours.

A. albopictus cells were also infected with an input multiplicity of 50 pfu/cell, but in this case the medium was changed several times so as to collect the virus yields between 1 and 18 hours, 22 and 40 hours, and 54 and 72 hours.

All virus samples were extracted once with Genetron (trichlorotrifluoroethane) and clarified by low speed centrifugation.

A sample (0.8 ml) of each preparation was then layered over 11.2 ml 14-40% sucrose (w/w) in TNE buffer made up in D_2O (Shenk and Stollar, 1973). Centrifugation was in the SW-41 rotor at 10° for 15 hours at 31,000 rpm. Fractions were then collected (0.6 ml) and assayed for plaque formation on BHK-21 cell monolayers.

with neuraminidase (Table 1). Although these results were consistent with the absence of sialic acid in mosquito cells, for confirmation we turned to a more direct chemical assay.

BHK-21 and *A. albopictus* cells were harvested, washed three times with PBS, and then assayed for sialic acid using the thiobarbituric acid method of Warren (1959). BHK cells contained 0.37 μm/ml packed cells (Table 2), a value similar to that obtained by others with these cells. Turning to the *A. albopictus* cells, we found that even after three washings with PBS, sialic acid was usually present when the cells were grown in MM medium containing 10% fetal calf serum. Since the serum component of the medium contains high levels of sialic acid, we thought it likely that the "cellular" sialic acid we were measuring was derived from the serum and was perhaps tightly bound to the cells. Consistent with this idea was the observa-

TABLE 1

Release of acid soluble radioactivity from glucosamine labeled cells

	TCA soluble cpm released from	
Enzyme used	*BHK cells*	*A. albopictus cells*
Cl. perfringens neuraminidase	63,696	159,077
	60,384	157,378
Vibrio cholerae neuraminidase	48,494	160,994
	52,502	152,539
Buffer only	22,036	163,922
	22,903	156,509

BHK cells and *A. albopictus* cells were labeled with ^{14}C-1-glucosamine medium for 48 hours. The ^{14}C-1-glucosamine (56.5 mc/mM) was added to a final concentration of 1 μc/ml. Cells were scraped from petri plates and washed 5 times with cold phosphate buffered saline, until the counts in the supernatant did not decrease further. Cells were then suspended in 8 ml of buffer A (0.5M sodium acetate pH 5.5 .15M NaCl, .009M $CaCl_2$) and duplicate 1 ml samples were incubated with *Cl. perfringens* neuraminidase (Worthington Biochemical Corp.) (0.1 mg), *V. cholerae* neuraminidase (Behring Diagnostics) (0.1 ml stock solution), or buffer alone. As measured by TCA precipitable cpm, the *A. albopictus* cell suspension contained approximately 4 times as much radioactivity as the syspension of BHK21 cells. After 30 minutes at 37° the cells were pelleted, and TCA added to the supernatant (final concentration 5%). The TCA treated supernatant was applied to a glass fiber filter and the filtrate collected. One ml of filtrate was added to 10 ml of counting liquid (1 vol. Triton-X and 2 vol. Permafluor-Packard) and counted in a scintillation counter.

tion that if the serum concentration in the medium was reduced, the level of the "cellular" sialic acid also fell. This was not so in the case of the BHK-21 cells. The situation was further clarified after it was found that the *A. albopictus* cells could be grown in MM medium without any serum. Under such conditions, the rate of growth was slower, and there was a marked tendency of the cells to aggregate; nevertheless, the cells did continue to grow for several weeks which was as long as we attempted to maintain them. Analysis of these cells showed no detectable sialic acid.

From these experiments, and those in which the cells were labeled with radioactive glucosamine, we conclude that the cultured *A. albopictus* cells like nearly all invertebrate cells do not contain sialic acid.

If cells do not contain sialic acid, it seemed probably that they would also lack sialyl transferase, the enzyme which attaches sialic acid moieties covalently onto glycoprotein. This was indeed the case. In contrast to chick cells no sialyl transferase activity was found in cultured *A. albopictus* cells. This was so whether the enzyme incubation was carried out at 28°C or at 37°C (data not shown).

We next proceeded to see whether Sindbis virus grown in mosquito cells contained sialic acid. This was done by growing virus in the presence of ^{14}C-labeled glucosamine and testing for the release by neuraminidase of TCA-soluble radioactivity. In the case of virus grown in chick or hamster cells between 9% and 11% of the input TCA precipitable radioactivity was released in TCA-soluble form (Table 3).

TABLE 2

Sialic acid content of media and cells

Materials assayed	μM/ml medium or packed cells
Fetal calf serum	
not dialysed	5.5
dialysed	4.8
4% bovine serum albumin	.01
MM medium with 5% fetal calf serum	.32 .37
MM medium without serum	< .01
BHK cells grown with 10% serum	0.37
overnight with 0.1% serum	0.37
overnight with 0.2% BSA	0.32
A. albopictus cells grown with 5% FCS	0.17
overnight with 0.2% FCS	0.09
A. albopictus cells grown without serum	
in suspension cultures	< .01
as monolayers	< .01

Cells to be assayed for sialic acid were harvested by scraping, washed 3 to 5 times in PBS and finally pelleted at 1200 rpm for 5 minutes.

In sharp contrast, no significant counts were released from Sindbis virus grown in mosquito cells. Thus, glucosamine did not label any moieties in the mosquito grown virus which were hydrolyzable by either of the 2 neuraminidase proportions tested. These results taken together are strong evidence that the presence or absence of sialic acid in the Sindbis virion, and by analogy that of other sugars as well, is dictated by the capabilities of the host cell.

3. Heterogeneity of Sindbis virus produced in A. albopictus cells.

What of the physical properties of the virus produced in mosquito cells? Previous experiments in our laboratory have shown that the progeny virus from hamster or CEF infected with our wild type virus banded sharply with a density close to 1.20 g/cc in sucrose-D_2O equilibrium gradients. Such virus could be monitored by radioactive label (e.g. ^{3}H-uridine) hemagglutinating activity or infectivity. When Sindbis virus grown in mosquito cells was examined in a similar way, a rather different picture was seen. Figure 2 shows sucrose-D_2O gradients of Sindbis virus, in which infectivity was used to identify virus. As noted above, virus from chick cells gave a sharp homogeneous peak. Virus taken from mosquito cells showed increasing heterogeneity with time after infection. At 18 hours, the main infective peak was at 1.20 g/cc, with lesser peaks of both greater and lesser density. Much the same picture was seen at 40 hours. By 72 hours, the predominant peak was at 1.17 g/cc with only small peaks at densities of 1.20 g/cc and 1.22 g/cc.

Observations by Brown and Gliedman (1973) describing size heterogeneity of virus particles from *A. albopictus* cells infected with Sindbis virus may be related

TABLE 3

Release of acid soluble radioactivity from glucosamine labeled sindbis virus

Origin of Virus	BHK Cells		Chick Embryo Fibroblasts		A. albopictus Cells	
Neuraminidase used	C[(a)]	P[(b)]	C	P	C	P
TCA precipitable cpm added	3014	3147	3074	3354	5623	6212
Acid soluble +Enz.	307	422	312	378	38	86
CPM released −Enz.	50	49	42	47	35	31
Δ	257	373	270	331	3	55
% of input cpm released	8.3	11.9	8.8	9.9	< 0.1	0.9

(a) Vibrio cholerae neuraminidase

(b) *Clostridium perfringens* neuraminidase

Sindbis virus was gorwn in BHK, CEF, and mosquito cells in the presence of ^{14}C-1-glucosamine (56.5 mc/mm) (5 μc/ml medium). Virus was harvested after 24 hours in the case of BHK and chick cells and 48 hours in the case of mosquito cells. Unlabeled carrier virus (from BHK cells) was added and the virus purified by equilibrium centrifugation in sucrose-D_2O gradients. Fractions containing virus (density about 1.20 g/cc) were pooled and used as substrate for neuraminidase. Incubation mixtures contained 0.1 ml of ^{14}C-1-glucosamine labeled virus, neuraminidase (50 μl of the cholera vibrio enzyme, or 50 μg of the *Cl. perfringens* enzyme) in a total volume of 1 ml containing buffer A (table 1) in the case of cholera vibrio enzyme or 0.1M sodium acetate pH 5.1 in the case of the *Cl. perfringens* enzyme. After incubation for 30 minutes at 37° 0.2 ml 30% TCA was added and the TCA filtrate collected. One ml of filtrate was added to 10 ml of counting liquid (table 1).

to our findings of density heterogeneity. In addition to normal sized particles (62 nm in diameter) they found 2 smaller class particles (52 nm and 39 nm in diameter); the latter were also noted to increase with time after infection. Physical separation and the biological properties of the smaller classes of particles have not yet been described. The basis of the density-heterogeneity of infective particles which we have seen is also unclear, but could possibly be due to factors such as aggregation, multiploid particles, or attachment of cellular membrane fragments to virions.

III. Aedes albopictus cells persistently infected with Sindbis virus.

Because of the lack of CPE after infection of mosquito cells by Sindbis virus, infection is invariably followed by the establishment of chronically infected cultures, cultures which will be referred to as *A. albopictus* (SV-C). After the initial infection, peak titers are usually reached between 24 and 72 hours, depending on the initial multiplicity of infection. Thereafter the titer slowly drops over a period of days to weeks usually leveling off at about 10^5-10^6 pfu/ml. depending also on how the chronically infected cells are maintained and split.

TABLE 4

Temperature-sensitive plaque formation on BHK cell monolayers by sindbis virus from chronically infected cultures of A. albopictus (a)

Incubation Temp. (C)	SV-C (PFU/ml)	EOP(b)	SV-W(b) (PFU/ml)	EOP(b)
28	4.1×10^6	1.0	1.6×10^8	1.0
34	4.4×10^6	1.1	1.7×10^8	1.1
37	8.2×10^4	2.0×10^{-2}	1.7×10^8	1.1
39.5	<10	$< 2.4 \times 10^{-6}$	1.6×10^8	1.0

(a) Adsorption was at 28 °C, after which the cultures were incubated at the temperatures indicated for 48 hours.

(b) Values represent the titer at the indicated temperature divided by the titer at 28 °C.

1. Generation of temperature-sensitive small plaque mutant virus.

While periodically monitoring virus levels in the *A. albopictus* (SV-C) cultures it was noted after a period of several months that the progeny virus gave rise predominantly and then exclusively to small plaques. Such plaques are illustrated in Fig. 3 and are to be contrasted with plaques formed by the wild type virus.

Further characterization of the virus (SV-C) revealed that it was temperature-sensitive both with respect to virus yield (Fig. 4) and plaque formation (Table 4). Twenty viral plaques were picked from BHK cell monolayers infected with SV-C, and were grown into small stocks on BHK cell monolayers. Both procedures were carried out at 34°C.

The titers of some of the SV-C clones and their efficiencies of plating at 34°C and 39.5°C are shown in Table 5. In each case the frequency of revertants (EOP at 39.5°/EOP at 34°) was less than 7×10^{-5}, and in some cases was not detectable.

Of further interest is the observation that with one exception the SV-C clone (19 out of 20) were RNA$^+$ at the non-permissive temperature. Experimental values for selected SV-C clones are shown in Table 6.

One additional property found in all the ts virus clones tested was increased thermal lability when incubated at 60°. Such an increase in thermal lability was also seen with the RNA$^+$ ts mutants of Sindbis virus described by Burge and Pfefferkorn (1966), and is usually taken as signifying and alteration in one of the viral structural proteins.

Selected viral clones were tested for their ability to complement one another (data not shown), but no complementation was observed. Under the same conditions Sindbis virus mutants ts-4 and ts-10 (Kindly provided by Dr. Elmer Pfefferkorn) did complement each other as expected. The failure to complement could be due to the presence of multiple genetic lesions, or could be due to the fact that the RNA$^+$ clones all have lesions in the same cistron. In the particular case of SV-C-2 which is RNA$^-$ (implying a defect in the RNA synthesizing enzymes) and also thermally labile (implying a defect in one of the structural proteins), there are probably at least 2 genetic lesions.

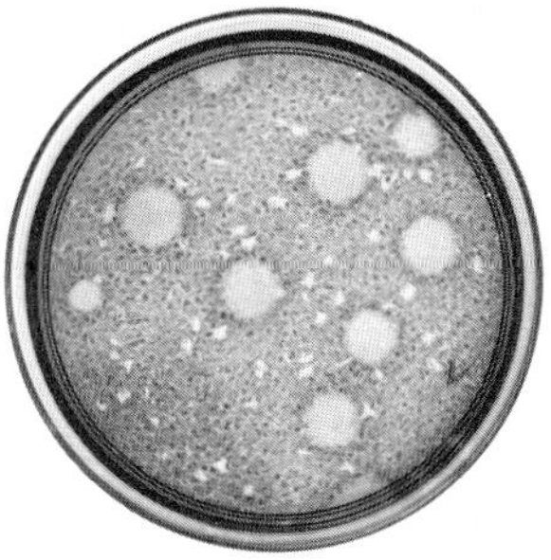

Fig. 3. Plaques produced on BHK-21 cell monolayer by SV-W and by Sindbis virus from a chronically infected culture of *A. albopictus* (SV-C). SV-W produces large, round plaques averaging 8 mm in diameter. SV-C produces irregularly shaped plaques measuring 1-2 mm. Monolayers were stained with neutral red 72 hours after addition of virus.

The evolution of temperature-sensitive virus and the properties of derivative plaque-purified clones are now being examined in greater detail in our laboratory.

2. *Resistance to superinfection with the homologous virus.*

By the time several weeks or months have passed after the initial infection of *A. albopictus* cells with Sindbis virus the level of infective virus may be only 10^{-3} or 10^{-4} of the peak titer achieved within the first few days. As the virus titer declines the proportion of cells containing viral antigen (as detected by fluorescent antibody) also drops, from 80-100% to a very low level of only 1% or 2%. In spite of this marked decline both in viral titer and in antigen-positive cells, the SV-C cultures as a whole are resistant to superinfection by the homologous virus (i.e. Sindbis), but not to a related by heterologous virus (e.g. Eastern equine encephalitis virus). Fig. 5A shows that in normal *A. albopictus* cells SV and EEE-V both replicate rapidly and to high titer. In the case of the *A. albopictus* (SV-C) cultures the small plaques produced by SV-C enabled us to differentiate the endogenously produced virus from the superinfecting EEEV or SV. Thus, it could be clearly shown that whereas EEEV could replicate to almost normal titers in the chronically infected cultures, superinfecting SV did not increase in titer at all.

At which stage in viral infection does this inhibition of the homologous virus occur? It is known that vertebrate cells infected with SV synthesize viral dsRNA with a sedimentation constant of about 22S. As shown in Fig. 6, the same is true of normal *A. albopictus* cells infected with SV or EEEV. Normal uninfected cells contain no detectable dsRNA by this method. When the *A. albopictus* (SV-C) cells were examined, they were found to contain much less dsRNA than the acutely infected cells; but instead of only a 22S peak there was also a significant peak which sedimented more slowly (about 15S) (Fig. 7). This pattern of dsRNA is reminiscent of that seen when chick or hamster cells were infected with virus stocks containing defective-interfering particles.

Superinfection with wild-type SV failed to alter this RNA pattern suggesting that the block to superinfection occurred at an early stage, before the superinfecting SV was able to induce any viral RNA synthesis. Whether or not the superinfecting virus was absorbed normally to the chronically infected cells has not been determined.

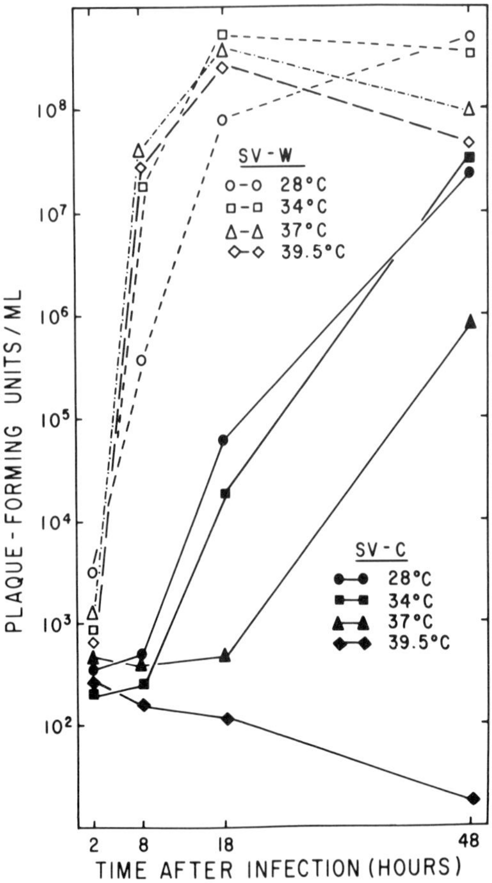

Fig. 4. Growth curves of SV-W and SV-C in chick cells incubated at 28, 34, 37, and 39.5 C. Prior to infection, phosphate-buffered saline and viral inocula were heated to 42 C. The virus inocula were warmed only for 2 min to prevent inactivation. The CEF monolayers were washed once with PBS and then were infected with either SV-C or SV-W at a multiplicity of infection of 1 PFU/cell. After adsorption for 1 h at 39.5 C, the inocula were removed, the plates were washed three times with prewarmed PBS, and prewarmed medium was added. Plates then were incubated at the appropriate temperature, and samples for plaque assay were taken at the times indicated. SV-W 0, 28 C; □, 34 C; ▵, 37 C; ◇, 39.5 C. SV-C: ●, 28 C; ■, 34 C; ▲, 37 C; ♦, 39.5 C.

What is the mechanism of the homologous interference to superinfection in the *A. albopictus* (SVOC) cultures?

Two possibilities can be suggested: (i) interference by temperature-sensitive viral mutants. In favor of this is the report describing efficient interference by SV-S (a plaque-purified ts mutant of SV) derived from cultures of *Aedes aegypti* cells chronically infected with Sindbis virus, (Peleg and Stollar, 1974) and (ii) interference

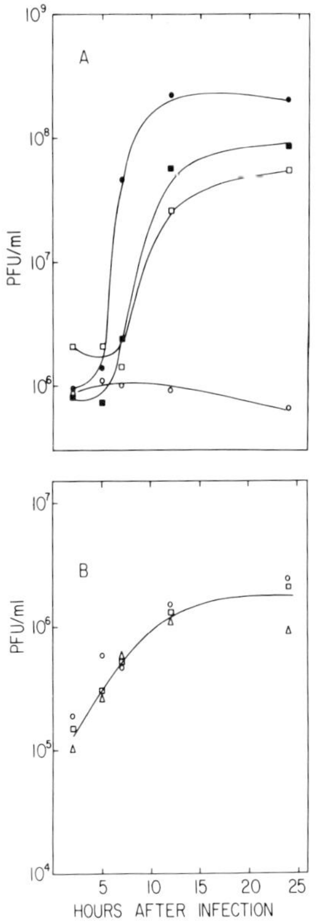

Fig. 5. Growth of SV-W and EEE in normal and SV-C cultures of *A. albopictus* cells and the effect of superinfection on the production of SV-C. A, Monolayer cultures of normal *A. albopictus* cells were infected with SV-W (●) or EEE (■) at input multiplicities of 20 and 80 PFU/cell respectively. Chronically infected cultures were infected with either SV-W (0) or EEE (□) at the same input multiplicities. Virus was adsorbed for 1 h at room temperature, after which the cultures were washed three times with phosphate-buffered saline. Medium was added, the cultures were incubated at 28 C, and samples were taken at the indicated times for plaque assay. Plaque assays were perfomed at 34 C on monolayers of BHK-21 cells. B, The production of SV-C in SV-C cultures of *A. albopictus* cells was monitored. Symbols: 0, not superinfected. □, superinfected with SV-W. Δ, superinfected with EEE; at input multiplicities of 0.02 and 0.08 PFU/cell, respectively.

by defective-interfering viral particles (or possibly by defective but replicating viral genomes in the cells of the chronically-infected cultures). As noted above, by analogy with our findings in vertebrate cell systems, the presence of a new smaller viral ds RNA species is quite suggestive of defectiveness.

But is there more direct evidence that DI particles of SV can be generated in mosquito cells? In recent experiments in our laboratory Igarashi serially passaged SV undiluted through *A. albopictus* cells in a manner similar to that used to generate DI particles in chick or hamster cells. After 30 passages no obvious cyclic variation or marked decrease in viral yield was noted. Further, cells infected with

TABLE 5

Frequency of revertants in stocks of sindbis virus mutant clones (a)

SV-C clone	Frequency of revertants (b)
SV-C-2	$< 10^{-7}$
SV-C-4	$< 10^{-5}$
SV-C-8	6.7×10^{-5}
SV-C-13	3×10^{-6}
SV-C-16	$< 10^{-6}$
SV-C-19	$< 10^{-6}$

(a) Cloned virus was obtained from well-isolated plaques, and a portion of this virus then was grown to a stock. Both the initial plaque purification and the growth of the stock were in BHK-21 cells at 34° C.

(b) Frequency of ts^+ revertants is the ratio of the titer measured at 39.5 °C to that measured at 34° C.

virus taken at various passage levels contained only 22S ds viral RNA and very little if any of smaller species.

In other experiments the response of *A. albopictus* cells to SV DI particles produced in vertebrate cells was examined. As noted above, when wild type virus (SV-W) was used to infect BHK or *A. albopictus* cells, only 22S ds viral RNA was produced (Fig. 8). However, SV-BP-18 (a stock produced in BHK-21 cells, and containing DI particles) which in BHK cells led to the synthesis of 22S and 12S dsRNA, induced only 22 dsRNA in the *A. albopictus* cells. Furthermore, whereas SV-BP-18 interfered well with SV-W in BHK cells no interference with SV-W could be demonstrated in *A. albopictus* cells. In other words, the DI particles in SV-BP-18 which are easily recognized by their effects in vertebrate cells appeared silent or inactive in mosquito cells.

IV. Isolation of a cell-fusing agent (CFA) from cultures of Aedes aegypti (Peleg)

It is becoming increasingly apparent that like established vertebrate cell lines, insect cell cultures may also contain "contaminating" viruses (Cunningham *et al.*, 1975). I would like briefly to summarize our findings with respect to an agent found in the *Aedes aegypti* cell line of Peleg (Stollar and Thomas, 1975).

In the course of experiments studying interference by SV-S (a small plaque ts mutant of Sindbis virus, derived from chronically infected cultures of *Aedes aegypti* cells) with the replication of SW-W, it was discovered that medium from "normal" *Aedes aegypti* cell cultures (Peleg) when applied to *Aedes albopictus* cells (Singh) led within 72 hours to marked syncytial formation (Fig. 9), and eventual death of most of the cell monolayers.

A plaque assay was developed for the CFA on *A. albopictus* cells (Fig. 10) thus facilitating its quantitation and enabling us to perform growth curve experiments. The CFA plaques may sometimes begin as "red plaques" with the cells in the early syncytia showing an increased uptake of neutral red. After another day or two such plaques become clear but may still be surrounded by a hyperstained red halo.

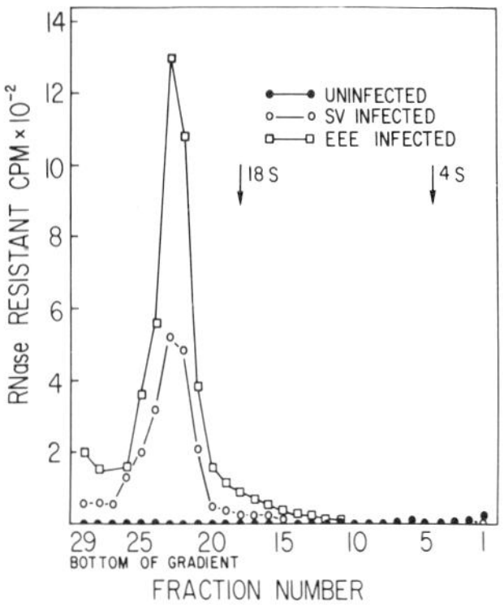

Fig. 6. Double-stranded RNA in normal cultures of *A. albopictus* cells infected with SV-W or EEE. Cultures which were uninfected or cultures infected with SV-W or EEE (input multiplicities of 20 and 80 PFU/cell, respectively) were labeled with uridine-5 ^3H (26.5 Ci/mM, 10μCi/ml) in the presence of actinomycin D(4μg/ml) from 6 to 11 h after infection. Cells were then harvested, and RNA was extracted. The RNA species were resolved by centrifugation in 5 to 20% linear sucrose gradients (27,000 rpm, 18.5 h, 4 C, Spinco SW-27 rotor). Each fraction was then treated with ribonuclease A, and the residual acid-precipitable radioactivity was measured. These preparations were centrifuged at the same time in different gradients and were drawn together by superimposing their marker RNA.

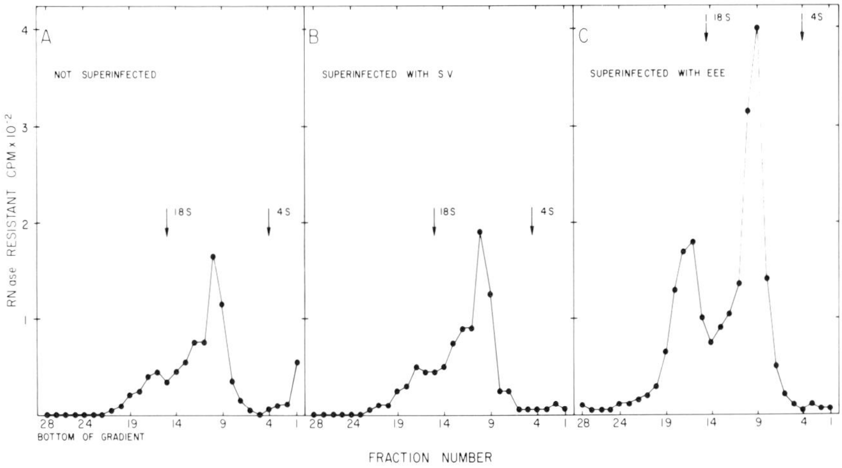

Fig. 7. Double-stranded RNA in SV-C cultures not superinfected (A) or superinfected with SV-W (B) or EEE virus (C). All procedures were as described in the legend to Fig. 6.

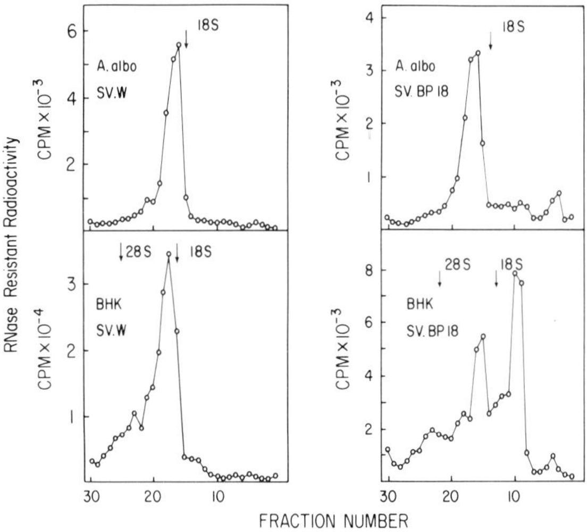

Fig. 8. DS RNA in BHK-21 or *A. albopictus* cells infected with SV-W or SV-BP-18:

Cultures were infected with the indicated virus stock at an input multiplicity of approximately 25 PFU/cell. BHK-21 cells were maintained at 37° and *A. albopictus* cells at 28°. Cultures were labeled with ^{3}H-uridine from 3 to 8 hrs after infection and in the presence of actinomycin D (1 µg/ml). RNA was extracted from the infected cultures and fractionated on sucrose velocity gradients. Individual fractions were then assayed for RNase resistant TCA precipitable radioactivity.

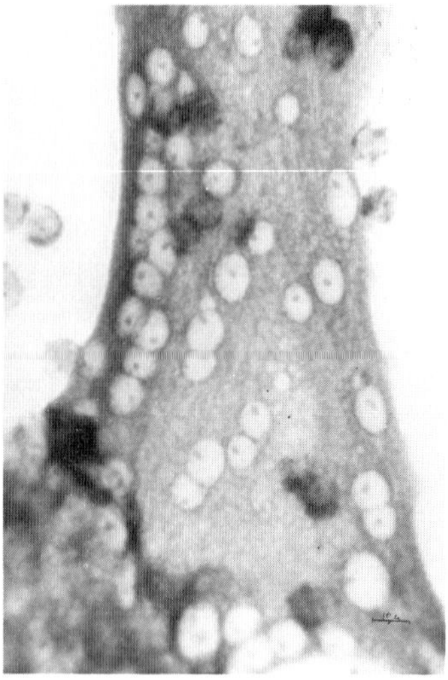

Fig. 9. Aedes albopictus cells infected with CFA 76 hours previously (X576).

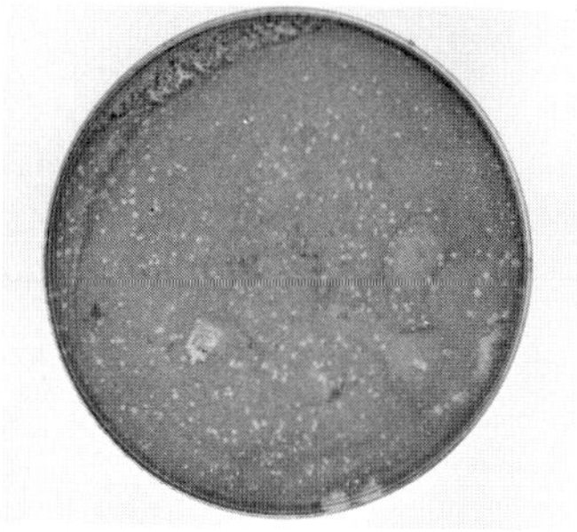

Fig. 10. Plaque assay of CFA on monolayer cultures of *Aedes albopictus* cells (Stollar and Thomas, 1975). The plates shown are 60 mm in diameter.

The growth curve shows that the CFA is readily found both in the medium and in cell extracts, in approximately equal amonts (Fig. 11). The latent period appeared to be about 12 hours, considerably longer than that of alphaviruses, but similar to that described for flaviviruses in mammalian cells (Stollar *et al.*, 1966).

Table 7 shows the effect of the CFA on various insect and mammalian cell lines. It was unable to replicate or cause fusion in any of the 3 mammalian cell lines tested (Vero, KB, and BHK). It is possible, however, that prolonged co-cultivation or blind passage would eventually produce or select for a variant capable of replication in one of these cell lines. Both replication and cell fusion occurred as described above in the normal *A. albopictus* cells and in *A. albopictus* cells chronically infected with Sindbis virus.

It was noted above when cell fusion occurred most of the cell monolayer was destroyed; in all experiments, however, there remained a small fraction of cells which were not fused. These "resistant" cells continued to grow normally and apparently accounted for the eventual recovery of the cultures several weeks after the marked cytopathic effect occurred. These cultures became carrier cultures of the CFA but differed very little in appearance from the normal *A. albopictus* cells. Of added interest is the observation that although these cultures always contained CFA they were resistant to fusion by superinfecting CFA. It is likely that, as in the case of cells chronically infected with Sindbis virus, these cells, *A. albo.* (CFA), are also resistant to superinfecting homologous virus. The *A. aegypti* cells from which the CFA was originally obtained, always contained CFA, but were never fused by superinfecting CFA.

Further observations lead us to believe that the CFA closely resembles in several respects the flaviviruses (group B togaviruses). Briefly, these observations are 1) the particle is spherical, has short projections, and measures about 50 nm in diameter, 2) the CFA is sensitive to ether and deoxycholate and thus appears to be an enveloped virus, and 3) the CFA sediments in a sucrose velocity gradient considerably more slowly than Sindbis virus, and at a rate expected for a flavivirus (Boulton and Westaway, 1972). However, serological tests in co-operation with Dr. Jordi Casals have failed to show cross reaction with a number of the flaviviruses. It is possible that the CFA may be one of the non-arthropod-borne togaviruses (Horzinek, 1973).

TABLE 6

RNA synthesis phenotype of sindbis virus variants (a)

Sindbis virus	SA at 39.5 °C: SA at 34 °C	RNA synthesis phenotype
SV-W	0.84	+
SV-C-2	< 0.01	−
SV-C-4	0.34	+
SV-C-8	0.39	+
SV-C-13	0.52	+
SV-C-16	0.53	+
SV-C-19	1.03	+

(a) Viral RNA synthesis (i.e., Actinomycin resistant) was measured by following ^{3}H-uridine incorporation into TCA precipitable material (Shenk *et al.*, 1974). Values are expressed as specific activities (counts per minute per absorbancy at 260 nm).

Our experience with the CFA underlines the importance of continual monitoring of the mosquito cell lines for unsuspected agents, and emphasizes the importance of a suitable indicator system for such monitoring. In this case, the *A. albopictus* cell line is the only sensitive indicator system presently available for the CFA.

V. Conclusions

Even though the study of togaviruses in the insect cell host is at an early stage, it is apparent that the use of mosquito cell cultures is both a necessary and productive means of studying the life cycle of arthropod-borne togaviruses.

In the acute infection (1-3 days after infection) evidence was presented showing the influence of the mosquito cell on the phenotype of Sindbis virus. Progeny virus from the *A. albopictus* cells in sharp contrast to virus from vertebrate cells 1) lacked sialic acid and 2) displayed marked heterogeneity in sucrose-D_2O gradients centrifuged to equilibrium.

In the stage of chronic infection (several months after the initial infection), it was noted (i) that the progeny virus gradually shifted to a small plaque temperature-sensitive genotype and (ii) that the cultures in this stage became resistant to superinfection with the homologous virus (Sindbis virus) but remained permissive with respect to the replication of a heterologous virus (Eastern equine encephalitis virus).

The apparent inability of defective-interfering particles generated in vertebrate cells to replicate or interfere in mosquito cells was noted with interest. If we might extrapolate from our observations in cell culture to the more complex level of the whole organism, we would propose that each of the alternate natural hosts, the insect, and the vertebrate can serve as a "biological filter" selecting for and against various of the viral forms generated in the alternate host. Thus, defective-interfering particles generated in the vertebrate host would be "silent" in the mosquito host; and temperature-sensitive viral mutants generated in the mosquito host would be unable to replicate at the higher body temperature of the avian or mammalian host.

TABLE 7

Response of different mammalian and mosquito cell lines to CFA (a)

Experiment	Cell line or subline	Infected with CFA[d]	Titer of CFA (PFU/ml)			CPE at	
			2 hr	5 hr	72 hr	55 hr	72 hr
I	BHK	+	4.7×10^3		$< 10^3$		0
	Vero	+	1.7×10^3		$< 10^3$		0
	KB	+	1.6×10^3		$< 10^3$		0
	A. albo. (normal)	+	5.5×10^2		2.6×10^7		4+
	A. albo. (SV)[b]	+	1.5×10^3		7.2×10^6		2+
	A. albo. (CFA)[c]	+	6.0×10^4		3.2×10^5		0
	A. aegypti (normal)	+	8.7×10^3		5.5×10^5		0
	A. aegypti (SV)	+	2.7×10^4		7.0×10^6		0
II	A. albo. (normal)	+	1.2×10^2	4.3×10^6		3+	4+
	A. albo. (normal)	–	N.D.[e]	N.D.		0	0
	A. albo. (SV)[f]	+	2.2×10^2	1.3×10^6		0–1+	3+
	A. albo. (SV)[f]	–	N.D.	N.D.		0	0
	A. albo. (CFA)[g]	+	4.3×10^3	4.6×10^6		0	0
	A. albo. (CFA)[g]	–	7.5×10^3	1.4×10^6		0	0
	A. aegypti (normal)	+	8.5×10^2	5.5×10^6		0	0
	A. aegypti (normal)	–	1.2×10^3	3.5×10^5		0	0

(a) CFA was added to cells (approximate MIO= 1 PFU/cell) for 60 minutes at room temperature and then removed; cells were fed with the appropriate medium and incubated at 34° in the case of vertebrate cells and at 28° in the case of mosquito cells. Cytopathic effect or syncytial fomation was graded on a 0 to 4+ scale. Samples of medium were taken at the times indicated and assayed by plaque formation on *A. albopictus* cells.

(b) These cells were initially infected 11 months previously with SV-W. Sindbis virus was usually present in the medium at levels near 10^5 PFU/ml and gave rise only to small plaques (Stollar and Shenk, 1973).

(c) These cells were initially infected 4 months previously with CFA.

(d) Cultures not infected with CFA were mock infected with PBS-2 containing 0.2% BSA.

(e) None detectable.

(f) Cultures were initially infected 17 months previously with SV.

(g) Cultures were initially infected 10 months previously with CFA.

Finally, the importance of adventitious viral agents in mosquito cell cultures was illustrated by our experience with the cell fusing agent found originally in the *A. aegypti* cells. How and when this agent first appeared in the *A. aegypti* cells is as yet unclear, as is its precise identification.

It is no exaggeration to say that the availability of insect cell cultures has opened new vistas for the study not only of the arthropodborne togaviruses but also undoubtedly for the identification and characterization of viruses still to be discovered.

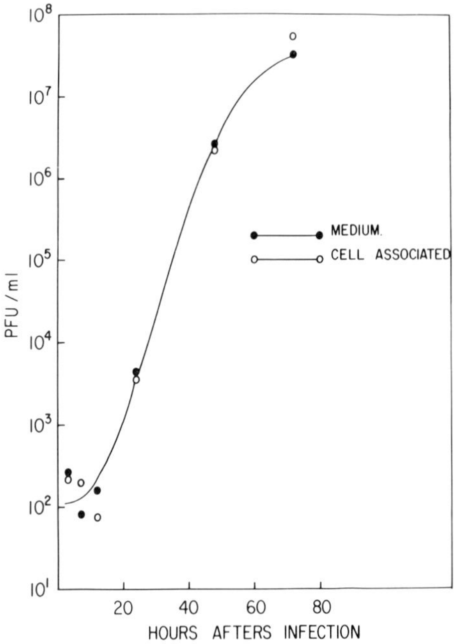

Fig. 11. Growth of CFA in *Aedes albopictus* cells. Monolayers of *A. albopictus* cells in 60 mm petri plates were infected with CFA at an input m.o.i. of 0.1 PFU/cell. After a 90 minutes adsorption period at room temperature the CFA was removed and cultures fed with 5 ml of MM medium per plate. Incubation was at 28°. At various times after infection, samples were then taken to measure both extracellular and cell-associated CFA. Assay was by plaque formation on *A. albopictus* monolayer cultures. To obtain cell-associated virus, the cells from one plate were washed, suspended in 2 ml of PBS-1 (containing 0.2% BSA), frozen and thawed twice, and then centrifuged at low speed. The supernatant was collected and taken to represent cell-associated virus.

Footnotes

1. Present Address: Department of Biochemistry
   Stanford University School of Medicine
   Stanford University Medical Center
   Stanford, California 94305

Acknowledgements

These investigations were supported by Grant No. GB37707 from the National Science Foundation, by Grant No. AI-11290 from the National Institute of Allergy and Infectious Diseases, and by the United States-Japan Cooperative Medical Science Program through Public Health Service Grant No. AI-05920.

VI. References

Boulton, R.W., and Westaway, E.G. (1972). *Virology 49,* 283.

Brown, D.T., and Gliedman, J.B. (1973). *J. Virol. 12,* 1534.

Burge, B.W., and Pfefferkorn, E.R. (1966). *Virology 30,* 204.

Cunningham, A., Buckley, S.M., Casals, J., and Webb, S.R. (1975). *J. Gen. Virol. 27,* 97.

Davey, M.W., Dennet, D.P., and Dalgarno. L. (1973). *J. Gen. Virol. 20,* 225.

Eagle, H. (1959). *Science 130,* 432.

Horzinek, M.C. (1973). *Prog. Med. Virol. 16,* 109.

Kraemer, P.M. (1967). *J. Cell. Phys. 69,* 199.

Mitsuhashi, J., and Maramorosch, K. (1964). *Contributions Boyce Thompson Institute 22,* 435.

Peleg, J. (1968). *Virology 35,* 617.

Peleg, J., and Stollar, V. (1974). *Archiv. fur die gesamte Virusforschung 45,* 309.

Pfefferkorn, E.R., and Shapiro, D. (1974). In: Comprehensive Virology Reproduction: Small and Intermediate RNA Viruses (Fraenkel-Conrat, H., and Wagner, R.R. eds.), 2, Plenum Publishing Corporation, New York.

Schlesinger, R.W., Stevens, T.M., and Miller, E.J. (1961). *Bact. Proc.*

Shenk, R.E., Koshelnyk, K.A., and Stollar, V. (1974). *J. Virol. 13,* 439.

Singh, K.R.P. (1967). *Curr. Sci. 36,* 506.

Stevens, T.M. (1970). *Proc. Soc. Exp. Biol. Med. 134,* 356.

Stollar, V., Stevens, T.M., and Schlesinger, R.W. (1966). *Virology 30,* 303.

Stollar, V., and Shenk, T.E. (1973). *J. Virol. 11,* 592.

Stollar, V., and Thomas, V.L. (1975). *Virology 64,* 367.

Warren, L. (1959). *J. Biol. Chem. 234,* 1971.

Warren, L. (1963). *Comp. Biochem. Physiol. 10,* 153.

Chapter 5

FURTHER STUDIES ON THE LATENT VIRUSES ISOLATED FROM SINGH'S *AEDES ALBOPICTUS* CELL LINE

H. Hirumi, K. Hirumi and G. Speyer

I. Introduction

Recently, two sublines (SL1 and SL2) of Singh's *Aedes albopictus* cell line (Singh, 1967) which undergo spontaneous syncytium formation were found to be contaminated with a viral complex which consisted of at least three different types of viral agents (Hirumi, 1975; Hirumi *et al.*, 1975). Earlier studies indicated that the morphology of these agents resembled that of the parvovirus, orbivirus and togavirus. One of them, presumably the togavirus-type agent, was serologically related to the alphavirus of the Togaviridae. In order to characterize the biological nature of these viral agents, further studies were made.

II. Materials and methods

Cell lines: A subline (SL3) of Singh's *Aedes albopictus* cell line was used as a "virus-free" cell line. The history of this subline was described earlier (Hirumi *et al.*, 1975). Two continuous cell lines of the *A. aegypti* mosquito, established by Singh (Singh, 1967) and Varma & Pudney (1969), were examined for their susceptibility to the viral agents.

Culture medium: Hirumi-Maramorosch Leafhopper (HML) medium (Hirumi and Maramorosch, 1964; Hirumi and Maramorosch, 1971) was used throughout the experiments.

Viral agents: SL3 cell cultures were inoculated with the *A. albopictus* syncytium-inducing agent (ALSA1) which was originally isolated from a subline (SL1) associated with spontaneous syncytium formation (Hirumi *et al.*, 1975). Viral agents used for this study were at passage 12 in the SL4 cells *in vitro.* Details concerning the history of the subline SL1 and the syncytium-inducing agent were also described earlier (Hirumi, 1975, Hirumi *et al.*, 1975). Infected cell cultures of the SL3 cells were subjected to three cycles of alternate freezing and thawing followed by removal of the cell debris by centrifugation for 10 min at 300 X *g.* The supernatant fluid was filtered through 0.45 μm Millipore filters and used as the original virus inoculum.

Chloroform treatment: The virus-containing supernatant fluid was mixed with chloroform (Hirumi and Maramorosch, 1964: Cunningham, *et al.*, 1975) and vigorously shaken for 2 hours at room temperature. After removing the chloroform, the syncytium-inducing ability of ALSA1 was tested by inoculating monolayer cultures (2- to 4-day-old) of SL3 cells in Disposo-Trays (13 mm diameter well, 24 wells/tray) (Linbro Scientific Co., New Haven, Conn.). The tray cultures were inoculated with serial dilutions (0.5 ml/well, 5 wells/dilution). Ten-fold serial dilutions (10^{-1} to 10^{-8}) of the chloroform-treated and non-treated virus inocula were prepared in HML culture media containing 5% heat-inactivated (56°C for 30 min.) fetal bovine serum. A total of 24 flask cultures (Falcon, plastic 30 cm^2 T-type flask) or SL3 cells were also inoculated with the treated and non-treated virus inocula (10^{-1} dilution, 1 ml/flask) and used for electron microscopic examination. After an absorption period of one hour on a rocker platform at room temperature, the inocula were removed from the cultures. After being rinsed with Earle's balanced salt solution (BSS), fresh medium (1 ml/well, 5 ml/flask) was added; cultures were then incubated at 27°C for up to 16 days. Three replicate experiments were made.

Susceptibility of A. aegypti cell lines: Flask cultures of *A. albopictus* SL3 cells, Singh's *A. aegypti* and Varma & Pudney's *A. aegypti* cells (10 flasks/cell line) were inoculated with the non-diluted ALSA1 in the same manner as described above. The same number of non-inoculated cultures were also prepared from each cell line as controls. At intervals of 24 hours for 10 days, the fluid phase of the inoculated and non-inoculated cultures (one culture/cell line, at 24 hour intervals) was removed, filtered through 0.45 μm Millipore filters; 10-fold serial dilutions of the culture fluid (10^{-1} to 10^{-8}) were then made in the HML culture medium. Newly prepared tray cultures of *A. albopictus* SL3 cells were inoculated with the serial dilutions as described above and incubated at 27°C for 10 days. After removing the media, flask cultures were used for electron microscopic examination.

Cytopathic effects: Syncytium formation in both tray and flask cultures was examined using an inverted phase microscope. The syncytium-inducing ability of ALSA1 was measured as the median tissue culture infectious dose ($TCID_{50}$/ml) 7 days after inoculation in the Disposo-Trays.

Electron microscopy: At intervals of 24 hours after inoculation, flask cultures of the infected and non-infected *A. albopictus* SL3 cells and two *A. aegypti* cell lines were prepared for thin section electron microscopy as described earlier (Hirumi *et al.*, 1975). Ultra-thin sections were stained with uranyl-magnesium acetate and lead citrate, and examined with a Siemens Elmiskop-I, modified to IA.

III. Results

Effects of chloroform treatments: The non-treated ALSA1 induced numerous syncytia of *A. albopictus* SL3 cells in all flask cultures 24 to 48 hours postinoculation (Fig. 1). The serial titration tests in the tray cultures indicated that the titer of the syncytium-inducing agent in the original inoculum was 2.7 x 10^5 $TCID_{50}$/ml.

Electron microscopic examination of SL3 cells inoculated with the non-treated ALSA1 revealed the presence of the three types of viral agents, the parvovirus (Fig. 2), orbivirus (Fig. 3) and togavirus (Fig. 4) type particles, in all cultures. No virus particles were seen in non-inoculated SL3 cells. The morphology of these viral agents and sites of their appearance in the host cells were very similar to those reported earlier (Hirumi, 1975; Hirumi *et al.*, 1975).

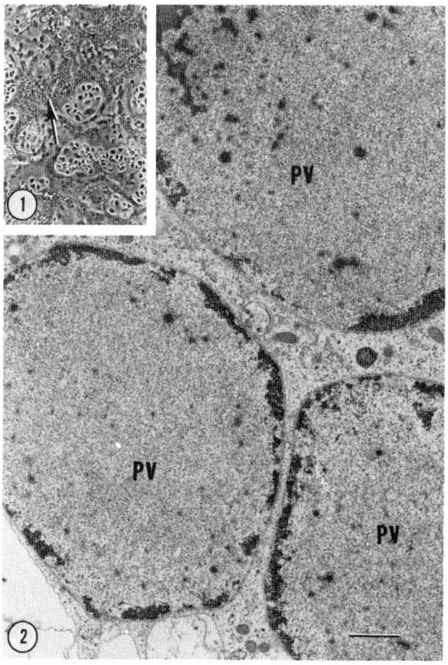

Fig. 1. Singh's *Aedes albopictus* SL3 cells inoculated with ALSA1 (2.7 X 10^5 $TCID_{50}$/ml), 48 hours postinoculation. Note large syncytia (arrow). Phase contrast micrograph. X 75.

Fig. 2. Portion of a large syncytium of *A. albopictus* SL3 cell, induced by ALSA1, 6 days after inoculation. Note nuclei contained numerous parvovirus-type particles (PV). Bar 1000 nm. X 12,600.

The chloroform-treated ALSA1, however, did not induce any syncytia in any flask, as well as tray cultures up to 16 days after inoculation, when the experiments were terminated. Preliminary experiments showed that chloroform treatments with lower chloroform concentrations, 5% and 10%, for shorter periods, 5 min. and 30 min. respectively, suppressed the syncytium-inducing ability of ALSA1 but did not completely inactivate it.

Electron microscopic examination of the syncytium-free SL3 cells inoculated with the chloroform-treated ALSA1 demonstrated the presence of many parvovirus- and orbivirus-type particles (Figs. 5 and 6, respectively). No togavirus-type particles were observed in any of the cultures (12 flasks) examined during the 16-day post-inoculation period.

Susceptibility of A. aegypti cells: The non-diluted ALSA1 (2.7 X 10^5 $TCID_{50}$/ml) induced numerous syncytia of SL3 cells in all cultures 24 hours after inoculation, whereas, no syncytia were seen in both inoculated and non-inoculated cultures of all *A. aegypti* cell cultures till 10 days after inoculation. Figures 7 and 11 illustrate ALSA1-inoculated Varma & Pudney's and Singh's *A. aegypti* cells, respectively, 10 days postinoculation. All cultures fluids, removed from the inoculated *A. aegypti* cell cultures at intervals of 24 hours after inoculation, induced syncytia of *A. albopictus* SL3 cells. Titers of syncytium-inducing agent in the culture fluid of the *A. aegypti* cells increased up to 5 X 10^8 $TCID_{50}$/ml during the 5 day postinoculation period. The culture fluid of non-infected *A. aegypti* cells did not induce any syncytia of *A. albopictus* SL3 cells.

Electron microscopic examination of the ALSA1-inoculated Varma & Pudney's *A. aegypti* cells revealed the presence of the parvovirus-, orbivirus- and togavirus-type particles (Figs. 8, 9 and 10, respectively). At the early stages of infection (24 to 48 hours postinoculation), only a few cells contained a small number of these virus particles. However, the numbers of infected cells, as well as of virus particles per cell, increased considerably during the later stages of infection. In the ALSA1-inoculated Singh's *A. aegypti* cells, all three types of viral agents were also observed (Figs. 12 - 14). Dual and/or triple infection with these virus particles were often seen in infected *A. aegypti* cells of both lines (Figs. 10 and 14). No significant differences concerning the replication of these three viral agents in *A. aegypti* cells were detected between the two cell lines. No virus particles were observed in the non-inoculated *A. aegypti* cells in all cultures.

IV. Discussion

The findings reported here indicate that the chloroform treatment (25% for 2 hours) completely inactivated the syncytium-inducing ability of ALSA1 in *A. albopictus* cell cultures. The presence of numerous non-enveloped virus particles (parvovirus-

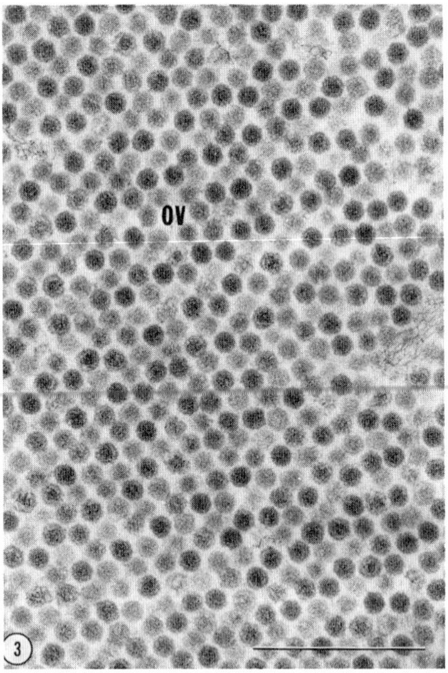

Fig. 3. Large accumulation of the orbivurus-type particles (OV) in an *A. albopictus* SL3 cell inoculated with ALSA1, 8 days post-inoculation. Bar: 500 nm. X 85,750.

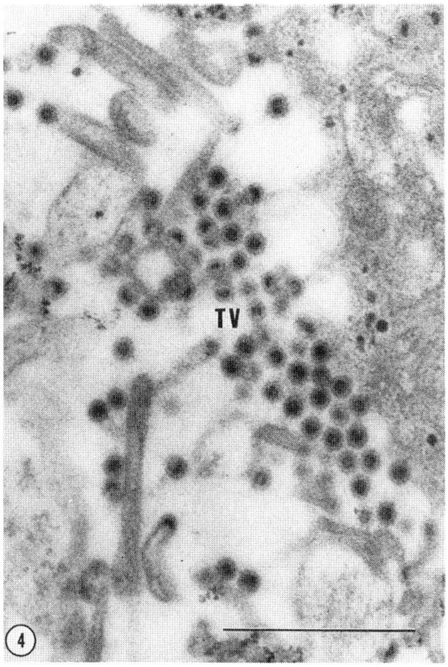

Fig. 4. Portion of the *A. albopictus* SL3 cell surface, inoculated with ALSA1, 3 days after inoculation. Note togavirus-type particles (TV) budding from the plasma membrane. Bar: 500 nm. X 95,700.

and orbivirus-type agents) and the absence of the enveloped particles (togavirus-type agent) in the *A. albopictus* cells inoculated with the chloroform-treated ALSA1 indicate that the treatment affected the infectivity of the enveloped agent, whereas, the non-enveloped ones were not affected. These results strongly suggest that the togavirus-type agent is indeed responsible for the syncytium induction of *A. albopictus* cells.

Recently, Cunningham *et al.* (1975) also independently detected the presence of a viral contaminant associated with syncytium formation in a subline of Singh's *A. albopictus* cells and identified it by serological tests as Chikungunya virus (an alphavirus of the Togaviridae). However, the virus particles illustrated in the publication appear to be non-enveloped virus particles whose morphology is identical to that of the orbivirus-type particles reported earlier (Hirumi, 1975, Hirumi *et al.*, 1975). Present studies indicated that the orbivirus-type agent detected in the *A. albopictus* cells is very unlikely to be the syncytial agent. In addition, all togaviruses are known to be enveloped (Horzinek, 1973). Thus, the subline may be contaminated with at least two viral agents, namely the "Chikungunya" virus and the orbivirus-type agent.

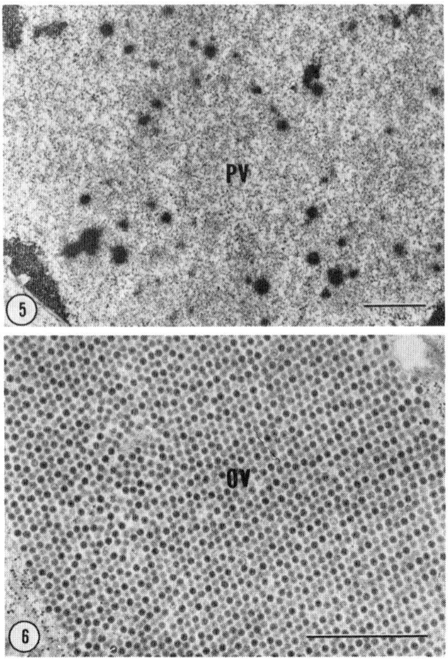

Fig. 5 and 6. Portion of *A. albopictus* SL3 cells inoculated with the chloroform-treated ALSA1. Bars: 1000 nm. *Fig. 5:* nucleus containing numerous parvovirus-type particles (PV), 2 days postinoculation. X 15,750. *Fig. 6:* paracrystalline arrays of the orbivirus-type particules (OV) in the cytoplasm of the host cell, 5 days postinoculation. X 30,900.

The present study clearly demonstrated that the togavirus-type agent could propagate in *A. aegypti* cells of both Singh's and Varma & Pudney's lines without inducing syncytia of the host cells. The agent propagated in the *A. aegypti* cells maintained the syncytium-inducing ability in *A. albopictus* cell cultures. Stollar and Thomas (1975) recently isolated a viral contaminant from Peleg's *A. aegypti* cell line (Peleg, 1969) and suggested that the agent may be similar to the group B togaviruses (flavoviruses of the Togaviridae). The agent caused cell fusions of Singh's *A. albopictus* cells although it did not induce syncytia of the *A. aegypti* cells. Further studies concerning serological relations between the two togavirus-type agents isolated from the *S. albopictus* and the *A. aegypti* cell lines would be of general interest.

The finding of cross-infection of the parvovirus- and orbivirus-type agents without cytopathic effects in the *A. albopictus* and the *A. aegypti* cells strongly indicates need for precaution in preventing viral contamination of these cell lines where the cell lines are currently in use.

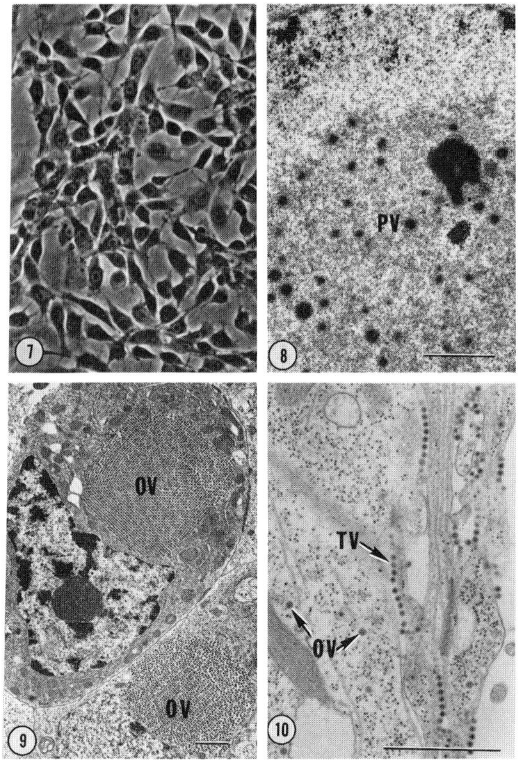

Fig. 7 - 10. Varma & Pudney's *A. aegypti* cells inoculated with ALSA1, Bar: 1000 nm. *Fig. 7:* no syncytium formation is seen, 10 days after inoculation. Phase contrast micrograph. X 300. *Fig. 8:* intranuclear parvovirus-type particles (PV), 6 days after inoculation. X 18,800. *Fig. 9:* paracyrstalline arrays of the orbivirus-type particles (OV) in the host cell cytoplasm, 5 days postinoculation. X 8,700. *Fig. 10:* togavirus-type particles (TV) budding from the host cell surface and orbivirus-type particles (OV), scattered in the cytoplasm, 8 days postinoculation. X 30,900.

V. References

Cunningham, A., Buckley, S.M., Casals, J. and Webb, S.R. (1975). *J. gen. Virol. 27,* 97.

Hirumi, H. (1975). *In:* Invertebrate Tissue Culture (K. Maramorosch, ed.). Academic Press, New York. *In press.*

Hirumi, H., and Maramorosch, K. (1964). *Science 144,* 1465.

Hirumi, H., and Maramorosch, K. (1971). *In:* Invertebrate Tissue Culture (C. Vago, ed.), V. 1, pp. 307-339. Academic Press, New York.

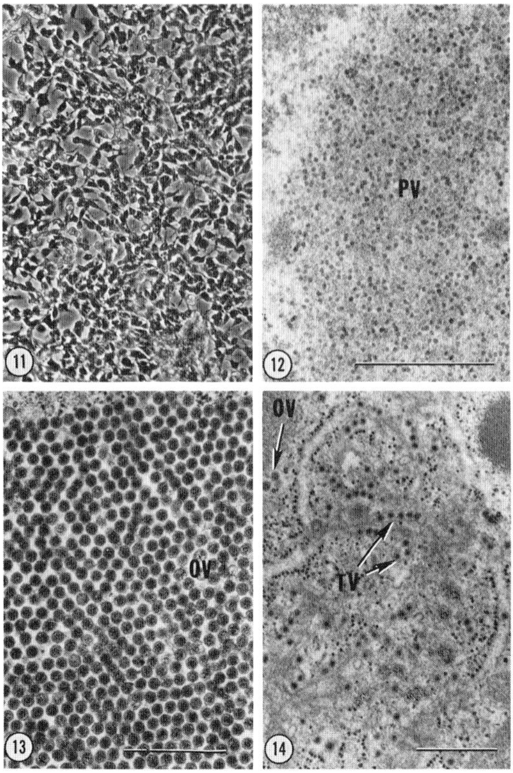

Fig. 11 - 14. Singh's *A. aegypti* cells inoculated with ALSA1. Bars: 500 nm. *Fig. 11:* no syncytium formation is seen, 10 days after inoculation. X 150. *Fig. 12:* parvovirus-type particles (PV) in the host cell nucleus, 6 days postinoculation. X 74,400. *Fig. 13:* paracrystalline arrays of orbivirus-type particles (OV) in the host cell cytoplasm, 8 days postinoculation. X 54,600. *Fig. 14:* togavirus-type particles (TV) in the irregularly elongated cisternae of the endoplasmic reticulum and orbivirus-type particles (OV) scattered in the host cell cytoplasm, 8 days postinoculation. X 42,600.

Hirumi, H., Hirumi, K., Speyer, G., Yunker, C.E., Thomas, L.A., Cory, J. and Sweet, B.H. (1975). *In Vitro. Submitted for publication.*

Horzinek, M.C. (1973). *J. gen. Virol. 20,* 87.

Peleg, J. (1969). *J. gen. Virol. 5,* 463.

Singh, K.R.P. (1967). *Current Sci.* (India) *36,* 506.

Stollar, V., and Thomas, V.L. (1975). *Virology 64,* 367.

Varma, M.G.R., and Pudney, M. (1969). *J. Med. Entomol. 6,* 432.

Chapter 6

APPLICATION OF TISSUE CULTURE TO PROBLEMS IN MALARIOLOGY

M.C. Rosales-Sharp and P.H. Silverman

I. Introduction

The methodology of tissue culture has been utilized in experimental malariology as a result of much need information on the metabolic requirements of the parasites during growth and development, on the host-parasite relationship with reference to host specificity, on the effectiveness of chemotherapeutic agents, and on the establishment of an efficient cell substrate or cell-free substrate that can support the developmental stages of the parasite. Evidently, the latter could be of great value in attempts to collect large quantities of the parasites, free of host tissue, for immunological studies. In recent years studies by several workers (Simpson *et al.*, 1974; Mitchell *et al.*, 1974; Nussenzweig *et al.*, 1972) actively engaged in basic malaria research, have shown that it is feasible to confer immunity against malaria using either the vertebrate forms (trophozoites and merozoites) or invertebrate forms (sporozoites) of the malarial parasite as a vaccine. However, the most foreseeable difficulty that must be resolved before a practical anti-malarial vaccine is developed is the methodology of harvesting large numbers of materials for vaccination purposes. Presently, sporozoites are collected from female mosquitoes either by laborious manual dissection of the salivary glands or the use of continuous gradient centrifugation techniques. On the other hand, the blood stages (trophozoites and merozoites) are harvested from blood collected from infected rodents and primates. The methods employed, although adequate for experimental purposes, are not practical systems in obtaining large quantities of antigenic materials for immunization purposes. Conceivably, the future practicality of malaria vaccine depends to a large measure on major technological breakthroughs in developing *in vitro* systems to cultivate sufficient antigenic materials. The potential value of the *in vitro* systems has long been recognized and several attempts have been made. So far, in spite of some of the promising results, success has been limited. An adequate *in vitro* system has yet to be established in order to accelerate studies on immunology of malarial infection and which would in all practicality be directed toward provision of antigenic materials for vaccination. This chapter will deal with efforts made towards the cultivation of the sporogonic forms or insect forms of the rodent malarial parasites, *Plasmodium berghei.*

II. Establishment of vector cell substrate

Considerable efforts have been made in the establishment of an *in vitro* vector cell substrate that hopefully will approximate the physiological and biochemical *in vivo* environment of the host tissue. The development of cell substrates taken from newly hatched larvae of *Anopheles stephensi* was initiated and established (Rosales-Ronquillo *et al.,* 1972). The cell substrate (primary cell culture of mosquito tissue, PCC) consisted of 9 different outgrowths: large and small epithelial cells, fibroblastlike cells, neuron-like cells, polygonal cells, 2 types of myoblast-like cells, and noncellular and cellular vesicles. They remained viable for a period of about 60 days without change of nutrient medium (Table 1). The medium was a modification of Schneider's medium (Schneider, 1969).

A cell line similar to that established by Schneider (1969) was also initiated. Cell monolayers were initiated from several trypsinized cellular vesicles and grown in culture medium as described by Schneider (1969).

One of the growth patterns observed in the primary cell culture (PCC) was the contractile, myoblast-like structures from gut fragments (Rosales-Ronquillo *et al.,* 1972). Preliminary histochemical studies revealed that these outgrowths are myogenic structures characteristic of gut muscles. *In vivo* it has been shown that ookinetes require an intercellular or extracellular association with the gut cells prior to their development in the next stage, the oocysts. Hoping that the myoblast structures *in vitro* will serve as a potential substrate for sporogonic development, a substrate consisting solely of large numbers of myoblast structure was established (unpublished). Addition of an insect hormone, ecdysone (10-20 µg per ml of the medium) was shown to enhance myoblast outgrowths. The myoblast outgrowths remain viable and contractile for about 20 days without change of ecdysone-treated medium (Figs. 1a and b; Figs. 2a and b).

Ancillary to the development of the vector cell substrate, the production of germfree adult *Anopheles stephensi* as a source of bacteria-free organs to serve as supplement to the cultivation substrate was initiated and successfully established (Rosales-Ronquillo *et al.,* 1973). An aseptic diet has been formulated and lately improved that yields 90% pupation and 100% of pupae emerging successfully into adults in about 3 days (Table 2). The procedure for preparing the improved diet is essentially the same as those described in an earlier paper (Rosales-Ronquillo *et al.,* 1973).

III. Cultivation of sporogonic forms in vector cell substrate

Various cultivation procedures were designed to evaluate the receptiveness of the 3 established vector substrates: primary cell cultures (ranging from 10- to 35-days old); anopheline cell line (subculture every 7 days) and myoblast-like cell cultures (ranging from 10- to 20-days old). Findings based on *in vivo* studies such as the various factors and requirements that influence parasite development were taken into consideration. For example, development of *Plasmodium berghei* is temperature-dependent ranging from $18^{\circ}C$ to $21^{\circ}C$. Antibiotics and serum were found to have adverse effects on the development of the parasite. The pH requirement of each spo-

TABLE 1

Composition of medium for primary cell culture of mosquito tissue, Anopheles stephensi *Liston.*

1. Salts		
$NaHCO_3$	0.408	g
KCl	1.283	g
$MgCl_2.6H_2O$	1.330	g
$MgSO_4.7H_2O$	0.466	g
NaCl	3.5	g
2. Amino acids		
L-arginine HCl	0.816	g
L-aspartic acid	0.408	g
L-asparagine	0.408	g
L-alanine	0.262	g
β-alanine	0.233	g
L-cystine HCl	0.029 (dissolve in 20 drops 1.0 N HCl)	
L-glutamic acid	0.700	g
L-glutamine	0.700	g
L-glycine	0.758	g
L-histidine	2.920	g
L-isoleucine	0.0584	g
L-leucine	0.0875	g
L-lysine HCl	0.729	g
L-methionine	0.0584	g
L-proline	0.408	g
L-phenylalanine	0.175	g
D,L-serine	1.283	g
L-tryptophan	0.116	g
L-threonine	0.204	g
L-valine	0.116	g
3. Sugars		
Sucrose	18.65	g
D-trehalose	0.584	g
D-glucose	1.16	g
4. $CaCl_2.H_2O$	2.29	g
5. E-cholesterol	2.3	ml
6. Fetal bovine serum (FBS)++	150	ml
7. NCTC 135+++	0.4	ml

Salts, amino acids, sugars and calcium were all dissolved in 750 ml of distilled water. After adding lipids, serum and NCTC 135, the amount was raised to 1000 cc volume.

+ Procedure for working solution was taken from Cancer Research, 1956, Vol. 16: 77-94.

++ Fetal bovine serum was inactivated for 30-45 min at 56°C.

+++ Special NCTC 135 (without amino acids, glucose and salts) adopted by Dr. Imogene Schneider (personal communication) was incorporated in the nutrient medium.

TABLE 2

Improved diet composition and formulation for rearing aseptic Anopheles stephensi *Liston.*

Cholesterol	20
Tween 80	10
$MgSO_4.7H_2O$	80
$FeSO_4.7H_2O$	18.5
$MnSO_4.7H_2O$	18.5
NaCl	18.5
CaCl	18.5
Liver Extract	500
Bacto Liver	500
Sucrose	10,000
RNA	1,000
Pyridoxine hydrochloride	4.59
Riboflavin	2.3
Thiamine hydrochloride	2.3
Calcium pantothenate	11.4
Pteroylglutamic acid	20-30
Niacinamide	11.4
Biotin	1
Choline Chloride	114
KH_2PO_4	685
K_2HPO_4	685
Casein *	

The amounts of each ingredient are shown in milligrams per 1000 ml of distilled water.
* 50 mg of casein were put in each rearing flask

rogonic phase is exacting: Ball and Chao (1964) showed that young *Plasmodium relictum* oocyst developed at a pH of 7 and a pH of 7.9 is favorable to sporozoite production. Finally after several attempts, a working procedure described by Rosales-Ronquillo and Silverman (1974) employing the 3 vector cell-substrates was found to be adequate to the cultivation of the early sporogonic forms (Rosales-Ronquillo and Silverman, 1974). The successful *in vitro* induction of the exflagellation of microgametes, fertilization of macrogametocytes, and formation of viable motile ookinetes of the rodent malarial parasite was demonstrated in the vector cell substrates. Quantitative analysis obtained from stained preparations showed that one drop of inoculum (plasmodially-infected blood of hamster) contains approximately 281 macrogametocytes. Eight drops of blood employed as inoculum per culture tube yield about 1800 zygotes which developed into viable, motile ookinetes about 24 to 30 hours into the culture period. The findings on ookinete movement *in vitro* and described by Speer, Rosales-Ronquillo and Silverman (1975) reaffirmed the knowledge (which is questioned by some) that the ookinete is motile.

IV. Cultivation of sporogonic forms in non-vector cell substrate

Successful cultivation of the early stages of the sporogonic forms of the malarial parasite was demonstrated in two vector cell substrates: primary cell cultures (PCC)

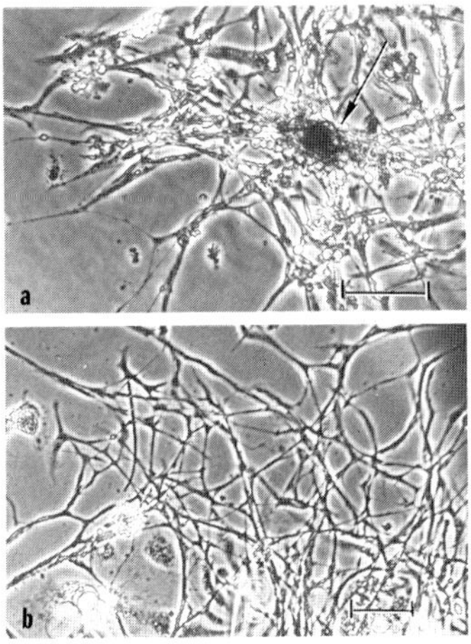

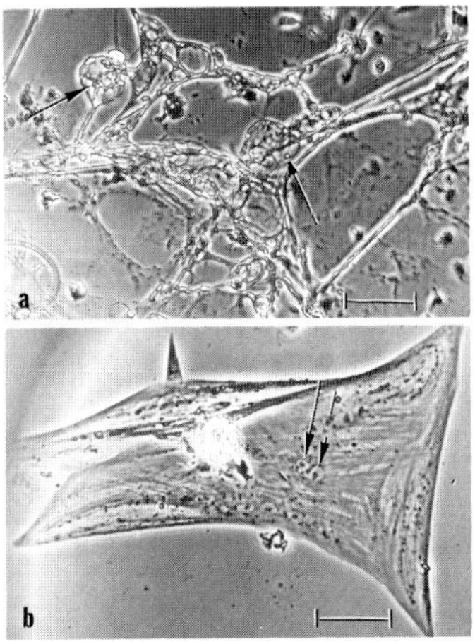

Fig. 1. A. stephensi primary tissue culture treated with ecdysone. Phase contrast of living cells. Appearance of myoblast cells (a) 4 days into the culture period. Note the gut fragment (arrow). Growth pattern of the myoblast cells (b) 15 days into the culture period. Scale lines represent 50 μ.

Fig. 2. A. stephensi primary tissue culture treated with ecdysone. Phase contrast of living cells. Appearance of myoblast cells (a) 25 days into the culture period. Degeneration is characterized by the presence of cytoplasmic blebs. *Figure 2b* shows multinucleated thin myoblast outgrowths. Arrows point to nuclei. Scale lines represent 50μ.

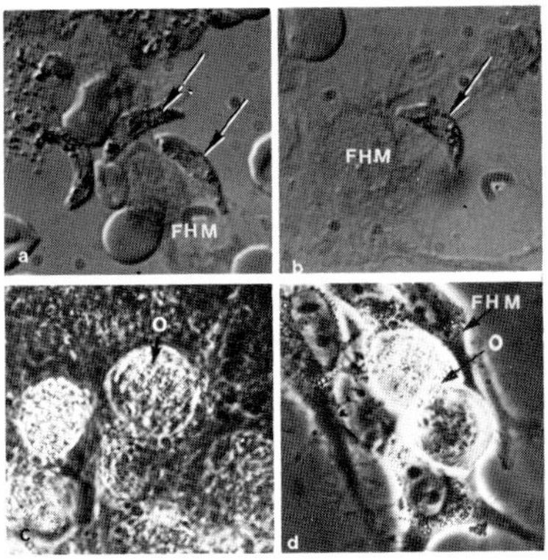

Fig. 3. Early sporogonic development in FHM (fat head minnow) epithelial cell line. Phase contrast of living cells. Appearance of ookinetes near the vicinity of FHM epithelial cells (a, b). Arrows point to ookinetes. Oocyst-like structures (O) 8 days into the culture period (c and d).

from mosquito tissues and an established anopheline cell line (Rosales-Ronquillo and Silverman, 1974). An ecdysone-treated primary cell culture (PCC) consisting of myoblast-like outgrowths also supported *in vitro* ookinete formation starting with erythrocytic gametocytes obtained from blood of plasmodially-infected hamsters (unpublished). However, all vector culture systems were not efficient in supporting development of the parasite beyond the ookinete stage. The capability of non-vector cell lines as culture substrates for malarial parasites was suggested by Dutky (1964). The capability of fish cell lines which were known to support growth of viruses was employed as culture substrate. A detailed culture procedure was described earlier by Rosales-Ronquillo and Silverman (1974). The results shows that Fat Head Minnow (FHM) epithelial cell lines supported ookinete formation similar to that occurring *in vivo.* In some FHM cultures, ookinetes were found within the intracytoplasmic vacuoles of the host cells 30 hours into the culture period (Rosales-Ronquillo and Silverman, 1974). Oocystlike structures were first seen in an 8-day-old culture (Fig. 3). However, they tend to lose their morphological integrity as the culture ages. On the other hand, the host cells were found to be healthy and capable of dividing for as long as 16 days into the culture period. It appears that the constituents of the inoculum (infected blood) have no adverse effect on the viability of the host cells. The capability of the FHM cell line to support further development of the sporogonic forms may be due to the ability of the cells to survive at 21°C for longer periods of time and their tolerance of a pH range of 7-7.8. It was also observed that FHM cells synthesize large amounts of lipids as the cell ages (unpublished). Zuckerman (1964) reported that sudanophil granules of oocysts of *P. gallinacium* and *P. cynomolgi* increase in size as parasites develop. These findings suggest that the developing oocyst has a high requirement for lipids during development. It appears then that the ability of the FHM cells to synthesize lipids in large amounts as the cell ages might just be the condition necessary to support development of *P. berghei* ookinete into oocysts.

No further development of the malarial parasite occurs beyond the oocyst-like formation. It appears that the host cells, although actually dividing, are not nutritionally capable of supporting further sporogonic development. This study demonstrated that the cultural system employed can approximate the normal host for only a limited period of time. Determination of the oocyst-like structures in culture has to be confirmed by electron microscopy and specific histochemical staining.

Since it was shown that a non-vector cell line, the fat head minnow (FHM) epithelial cell line, seem to support development beyond ookinete stage, *in vitro* cultivation using other heterologous cell lines was initiated. Other established cell lines employed in an attempt to find a cell type capable of supporting malarial parasite development *in vitro* were: Rainbow Trout Gonadal (RTG-2); Diploid Grass Frog Embryo (ICR-134); Bull Frog Tongue (FT); Blue Gill Fry (BR-2); and African Clawed Toad Kidney (A6). Preliminary results employing these cell lines other than fat head minnow (FHM) are shown in Table 3.

Of the unusual host employed, aside from the fat head minnow (FHM), the African clawed toad kidney (A6) also supported few ookinete formation. More work along this line of investigation needs to be done.

Previous studies have shown that oocysts of varying ages continued their development *in vitro* for an additional 4 to 5 days when imaginal organs were added as supplements to either cell-cultured substrates or cell-free substrates (Chao and Ball, 1964; Schneider, 1968; Walliker and Robertson, 1970). However, the presence of bacterial contamination sets limitations to the cultivation. Attempts to overcome

TABLE 3

Cultivation of mosquito phase of Plasmodium berghei *in non-vector cell lines.*

Cell Lines	Maturation of Gamete	Zygote Formation	Ookinete Formation	Oocyst-like Structures
Fish Cell Lines:				
Fat Head Minnow (FHM)	++++	++++	++++	+++
Rainbow Trout Gonadal	-	-	-	-
Blue Gill Fry (BF-2)	-	-	-	-
Frog Cell Line				
Bull Frog Tongue (FT)	-	-	-	-
Diploid Grass Frog Embryo (ICR-134)	-	-	-	--
Toad Cell Line				
African Clawed Toad Kidney (A6)	+++	++	+	-

Legend (+) as to relative degree of development:
++++ indicates large number
+++ indicates slight number
++ indicates few number

this difficulty have been successful by rearing germ-free mosquitoes (Rosales-Ronquillo *et al.,* 1973). Aseptic guts, salivary glands and ovaries were added as supplements to cell substrates in *in vitro* culturing of malarial parasites. Viability of both the growing cell substrate and organ supplements persisted for as long as 20 days as indicated by their morphological and behavioral activities. Presently, the capability of this culture system to support sporogonic development is being investigated. Preliminary results showed the presence of substantial numbers of fertilized macrogametocytes and a few ookinetes 24 hours into the culture period. Further determination of the adequacy of such a system as culture substrate have yet to be made.

In addition to all efforts to cultivate the mosquito phase of *P. berghei* employing cell culture substrates, experiments using organs were initiated. The techniques involve the incubation of dissected guts (taken from females after blood-feeding on plasmodially-infected hamsters) in culture medium. Earlier attempts by Ball and Chao (1960) and Schneider (1968) using guts taken from conventionally reared females produced limited success due to the presence of bacteria and yeast in the guts. The use of heavy doses of antibiotics was found to be detrimental to the parasite. With the availability of aseptically reared female mosquitoes, the use of organ culture as a means to produce bacteria-free sporozoites was pursued (Rosales-Ronquillo *et al.*, 1973). The viability of the dissected aseptic guts was tested in different cell-free media. Guts remained active and pulsatile for long periods (ranging from 15-30 days) in all media at 21°C. Preliminary results of current attempts to cultivate the parasite by route of organ culture are shown in Table 4.

The best medium so far employed is medium 539 supplemented with fetal bovine serum (FBS). Giemsa-stained guts showed formation of young oocysts. A similar result was obtained when midguts were incubated in 5-days-old primary cell substrate (Rosales-Ronquillo *et al.*, 1973). So far, no development beyond the young oocyst stage occurred.

TABLE 4

Use of organ culture to cultivate sporogonic forms of Plasmodium berghei.

Media	Non-Infected Gut	Infected Gut				
		Developmental Stages of the Parasite				
	Viability* (days)	Viability* (days)	Maturation of Gamete	Zygote Formation	Ookinete Formation	Oocyst Formation
Medium 539	20	20	+++	+	+	-
Medium 539 + Fetal Bovine Serum	30	25	++++	++	++	+
Minimum Essential Medium	20	15-20	+++	-	-	-
Grace Insect Medium	20	15	-	-	-	-
NCTC 135	30	20	+	+	-	-

* Viability appraised by the Contractility in culture

** Media not changed throughout cultivation

Legend (+) as to relative degree of development

++++ indicates large number

+++ indicates slight number

++ indicates few number

\+ indicates trace

V. Conclusion

So far there is no known methodology for continuous *in vitro* cultivation of the sporogonic forms of the malarial parasite from gametocytes to sporozoite stage. Despite the limited progress made in the last 20 years by Ball and Chao and Dr. Imogene Schneider and presently by Rosales-Ronquillo and Silverman (1974), the attainment of the objective indeed seems quite remote. Perhaps it is of utmost importance that one must recognize the complexity of the biochemical and physiological nature of the parasite and include such in all workable hypotheses. For instance, the sporogonic forms of the malarial parasite have a complex life history during their development requiring successive sites in a variety of mosquito tissue under different conditions. To approximate such changing and exacting *in situ* requirements of the parasites in an *in vitro* milieu in attempts to cultivate the entire sporogonic form in a single preparation would involve extensive experimentation. In spite of all the tremendous amount of work directed towards such a goal, one is not close to a system for continuous *in vitro* single culture of the whole life cycle. Work along this line should be encouraged in order to augment and perhaps optimize some studies on immunology of malaria infection.

VI. References

Ball, G.H. and Chao, J. (1960) *Exp. Parasitol. 9:* 47.

Ball, G.H. and Chao, J. (1964) *Amer. J. trop. Med. Hyg. 13:* 181.

Dutky, S. R. (1964) *Amer. J. trop. Med. Hyg. 13:* 193.

Mitchell, G. H., Butcher, G. A. and Cohen, S. (1974) *Nature 252:* 311.

Nussenzweig, R. S., Vanderberg, J., Spitalny, G. L., Rivera, C. I. O., Orton, C. and Most, H. (1972) *Amer. J. trop Med. Hyg. 21:* 722.

Rosales-Ronquillo, M. C., Simons, R. W. and Silverman, P. H. (1973). *Ann. Entomol. Soc. Amer. 66;* 949.

Rosales-Ronquillo, M. C., Simons, R. W., and Silverman, P. H. (1972). *Ann. Entomol. Soc. Amer. 65,* 721.

Rosales-Ronquillo, M. C., Simons, R. W. and Silverman, P. H. (1973). *J. Parasitol. 60;* 819.

Rosales-Ronquillo, M. C., Nienaber, G. and Silverman, P. H. (1974) *J. Parasitol. 60:* 1039.

Schneider, I. (1968) *Exp. Parasitol. 22:* 178.

Schneider, I. (1969) *J. Cell Biol. 42:* 603.

Simpson, G. L., Schenkel, R. H. and Silverman, P. H. (1974) *Nature 247:* 304.

Speer, C. A., Rosales-Ronquillo, M. C. and Silverman, P. H. (1974) *J. Invert. Path. 25:* 73.

Walliker, D., and Robertson, E. (1970) *Trans. Roy. Soc. trop. Med. Hyg. 64:* 5.

Zuckerman, A. (1964) *Amer. J. trop. Med. Hyg. 13:* 209.

Chapter 7

APPLICATION OF TISSUE CULTURE OF A PULMONATE SNAIL TO CULTURE OF LARVAL *Schistosoma mansoni*

E.L. Hansen

I. Introduction

The use of tissue culture for cultivation of parasites is a logical application of the recognition of the nature of parasitism. Over a long period of evolution parasites have become adapted to and dependent upon their hosts and are unable to exist outside of these hosts except for limited periods or in a special form. Parasites are, therefore, difficult to cultivate *in vitro*. Cultures are characterized by slow and limited growth and usually end at some critical stage in development. Such a limitation indicates deficiencies in nutrition or in the cultural environment. It is possible that these deficiencies can be met by an associated culture of host tissue.

Cultivation of parasites synxenically with tissue cultures was reviewed by Chao in 1973. Most studies had been made with protozoa having intracellular parasitic stages. Among metazoa, only microfilaria have been extensively studied. Microfilaria are the intermediate stage, occurring in insects, of nematodes parasitic in man and other animals. Extension of the synxenic technique to culture of the intermediate stage of a trematode, specifically *Schistosoma mansoni*, has only recently been attempted. The technique has been made the more difficult by the paucity of information on the physiology of the intermediate host, *Biomphalaria glabrata*, a pulmonate snail, and also by the complexity of the trematode developmental stages.

II. Larval development of *Schistosoma mansoni*

In the course of development in the snail, larval *S. mansoni* pass through several asexual generations. Infection begins with a miracidium which, remaining close to the site of penetration, converts to a mother, or primary sporocyst. The mother sporocyst gives rise to second generation, or daughter sporocysts which migrate to the digestive gland, gonad, and rectal ridge of the snail. These second generation sporocysts give rise to tertiary, etc., sporocyst generations and to cercariae. An individual sporocyst may go direct to cercarial production or may first produce another sporocyst generation and then start cercarial production. The sequence takes 20 to 30 days before appearance of the first cercaria. Cercariae continue to emerge from the snail until it eventually dies. By infecting man the cercariae initiate the sexual phase of the trematode life cycle.

The host-parasite relationship thus differs in the successive larval stages. Miracidia and cercariae are active and free living. Sporocysts occur only within the snail. While small they have sluggish motility, their movement in the tissues being aided by the open hemolymph circulation. They become sessile and grow in two entirely different sites, mother sporocysts in the peripheral tissues, and daughter sporocysts in the internal organs. In both sites the sporocysts form close association with snail tissue by means of microvillous processes which can be seen also in cultured specimens (Krupa & Bogitsh 1972, DiConza & Basch 1974a) (Fig. 1). Nutriments must pass through the multilayered wall of the sporocyst to reach the embryos that lie within its brood chamber.

Development is rapid and continuous, several hundred cercariae being released each day from the snail. The rate of release from an individual sporocyst has not been determined for a normal infection where large numbers of sporocysts are present. However, some idea of the rate can be obtained from experiments in which single daughter sporocysts were implanted (Hansen 1973). In those cases where development proceeded directly to cercariae rather than to successive generations of sporocysts, the rate of cercarial emergence averaged 7 per day and continued for 55 days, at which time the snails were dissected to ascertain the sporocyst population.

The effect of the sporocysts on their snail host has been described as comparable to starvation (Meuleman 1972). Earlier, Cheng (1963) had reported a decrease in cellular glycogen and in hemolymph proteins. Indirectly this is a comment on the efficiency of intake and on the high nutritional requirement of the sporocysts. *In vitro* culture must thus provide for a rapid intake of nutrients as well as substitute for a complexity of interrelations between trematode and host tissues (Heyneman 1975).

III. Types of culture

1. Organ culture of *Biomphalaria glabrata*.

Organ cultures of *B. glabrata* have been of great value in providing basic data on media. This data has been used in design of media for snail tissue cultures and for

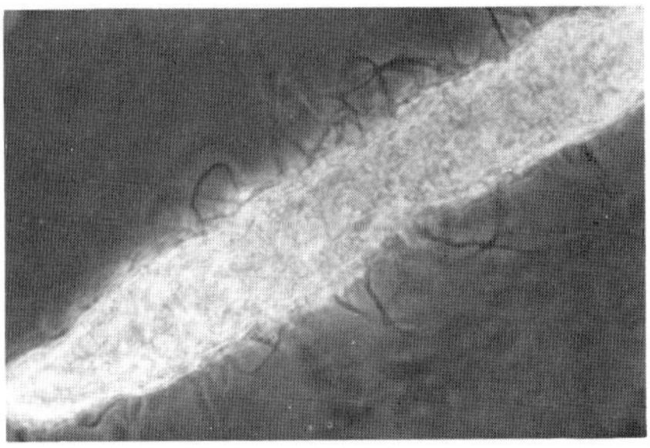

Fig. 1. Schistosoma mansoni, portion of a daughter sporocyst, 350 μm long, cultured axenically for 6 days, showing tegumental processes. Gluteraldehyde fixed. Phase contrast; bar 20 μm.

culture of sporocysts. The salt-sugar solution of Chernin (1963) has been used in most subsequent culture work. Chernin (1964) showed that maturation of cercariae continued in cultured fragments of parasitized digestive gland. Thus development of the parasite continued even in tissue removed from the influence of the intact host. Cercarial development in the fragments was affected by the composition of the medium. Benex (1967), using an entirely different medium showed that isolated tentacles could be cultured and were suitable for study of miracidial penetration. Other aspects of molluscan organ culture have recently been reviewed by Bayne (1975).

2. Tissue culture of *Biomphalaria glabrata.*

Only recently have tissue cultures of *B. glabrata* become available for synxenic culture of the trematode. The first reports on tissue cultures described the preparation of the cell suspensions and the morphology in primary cultures (Burch & Cuadros 1965, Cheng & Arndt 1973, Basch & DiConza 1973 & 1974, Hansen & Perez-Mendez 1973). In 1974 a cell line, "Bge", was established from embryo tissue (Hansen 1974 & 1975). A wide variety of formulations of culture media were used for the primary cultures and for the Bge cell line. These formulations fell within a narrow range of pH and osmolality but differed in ionic composition and in content of peptones, amino acids, vitamins, and minor constituents. There is presently no clear indication as to the criticality of the various components. The formulations usually included portions of media that were commercially available and designed for culture of other types of tissue. The original references should be consulted for details.

The effect of different media was conveniently tested by replacing the existing medium on established primary cultures. The cultures were then observed for differences in growth and morphology. This method avoided difficulties in starting primary cultures in a new medium. It was suitable for quickly eliminating toxic or inhibiting mixtures and for detecting improvements in growth. One such alternate medium produced repeated slow spasmodic cellular contractions which were reversible with change of medium (Hansen & Perez-Mendez 1973). Colony formation, the prerequisite to initiation of the Bge cell line, was observed in several types of media (Hansen 1975b). The Bge cell line, though started in a medium that had been based on analysis of hemolymph, was later transferred to one based on diluted

Schneider's *Drosophila* medium and grew equally well. Other medium variations have subsequently been successfully used in the 54 subcultures to date, indicating again that a specific medium was not required for culture of *B. glabrata* cells.

Procedures for maintaining the cell line have been described in detail (Hansen 1975b). To this two recent observations can be added: the complete medium can be stored at 4°C for at least 4 months without losing its effectiveness, and the special trypsinization solution can be dispensed with. A mixture of trypsin-EDTA 10x GIBCO 540 (Grand Island Biological Co. N.Y., 14072) diluted 1:50 in water was used with satisfactory results.

3. Culture of sporocysts of *Schistosoma mansoni*.

Both small motile sporocysts and large sessile forms have been used to initiate cultures. Motile second generation sporocysts were particularly convenient because they could be released by dissection of mother sporocysts with minimal mechanical damage and minimal exposure to possible inhibitory effects of antibiotics. The major sterilization was done on the tissue fragment containing the mother sporocyst, and the dissection was carried out in a very dilute antibiotic solution. Small motile tertiary generation sporocysts were obtained from infected tissue or were produced in cultures of second generation sporocysts (Hansen 1975a). Large sporocysts imbedded in snail tissue were liberated from the tissue by trypsinization and teasing before introducing into culture.

A variety of media have been used (DiConza & Basch 1974a, Hansen *et al.* 1974a & 1974b, Buecher *et al.* 1974). As with the molluscan cells, osmolality is important (Hansen *et al.* 1973). However, to date, only a minimal response has been obtained and the tolerance to different media is only in terms of survival and limited development. It is therefore possible that medium composition may be critical in obtaining full development of the sporocysts.

4. *Synxenic cultures.*

For synxenic cultures a tissue culture layer was first established and then the larval trematodes were added. They were placed directly on the cell layer or were contained within small chambers in which they were separated from the cells by a transparent and porous Nucleopore membrane (Hansen *et al.*, 1973). By being enclosed within the chamber the trematodes were easy to observe and did not become overgrown by tissue culture cells. The membrane permitted passage of fluid. The entire assembly could be quickly and easily moved to a new cell culture.

The tissue culture used initially was of an unidentified arthropod. This was subsequently replaced by cultures of *Aedes albopictus* (DiConza & Hansen 1973). As soon as cultures of *Biomphalaria glabrata* became available they were used, first as primary cultures and later as subcultures of the Bge cell line.

The favorable sporocyst response in synxenic culture with arthropod cells (DiConza & Hansen 1973) led immediately to application of *Aedes albopictus* tissue cultures. Sporocysts were maintained sufficiently well to produce a new generation of daughter sporocysts (Hansen *et al.* 1973). Since in these latter cultures the sporocysts were separated from the cell layer by the Nucleopore membrane, the effect of the associated cells was mediated by the medium. This conditioning effect was retained after the medium had been harvested and stored at 4°C, and this stored medium was successfully used for axenic cultures of sporocysts (Hansen *et al.* 1974b). It was then found that the conditioning effect could be replaced by controlling the redox potential of the medium (Hansen *et al.* 1974b). Sulfhydryl compounds were added and the gas phase adjusted to lower oxygen and increased carbon dioxide (Buecher *et al.* 1974). A partially reduced condition probably also existed in the sealed tubes containing deep medium in which DiConza and Basch (1974a) maintained sporocysts that retained ability to establish an infection when reintroduced into a snail 3 weeks later.

The sporocyst response in axenic cultures was at a minimal level. At best a single progeny sporocyst generation was produced. This is in marked contrast to the great reproductive potential of a sporocyst *in vivo*. Association with *Aedes albopictus* had the great disadvantage that in providing medium of low osmolality for sporocysts the decrease from 280 mOsm to 120 mOsm made conditions unfavorable for the tissue culture. The *Aedes* cells did not divide but gradually rounded and detached.

Primary tissue cultures of *B. glabrata* obviated this disparity in osmolality and the medium used was suitable for sporocysts as well as for the cell cultures. Good sporocyst growth was obtained even though the primary tissue cultures were gradually declining during the incubation period. The extent of germinal ball development at 20 days is shown in Fig. 2. These embryos, when examined later by phase contrast after Feulgen stain, had the characteristic fluorescence of cercarial embryos (Hansen 1975a).

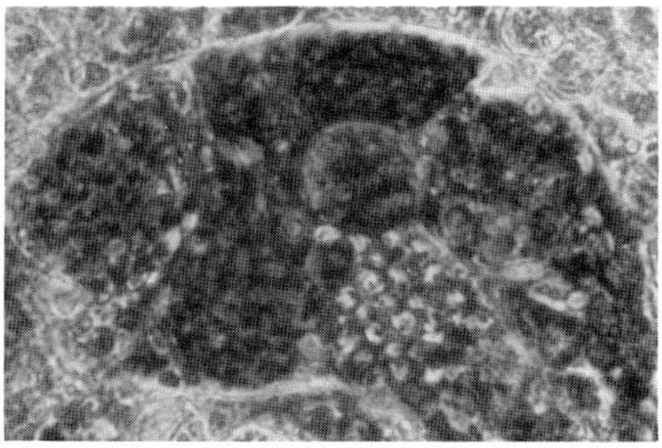

Fig. 2. Embryos in sporocyst of *Schistosoma mansoni* at 20 days grown in a primary culture of *Biomphalaria glabrata* cells; bar 20 μm.

A culture of the Bge cell line persisted in good condition for at least 19 days if the medium was replenished at intervals of 2-3 days. Subcultures in the twentieth passage were used synxenically with tertiary daughter sporocysts which had emerged in other cultures. They were placed on the cell layer (Fig. 3a). During incubation for 12 days (in medium 315, see below), the sporocysts gradually became covered with cells from the tissue culture as shown in Fig. 3b. After the culture was terminated to permit detailed examination, the sporocysts were found to have grown from 120 μm x 25 μm to 370 μm x 60 μm. They retained wall motility and were packed with large clear germinal cells as well as small embryos.

In other synxenic cultures the sporocysts were held in Nucleopore chambers; growth response was not improved over that which had been obtained in earlier synxenic cultures and there was only a minimal emergence of tertiary generation sporocysts.

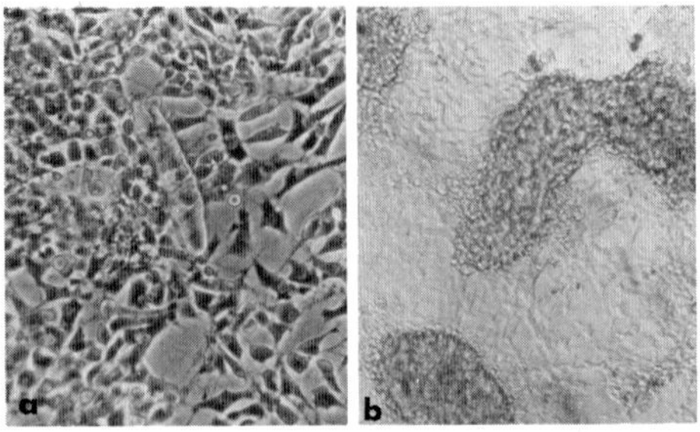

Fig. 3. Synxenic culture, *Schistosoma mansoni* sporocyst in culture of Bge cell line; phase contrast.

a) Sporocyst on 48 hours culture of Bge cells in 20th subculture: bar 50 µm.

b) Sporocyst grown and covered by cells 12 days later; bar 100 µm.

The large second generation sporocysts also contained cercarial embryos, but these showed very little development in culture. This raised the question of whether the degree of rapid cell multiplication and subsequent differentiation that occured in cercarial maturation could be supported by *in vitro* culture conditions. Small cercarial embryos, after formation of the differentiated wall (Meuleman & Holzmann 1975), lie free in the brood chamber of the sporocyst and can be easily dissected out. We found that development continued when these embryos were introduced into synxenic culture with Bge cells. Round embryos 30-40 µm in diameter, Stage III as described by Cheng & Bier (1972), developed to tailed cercariae 150-200 µm long in 4 to 7 days (Hansen 1974a). These cercariae retained motility for a further 5 days, and showed schistosomule-like movement. In contrast, in axenic culture, even under reducing conditions (Buecher *et al.* 1974), there was only partial tail growth. These experiments have now been repeated in another laboratory and the cercariae are being tested for infectivity (J. J. DiConza, personal communication).

The media for synxenic cultures were formulated by diluting the Bge culture medium (Hansen 1974) and adding components from media that were under study for culture of the sporocysts. The final medium (170 mOsm) contained 15% Schneider's *Drosophila* medium, (Schneider 1964, 1966) (GIBCO 172), lactalbumin hydrolysate 2.5 g/l, galactose 1.5 g/l, nucleic acid precursors 50 µm, choline 0.65 mg/l, inositol 0.46 mg/l, sodium pyruvate 5 mg/l, l-alanine 6 mg/l, l-asparagine 2 mg/l, Chernin's salts at one-half fold, HEPES buffer 1.5 mM, and 5x MEM vitamins. Inactivated fetal calf serum was included at 10% except in medium 315 (130 mOsm) in which it was decreased to 0.5% and 4.5% of hemolymph from parasitized snails added. The hemolymph had been aspirated from the cardiac region, diluted with an equal volume of saline containing 0.1 mM dithiothreitol, filtered for sterility through a Millipore membrane of 0.3 µm porosity, and stored in frozen aliquots.

The particular characteristics of the medium are the low osmolality and content of galactose. The osmolality of *B. glabrata* hemolymph is 120 mOsm. Galactose was included because it is present in snail hemolymph (Liu & Hansen 1975). It increased growth of Bge cells when added to media that already contained glucose and trehalose. Its utilization *in vitro* by sporocysts of other trematodes, *Microphalus* spp., has recently been shown by McManus & James (1975). Requirements for the other components were not tested.

IV. Application of tissue culture

1. *Design of media for culture of larval schistosomes.*

From the onset, in the attempt to cultivate larval *S. mansoni*, it has been valuable to consider parallels between requirements of the trematode and those of snail tissue culture. With both systems a wide variety of basal media can be tolerated. Both systems showed great sensitivity to serum. While growth of tissue cultures was equally good in a number of different media, the requirements for optimal response of sporocysts has not been achieved in any medium formulated to date. Peptone hydrolysates improved tissue culture growth but sporocysts showed a differential response to certain peptones, development being better with lactalbumin hydrolysates and the overriding effect of serum has made it difficult to assess the effect of specific components of the basal media. All possible requirements have been included at rather high levels. Even when an approach to a rational medium had been attempted by formulating the amino acids accordingly to those in hemolymph, the levels included have usually been increased. The improved trematode response in synxenic culture with Bge cells will now make it possible to evaluate improved formulations of media.

2. *Serum and hemolymph.*

Because snail hemolymph initially appeared to be toxic, and in view of the successful use of vertebrate serum in cultures of insect tissue (Weiss 1971), media for culture of snail cells and larval trematodes have contained vertebrate serum. Most commercially available types of serum, including newborn calf, dialyzed calf, horse, goat, rabbit and chicken (Hansen 1975c) were toxic to snail tissue culture and to larval *S. mansoni*. This toxicity was decreased by heating the sera to 56°C for 30 min. Fetal calf serum proved least toxic, but only selected batches were suitable. Serum lots were evaluated by snail tissue cultures, initially using primary cultures and later the Bge cell line, before use in sporocyst cultures. The toxic effect of a poor lot of serum was manifest by clumping of the cells and low attachment. A typical response is shown in Fig. 4 together with a parallel culture in control medium. When the entire test medium which gave this poor response was replaced by control medium there was a rapid recovery. Outgrowths of cells developed and were sufficient for subculture.

Even "good" lots of serum seem to have some slight toxic effects and differ in the rate of Bge cell growth that they support. In the steps which led to the isolation of the snail cell line the introduction of short periods of cultivation in serum free medium appeared to have been essential for the appearance of the colonies from one of which the cell line was eventually derived (Hansen 1976b). Basch and Di-Conza resorted to using a selected human donor as their source of serum for sporocyst culture (personal communication).

Attempts to eliminate serum have not been successful. Sporocysts showed no growth in medium without serum (Hansen 1974a); tissue culture cells did not divide. Substitution of albumin was not successful.

The requirement for serum in culture medium has been attributed to both its physical effects and to its nutritional components (Taylor 1974). Vertebrate serum and snail hemolymph differ in many respects and furthermore the hemolymph composition is changed by parasitism (Lee & Cheng 1972b). There is a depletion of certain of the hemoglobin proteins (Anteson & Williams 1975). The selective nature of this

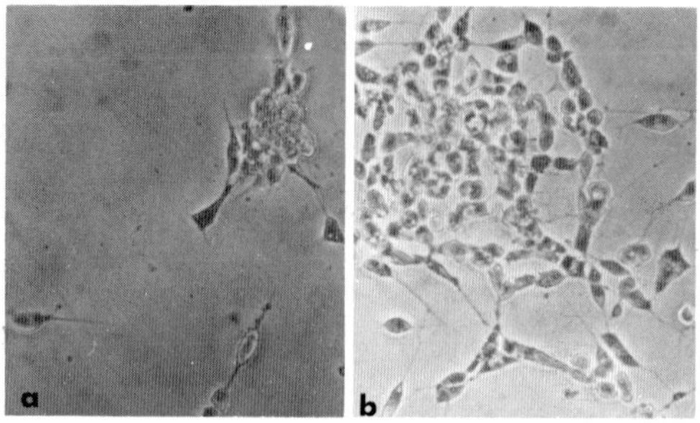

Fig. 4. Bge cell line in subculture 43, incubated 24 hours at 27°C in
a) medium with poor serum and
b) control medium. Phase contrast; bar 50 μm

depletion raises the question of whether the different proteins that occur in vertebrate serum can be utilized by sporocysts.

Hemolymph can be used in culture medium if it is first treated. It does not coagulate on withdrawal but rapidly changes from red to green as it is exposed to the air, becoming toxic to both tissue cultures and sporocysts. This change was prevented by the procedure of aspirating the hemolymph into 0.1 mM dithiothreitol in saline, thus maintaining a reduced condition. The response of tissue cultures and sporocysts in media containing hemolymph treated in this manner was excellent and warrants further investigation.

3. *Tissue culture of the digestive gland.*

There has been repeated speculation that trematode development is influenced by snail hormones (Coles 1970, Bayne 1972). Although each individual step in development can take place *in vitro*, i.e., miracidium to primary sporocyst (Voge & Seidel 1972, Basch & DiConza 1974a), secondary sporocyst to tertiary sporocyst (Hansen *et al.* 1974b), and sporocyst to cercariae (Chernin 1964, Hansen 1975a) the entire developmental cycle has not, as yet, been carried out as one step. It is therefore possible that, to achieve the entire sequence *in vitro*, additional factors from the snail might be required. One possible experimental approach is to culture tissue of the preferred organ, the digestive gland.

In preliminary experiments cells of trypsinized digestive gland attached within 24 hours. Cells migrated from the explants during 22 days of culture. Outgrowths at 46 days are shown in Fig. 5. They remained attached for 63 days but no further cell growth was observed and cultures were terminated. The cell growth was too sparse to attempt association with sporocysts in synxenic culture. However these initial findings suggest that difficulties with culture of the digestive gland can be overcome by frequent replacement of medium so that disintegrating cells are removed. A sequence of media was used for this culture as for the culture of embryo cells (Hansen 1975b).

4. *Culture of cercariae.*

Each successive trematode generation starts from a single germinal cell and necessitates great cell multiplication. Even an early embryo 18 μm in diameter already contains 42 cells. Cercarial growth particularly necessitates rapid cell

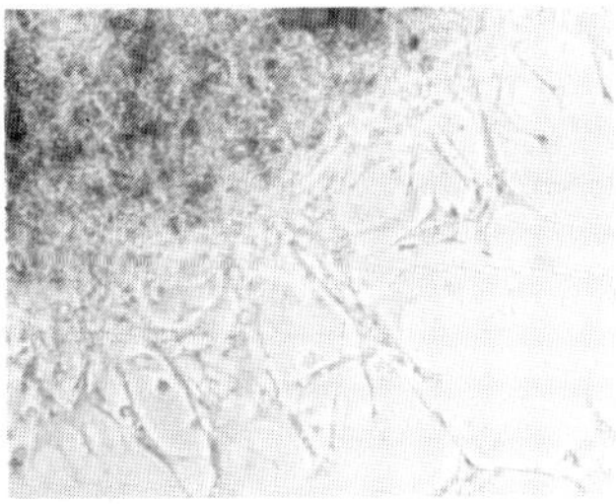

Fig. 5. Explant of digestive gland of *Biomphalaria glabrata,* cell outgrowths at 46 days in culture. Phase contrast; bar 100 μm.

multiplication. Since in culture cercariae did not develop until the embryos had been surgically liberated from the sporocyst, it is possible that under *in vitro* conditions the sporocyst wall impeded availability of nutrients to embryos within the brood chamber. During cultivation of second generation sporocysts from the digestive gland the walls noticeably increased in granulation and few tegumental processes were seen. Presumably it is essential also that the enzymatic activity in the sporocyst wall that has been demonstrated *in vivo* (Krupa & Bogitsh 1972) should be retained in cultured sporocysts.

Cercarial development in synxenic culture now makes it possible to evaluate the specific contribution of the tissue culture. In our experiments this contribution could not be replaced by addition of sulfhydryl compounds to axenic media. Undefined but beneficial response to conditioned medium has been reported in other types of tissue culture where the medium was suboptimal (Morton & Isaacs 1972). Trematode uptake of ^{59}Fe from snail hemolymph (Lee & Cheng 1972b) may indicate a special requirement for heme and a possible dependence on cytochromes from the associated tissue culture.

Development of cercariae from embryos in synxenic culture will provide a source of axenic cercariae for future study. The transformation from cercaria to schistosomule may be different in cultured cercariae that have not had to pass through penetration phases first to get out of the snail and then into the new host, with a period in between of existence in an unfavorable environment. The change of normal cercariae to schistosomule-like organisms in a vertebrate tissue culture was demonstrated by Jensen *et al.* (1965).

5. *Immunity.*

The Bge cell line offers possibilities for investigations of snail immunity to trematode infection. Immunity appears to be associated with production of macrophage-like cells. These hemocytes occur in the snail in numbers that are probably too few to initiate a culture, but is is quite probable that they could be maintained in association with cultures of the Bge cell line and thus be available for study.

6. *Developmental control.*

While admittedly synxenic cultures are more difficult to interpret than axenic cultures, it is possible that their complexity might be essential to the complexity of the successive larval stages. Development of sporocysts in synxenic cultures was sufficiently good to suggest that the reproductive potential of sporocysts could be realized *in vitro* and thus make it possible to investigate the nature of the factors controlling embryonic development to either cercariae or sporocysts from the genetically idential germinal cells in the sporocyst brood chamber.

7. *Cultivation of schistosome cells.*

The close relationship between snail and trematode was exploited for initiation of a tissue culture of *S. mansoni.* Using the same methods as for snail tissue we were able to maintain cercarial cells in culture. Tiny cells migrated from trypsinized cercarial embryos and remained attached for 27 days. Attempts to culture cells from trypsinized, comminuted sporocysts were not successful. Since the amount of tissue that can be obtained from the parasites is very small it is difficult to reach a suitable balance between cells and medium and it is possible that cultivation would be improved by using conditioned medium from the Bge cell line.

8. *Species considerations.*

The above *in vitro* studies with *S. mansoni* and snail tissue culture were made with a laboratory strain of *B. glabrata* that combined characteristics of high susceptibility and albinism (Newton 1955). The trematode, originally collected in Puerto Rico, has been maintained for more than 10 years in laboratory animals. However, in the field, the life cycles of human schistosomes involve several species of parasites and several species and strains of snails. The extent to which the present findings can be extended to cultivation of other snails and other larval schistosomes can now be tested. While some difficulty might be anticipated, as has been encountered in establishing cultures from related species of other invertebrates, such differences in requirements can contribute to our knowledge of host-parasite specificity.

V. Conclusion

Even though the use of synxenic cultures for study of *S. mansoni* is still in a preliminary stage, the results are an encouraging contribution toward the ultimate goal of maintaining the entire larval cycle of the parasite *in vitro.*

Acknowledgment

The author is indebted to Dr. E. Buecher who participated in portions of this work and to B. David, M. Woodmansee, G. Perez-Mendez and R. Yescott for assistance. Dr. D. Heyneman generously supplied to infected snails.

Supported by N.I.A.I.D. grant A1-07359 and contract N01-A1-22525 for work at the former Clinical Pharmacology Research Institute, Berkeley, California.

(Present address is 561 Santa Barbara Rd., Berkeley, CA 94707)

VI. References

Anteson, R. K. and Williams, J. F. (1975). *J. Parasit.* *61,* 149.

Basch, P. F. and DiConza, J. J. (1973). *Am. J. Trop. Med. Hyg.* *22,* 805.

Basch, P. F. and DiConza, J. J. (1974a). *J. Parasit.* *60,* 935.

Basch, P. F. and DiConza, J.J. (1974b). *J. Invert. Path.* *24,* 125.

Bayne, C. J. (1972). *Parasitology* *64,* 501.

Bayne, C. J. (1975). In: "Invertebrate Tissue Culture" (Maramorosch, K., ed.), Academic Press (in press).

Benex, J. (1967). *Ann. de Parasit. 42,* 493.

Buecher, E. J., Perez-Mendez, G., Hansen, E. L. and Yarwood, E. (1974). *Proc. Soc. Exp. Biol. Med. 146,* 1101.

Burch, J. B. and Cuadros, C. (1965). *Nature 206,* 637.

Chao, J. (1973). In: "Current Topics in Comparative Pathobiology" (Cheng, T. C. ed.), Vol. 2, pp. 107-144, Academic Press, New York.

Cheng, T. C. (1963). *Ann. N. Y. Acad. Sci. 113,* 289.

Cheng, T. C. and Bier, J. W. (1972). *Parasitology 66,* 129.

Cheng, T. C. and Arndt, R. J. (1973). *J. Invert. Path. 22,* 308.

Chernin, E. (1963). *J. Parasit. 49,* 353.

Chernin, E. (1964). *J. Parasit. 50,* 531.

Coles, G. C. (1970). *Comp, Biochem. Physiol. 34,* 213.

DiConza, J. J. and Hansen, E. L. (1973). *J. Parasit. 59,* 211.

DiConza, J. J. and Basch, P. F. (1974). *J. Parasit. 60,* 757.

Flandre, O. (1971). In: "Invertebrate Tissue Culture" (Vago, C. ed.), Vol. I, pp. 361-383, Academic Press, New York.

Hansen, E. L. (1973). *Int. J. Parasit. 3,* 267.

Hansen, E. L. (1974). *IRCS, Med. Sci. 2,* 1703.

Hansen, E. L. (1975a). Ann. N. Y. Acad. Sci. (Cheng, T. C. and Bulla, L. A. eds.) (in press).

Hansen, E. L. (1975b). In: "Invertebrate Tissue Culture". (Maramorosch, K. ed.), Academic Press, New York (in press).

Hansen, E. L. (1975c). Ann. Meeting Tissue Cult. Assoc. (abstract # 34).

Hansen, E. L. and Perez-Mendez, G. (1973). *IRCS, Med. Sci. 73-11,* 1-0-12.

Hansen, E. L., Perez-Mendez, G., Long, S. and Yarwood, E. (1973). *Exp. Parasit. 33,* 486.

Hansen, E. L., Perez-Mendez, G. and Yarwood, E. (1974a). *Exp. Parasit. 36,* 40.

Hansen, E. L., Perez-Mendez, G., Yarwood, E. and Buecher, E. J. (1974b). *J. Parasit. 60,* 371.

Heyneman, D. (1975). In: "Invertebrate Tissue Culture". (Maramorsch, K. ed.), Academic Press, New York, (in press).

Jensen, D. V., Stirewalt, M. H..and Walters, M. (1965). *Exp. Parasit. 17,* 15.

Krupa, P. L. and Bogitsh, B. J. (1972). *J. Parasit. 58,* 495.

Lee, F. O. and Cheng, T. C. (1972a). *Exp. Parasit. 31,* 203.

Lee, F. O. and Cheng, T. C. (1972b). *J. Parasit. 58,* 481.

Liu, C. L., and Hansen, E. L. (1975). *J. Invert. Path.* (in press).

McManus, D. P. and Jones, B. L. (1975). *Int. J. Parasit. 5,* 177.

Meuleman, E. (1972). *Neth. J. Zool. 22,* 355.

Meuleman, E. A. and Holzmann, P. J. (1975). *Z. Parasitenk. 45,* 307.

Morton, H. J. and Isaacs, R. J. (1972). *J. Natl. Cancer Inst. 49,* 1071.

Newton, W. L. (1955). *J. Parasit. 41,* 526.

Schneider, I. (1964). *J. Exp. Zool. 156,* 91.

Schneider, I. (1966). *J. Embryol. Exp. Morph. 15,* 271.

Taylor, W. G. (1974). *J. Natl. Cancer Inst. 53,* 1449.

Voge, M. and Seidel, J. S. (1972). *J. Parasit. 58,* 699.

Weiss, E. (1971). Current Topics in Microbiology and Immunology No. 55, pp. 1-288, (Weiss, E. ed.), Springer-Verlag, New York.

II

APPLICATION IN BIOLOGY

Chapter 8

INSECT CELL AND TISSUE CULTURE AS A TOOL FOR DEVELOPMENTAL BIOLOGY

J.C. Landureau

I. Introduction

Cell and tissue cultures have found a very broad use in various fields of biology, They could not but stimulate a renewed interest at the moment when, after the conceptual revolution which permitted the study of bacteria and viruses, contemporary biology is analysing the function of genetic material of eucaryotic cells.

The utilisation of liquid media facilitates easy cultivation of almost all tissue categories of higher vertebrates. Numerous cell strains have been established. They provide homogenous material, in a theoretically unlimited quantity, which can multiply in an easily regulated medium. Moreover, certain lines, aside from a basic program necessary for their maintenance *in vitro*, retain specialized activities. The best known example in mammals is the model of: hepatoma cells from a rat, secreting albumine and synthesizing tyrosine amino transferase by the action of steroid hormones. The results have stimulated a vast amount of research on the regulation of gene expression by somatic cell hybridization (Davidson, and de la Cruz, 1974).

We are far from these results in the case of invertebrates: the development of insect cell lines started only in 1962, when Grace established the first cell strain (Grace, 1962). The author recognized the very small percentage of cultures which have evolved and deplored the polyploid character of cell caryotypes (up to 64 n chromosomes).

After a prolonged study, I was able to establish conditions which permitted me to regularly obtain euploid cell lines (Landureau, 1968). These lines can be grown in a synthetic medium and therefore are completely comparable to those available for mammals.

I notice then that certain strains of hemocyte origin maintained *in vitro* performed specialized biosynthetic activities. These activities seemed to be related to physiological functions which are, in insect biology, regulated by the molts. Given the rather advanced knowledge which we possess of the neuroendocrine control of these phenomena of growth, it seemed to me that my cell lines could constitute an interesting material for studying the function and regulation of differentiated activities.

Moreover, the experience acquired over the years permitted me, to cultivate tissues and organs taken from defined stages of differentiation, so as to resolve problems of insect physiology.

The success of these techniques has removed real or imaginary obstacles which seemed to prevent the study of cell differentiation and of the biology of insect development *in vitro*.

II. Perfection of the *in vitro* techniques

The realization of insect cell cultures faced numerous difficulties: some were concerned with general methods of cell culture such as selection of flasks needed for the reduced volume of available tissues, pronounced cell sensitivity to the usual agents of dissociation, etc. Others were related to the establishment of the appropriate culture medium. The latter had to take into account the physico-chemical characteristics of insect hemolymph. Moreover, protein supplementation is needed as a general rule, and up to 1962 the only successful addiditive was provided by hemolymph from diapausing pupae (Grace, 1962).

A. Obtaining the first regular cell cultures

I chose cockroach embryos: the interest in these insects lies in their primitive characteristics (the order has changed very little since the permo-carboniferous period) and also in the fact that we are dealing with a material whose physiology is relatively well known.

1. *Primary cultures:*

The cultures were started from embryos in stage 26 (a total of 32 stages in our experimental conditions). In this material we notice the fixation of numerous cell categories. Many disappeared at the end of the first week.

It is necessary to wait 10 to 15 days to observe in almost all cultures new mitosis. This leads to a wave of second order migrations, the cells forming networks or laminae which bind together the small tissular fragments. The bottom of the flask is covered in a short time. Nevertheless, the transferring of these primary cultures remained difficult for a long time.

2. *EPa strain:*

The development of long term cultures greatly benefited from the progressive improvements of culture conditions. In all cases studied *(Blattella germanica, Blabera fusca* and *Periplaneta americana),* the culture of embryonic cells became theoretically unlimited and thus I succeeded in cultivating very different cellular categories for several years. The most interesting result concerns the establishment of a cell line which multiplies very rapidly: the EPa strain from embryonic cells of *P. americana* (Landureau, 1968). These cells are apparently homogeneous (Fig. 1), their caryotypes remain essentially diploid with less than 25% of tetraploid metaphases.

EPa cells do not show contact inhibition; on the contrary, they tend to form important aggregates. This property has been studied according to the techniques of Curtis (1969). Embryonic cells of cockroaches resemble by their characteristics insect hemocytes. Contrary to all cell lines of higher vertebrates, they retain their ability to aggregate in the absence of Mg^{2+} or Ca^{2+}, or both, in the culture medium.

3. *EPa clones:*

While the fibroblasts developed from *B. fusca* embryos allowed me to propagate *in vitro*, for the first time, cockroach intracellular bacteria, the same bacteria, brought into contact with other lines, were rapidly destroyed. The result was often the same, especially with EPa strain, in the course of accidental infections by bacteria, yeasts or molds.

To analyse this phenomenon, I studied the enzymatic activity in the metabolized culture medium and compared this with bacteriolytic agents described in the tissues and especially in the insect hemolymph (Bernier *et al.*, 1974).

The producing cells were isolated by cloning (Fig. 2). They enclose in their cytoplasm numerous granules which could correspond to the storage of the bacteriolytic enzymes. After an infection or simply by changing the metabolized medium, the cells react within less than 24 hours by releasing these proteins, and this increases or re-establishes the enzymatic activity.

Some granules are filled with microtubules which can be observed in cross section or longitudinal section. The interior of the granule tends to become impervious to electrons but the microtubules always remain distinct. I was able to find this

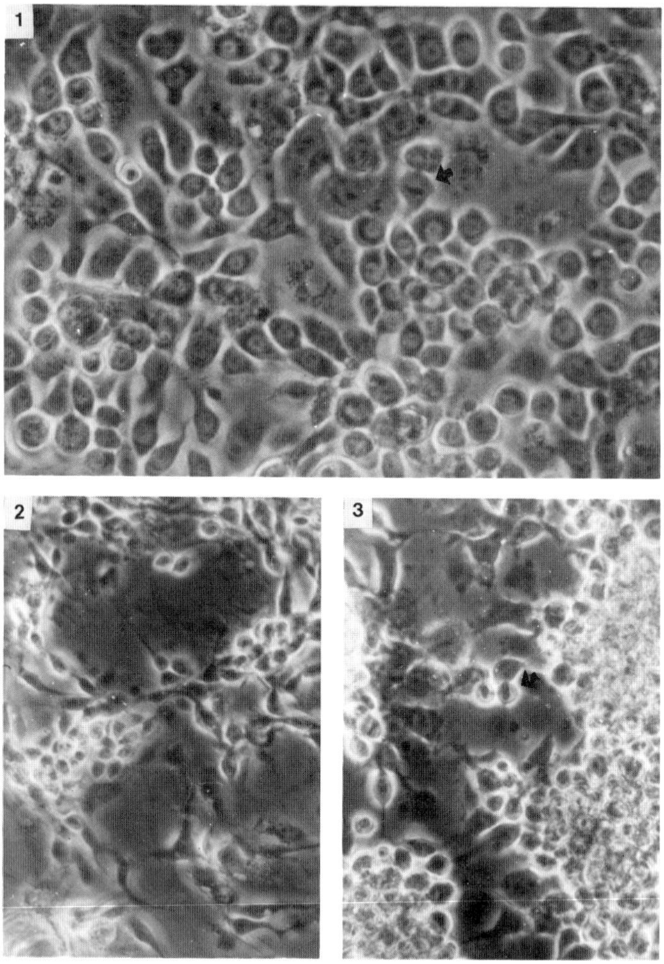

Insect cell culture: *Fig. 1:* embryonic *P. americana* cell line (EPa strain) x300. *Fig. 2:* embryonic *P. americana* hemocyte line (EPa clone) x200. *Fig. 3:* ♂ nymphal hemocyte line (Hpa 33 strain) x200.

peculiar ultrastructure on numerous electronic micrographs of insect hemocytes. (Sharder, 1972). On the basis of these observations I attributed a hemocyte origin to the EPa clones which produce bacteriolytic activity.

Other EPa clones do not show any of the above-mentioned characteristics. Some of them are responsible for the conditioning of the culture medium by the EPs strain. They release *in vitro* one or more factors capable of acting directly on cell multiplication. This property has been beneficial for the establishment of all the other cell lines.

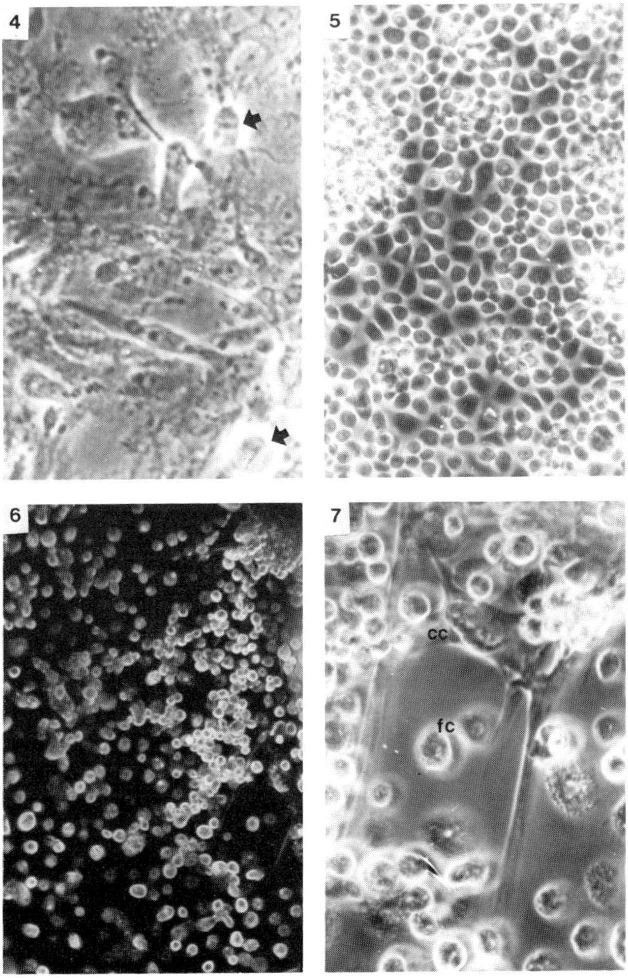

Fig. 4. cell strain from *P. americana* integument x200. *Fig. 5. Malacosoma disstria* cell strain grown in medium L21 x200. *Fig. 6-7.* primary cultures from *Samia cynthia* ovarioles.
fc = follicular cells, cc = contractile cells
x75 (Fig. 6), x200 (Fig. 7).

B. Studying cell metabolism

At present, the EPa line is the only insect cell strain whose nutritive requirements has been analyzed systematically because it can develope normally in a synthetic medium. This last step was difficult: it was necessary to elucidate the enigmatic role of the protein supplementation which seems indispensable to all cells *in vitro*.

1. *Protein supplementation:*

The search for the role played by the serum is a general problem of tissue culture and my interpretations may be extended further.

The favourable action of a mammaliam serum on the multiplication of cockroach cells takes place only when the serum comes from embryos. I managed to reproduce it by substituting purified fractions taken from an adult serum, but which were introduced in the same ratio as that of the embryonic serum. I was able to show the particular importance of the α-globulins and of the COHN fraction V.

Two complementary but distinct effects are involved: - one consists of a cell protection by serum antiproteases (Landureau and Steinbuch, 1970). This role has just been confirmed for mammals (Wallis, *et al.*, 1969). It would intervene especially after the cell transfer which brings about the appearance of measurable proteolytic activity with the destruction of a part of the cells. - the other action establishes the role of certain binding proteins and particularly those carrying vitamin B12 (Landureau and Steinbuch, 1969).

2. *Cell nutritive requirements:*

The growing of the EPa strain in a defined medium allowed me to study systematically the nutritive requirements of these insect cells.

As for amino acids (Landureau and Jollès, 1969), we completed the study of deficiencies by measuring their consumption by the cells (Technicon auto analyser). It was noticed, among other things, that it is impossible for the EPa cells to synthesize glycine even in the presence of sufficient quantities of serine, the immediate precursor of glycine. The synthesis of proline is not accomplished from glutamic acid nor from arginine via the ornithine-urea cycle. The existence of this cycle in insects remains controversial.

My conclusions concerning the sulphurous amino acids confirm the studies on cockroach physiology: deprived of their intracellular symbiotic bacteria by an appropriate process, the insects no longer manage to incorporate ^{35}S in the form of cysteine and methionine (Henry and Block, 1960).

The symptoms which follow hydrosoluble vitamin deficiency generally appear slowly (Landureau, 1969), with the exception of deficiencies in pantothenate and in vitamin B12. The removal of folic acid only weakly affects cell multiplication. This was confirmed later by the addition of a structural analogue of the vitamin (aminopterine) to the already deficient medium (Landureau, 1970).

3. *Cyanocobalamine in biosynthetic pathways:*

The biological role of vitamin B12 is still relatively unknown at least in Metazoans. The unusually strict requirement of the EPa cells with respect to vitamin B12 therefore offered an exceptional opportunity to analyse the function of this vitamin (Landureau, 1970).

I began by studying the pathways of interconversion of the main sulphurous amino acids by the EPa cultures, for it is well known that vitamin B12 is capable of interacting with folic acid for methionine biosynthesis. I proceeded to examine systematically the deficiencies in cysteine and in methionine and their correction by adding to the medium various known precursors (Table 1). Among these, only cystathionine manages to replace satisfactorily one or the other of sulphurous amino acids or even the two simultaneously.

TABLE 1

Utilization by the EPa strain of sulphurous amino acid precursors.

	Cysteine-Cystathionine-Homocysteine-Methionine
Cysteine deficiency	
+ methionine	– ←- +
+ methionine, + homocysteine	– ←———————————— + +
+ methionine, + cystathionine	– ⇐═════ + +
Methionine deficiency	
+ cysteine	+ ————————————————→ –
+ cysteine, + cystathionine	+ + ══════════⇒ –
+ cysteine, + homocysteine	+ + ————→ –
Double deficiency (cysteine and methionine)	
+ cystathionine	– ←——— + ——————————→ –
+ homocysteine	– ←- - - - - - - - - - - - + - - - - - - - → –

concentration of each precursor: 4mM.
+ - - - - - - - - - → uncorrected deficiency (limited survival)
+ ——————→ partially corrected deficiency (slow growth, difficult cell transfer)
+ ══════⇒ corrected deficiency.

The rapid death of cells, whether or not the medium includes methionine, proves however that the role of vitamin B12 can not be limited to its involvement in the biosynthesis of these amino acids.

In 1964, Sanford and Dupree (1964) proposed in the discussion of their results the hypothesis of a possible participation of vitamin B12 in nucleic acid synthesis of mammalian cells. I was able to demonstrate and determine this role with the EPa strain. The "sparing" effect manifested in the cultures by vitamin B12 and the improvement of the situation by puric and pyrimidic precursors suggested the existence of an alternative pathway in the synthesis of these precursors.

It is difficult to go beyond such affirmations. With Dr Abeles (Department of Biochemistry, Brandeis University), we checked the existence in the EPa cells of a ribonucleotide reductase which has as a coenzyme the 5'deoxyadenosyl-B12. This reaction concerns the reduction of ribonucleotides to deoxyribonucleotides. It had been found until now only in cells of *Lactobacillus leichmanii*, J. L. Becker (personal communication) has just confirmed it in the case of another insect cell strain (*D. melanogaster*).

C. Extending the results

Although the culture medium has been especially adapted to the requirements of a cockroach cell line, it permitted regular cultivation of tissues and organs taken during insect development. Only minor modifications of the medium were necessary.

1. *Cockroach tissue culture:*
We were able to develop embryonic cell cultures from any stage of the parthenogenetic development and even from a single individual (see below).

In order to confirm the hemocyte origin of some EPa clones, I introduced into culture the hemolymph taken from nymphal or imaginal cockroaches. Two new strains - HPa 33, for ♂ nymphal hemocytes of *P. americana* and HPa 34, for ♀ adult hemocytes of *P. americana* - were established. They are completely comparable to the embryonic hemocytes (Landureau and Grellet, 1975) (Fig. 3).

By cultivating *P. americana* integument, I isolated another cell strain (Fig. 4). These cells in no way resemble embryonic or hemocyte lines. They very rapidly bind ^{3}H-ecdysterone even at weak concentrations (10^{-9} M.) and shortly after, they seem to give rise to an intense biosynthetic activity. At the present time, the cuticular nature of the cell secretions is being studied.

2. *Tissue culture of other insects:*

The culture medium used for *P. americana* is perfectly suited for other orders of insects. It allows the culture of integuments and various endocrine glands of *L. migratoria, S. gregaria* and *T. molitor.*

On the other hand, in the case of Lepidoptera Saturnides for example, this medium had to be modified to correspond with the physico-chemical characteristics of the hemolymph. The pH and the osmotic pressure fall respectively to 6.5 and 320 mosmol. The saline fraction changes and moreover two amino acids must be added for these cells do not synthesize them.

This original medium, called L21 (Table 2), proved far superior to that of Grace for the cultivation of germinal cysts and the spermiduct of *S. cynthia,* as well as for Lepidoptera cells in primary cultures (Fig. 6-7) or in established strains (Fig. 5, *Malacosoma disstria* line, courtesy of Dr Sohi).

It is therefore reasonable to expect in the near future a greater uniformity of the *in vitro* techniques. The major ingredients of the culture medium would be identical irrespective of the insects studied, and the only variation would be in the osmotic pressure and the Na/K ration.

III. Cockroach cell lines and developmental biology

There are only a few insect cell strains maintaining a recognizable, reproducible and regular differentiated activity, in spite of their needs for *in vitro* proliferation. These cells of course must be genotypically unchanged or in conditions of minimal change.

I think that cockroach hemocyte lines fulfil the requirements: they perform their physiological functions for years, but their caryotypes are alwasy euploid with less than 25% of tetraploid metaphases.

For this reason, we studied by cell culture techniques the maintenance of haploidy and aneuploidy in cockroach parthenogenesis.

A. Characterization of hemocyte secretions.

EPa and HPa hemocytes resemble insect plasmatocytes, but their ultrastructure modulates considerably according to the culture conditions: their cytoplasm can either produce numerous granules (granulocytes, coagulocytes, adipohemocytes), or, can be devoid of them, while the nucleoplasm ratio increases (prohemocytes). Thus, the transition among these different types of hemocytes, not evident *in vivo,* is easily demonstrated *in vitro.*

TABLE 2

Culture media for insect cells (concentrations in mg/liter).
Medium S20 for Heterometabolous insects: cockroach, locust.
Medium L21 for Holometabolous insects: Lepidoptera.

A) Part of the medium common to S20 and L21:	
amino acids (L-form)	
arginine - HCl	750
aspartic acid	500
cysteine - HCl	375
glutamic acid	1500
glycine	750
histidine	200
isoleucine	100
leucine	200
lysine - HCl	200
methionine	250
phenylalanine	100
proline	650
threonine	100
tryptophan	100
tyrosine	100
valine	100
inorganic salts	
$CaCl_2$, $2H_2O$	500
$MnSO_4$, H_2O	35
buffer	
PO_3H_3	900
vitamins	
d-biotin	0,05
Ca-pantothenate	1
choline - HCl	250
cyanocobalamine	0,2
folic acid	0,05
inositol	0,5
nicotinamide	0,25
pyridoxine - HCl	0,25
riboflavin	0,5
thiamine - HCl	2
sugar	
glucose	4000
antibiotics	
penicillin -G	125
streptomycin, sulfate	50
B) Distinctive part: *S20*	
KCl	1050
$MgSO_4$, 7 H_2O	1250
NaCl	8500
glutamine	75
serine	35
L21	
KCl	3000
$MgSO_4$, 7 H_2O	4120
$MgCl_2$, 6 H_2O	1120
glutamine	500
serine	400
sucrose	5000

After a week's culture, the cells accumulate glycogen and lipid reserves. They contain two distinct categories of granules (Fig. 8-9). - The first, dense and homogeneous (Fig. 11) manifests a positive phosphatase acid reaction (Fig. 12). Granules appear through the condensation of material on the side of maturation of dictyosomes where the cisternae from the reticulum are found. In close proximity some vesicles and some agranular membrane system may be seen. There is therefore a very close resemblance between the genesis of lysosomes and that of this first category of granules. The content of dense granules is easily digested by proteolytic enzymes (pepsine and pronase) (Fig. 13). - The second category contains microtubules often grouped in juxtaposed bundles (Fig. 9). Unlike the first ones, these granules have no acid phosphatase and apparently are not digested by pepsin and pronase. On the contrary, they show a positive reaction for thiamine pyrophosphatase. The Golgi from which they originate (Fig. 10) sometimes possess close

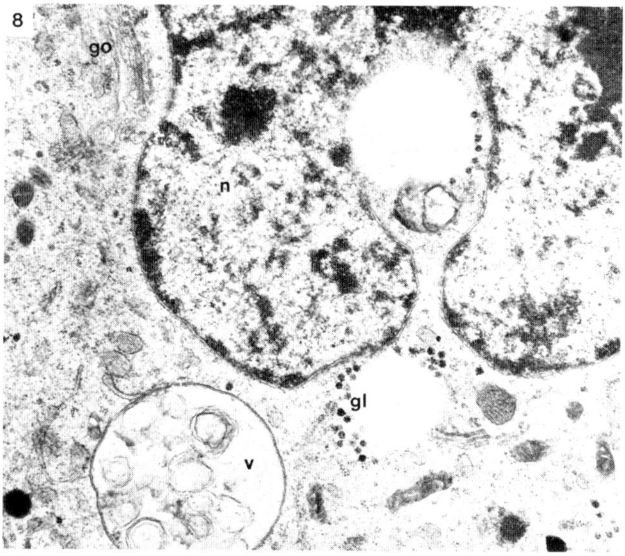

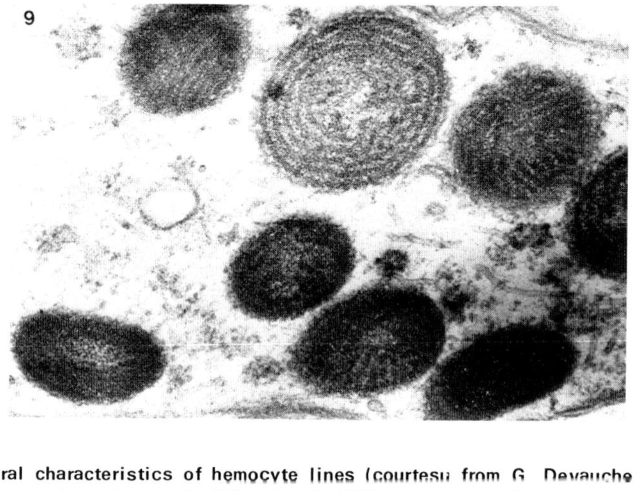

Ultrastructural characteristics of hemocyte lines (courtesu from G. Devauchelle). *Fig. 8.:* embryonic *P. americana* hemocyte (EPa clone) x4000 go = golgi, gl = glycogen, v = vacuole *Fig. 9:* multitubular granules in EPa hemocytes, 9 days after medium replacement. x 25000

relationship with the nuclear membrane. Producing large multitubular vesicles (Fig. 10). The older granules tend to become more opaque. Their content is directly released into the culture medium. It should be recalled that the same release takes place from granulocytes into the insect hemolymph (Devauchelli, 1972).

1. *Involvement in insect immunity and ecdysis:*

Phagocytosis is the first cell reaction implied in insect immunity. Like hemocytes, the EPa and HPa cells are capable of rapidly ingesting particules of various size and even entire cells (bacteria or other cellular categories).

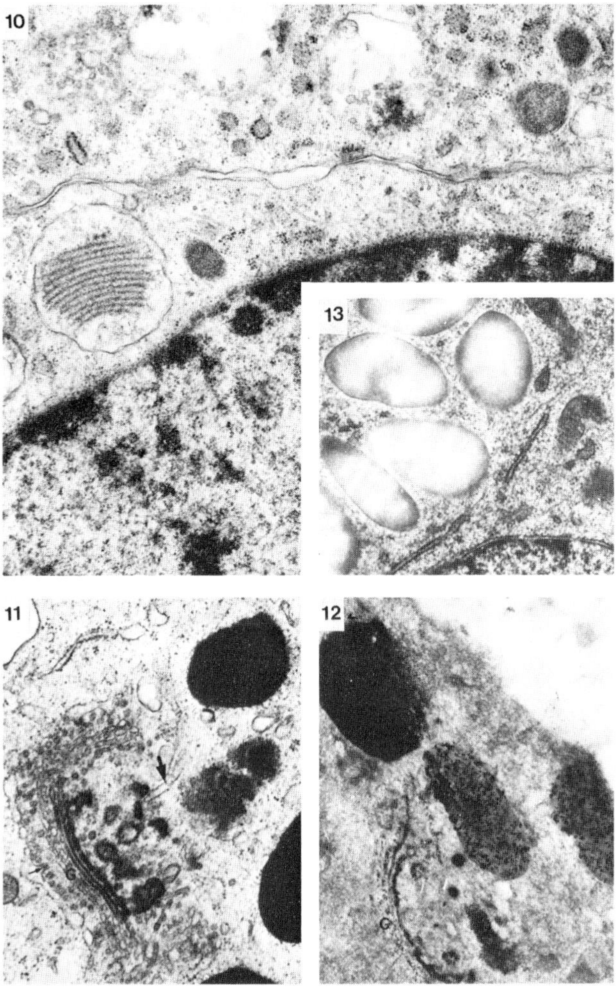

Fig. 10: genesis of multitubular granule from Golgi vesicle x 12000. *Fig. 11:* genesis of dense granules from Golgi x 25000. *Fig. 12:* acid phosphatase reaction in dense granules x 25000. *Fig. 13:* digestion of dense granules by pronase x 25000.

Moreover, their culture medium strongly inhibits the growth of different microorganisms. The lytic activity was determined by measuring the decrease of the optical density of a suspension of dead *Micrococcus lysodeikticus* cells at 583 nm with a Vitatron photometer. Hen egg white lysozyme was taken as the reference enzyme. Recovered values are similar to those of insect blood (Bernier *et al.*, 1974). They range from 10 to 50 μg/ml. The lytic activity increases by 100 or 1000 times if a defense reaction is induced *in vivo* or *in vitro* by the hemocytes (Boman *et al.*, 1974, Mohrig and Messner, 1968). This activity however should not be confused with a lysozymic activity nor can it be the result of a single biochemical entity. In effect, from a serum-free medium, we characterized after preliminary purification and starch gel electrophoresis, several proteins, of which only two correspond to a chitinase (M.W. 25,000 and 30,000). The broad spectrum of bacteriolytic activity due to the hemocytes probably results from the combined action of several enzymes (Welsch, M., 1966).

The regular production of these muralytic enzymes permits us to understand better the way in which the cockroach regulates the intense multiplication of its intracellular symbiotic bacteria. It is known, in fact, that the thin bacterial cell wall is composed of mureine (Daniel and Brooks, 1967) and that its destruction is induced in the living insect by a treatment of lysozyme (Malke, 1964)

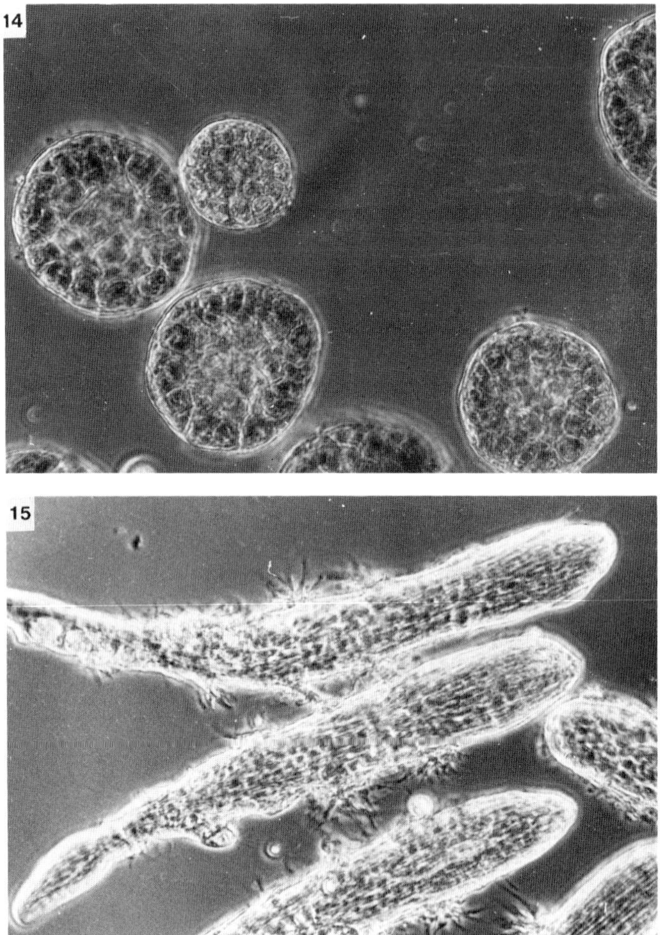

In vitro spermatogenesis of *Samia cynthia* **(courtesy from A. Szollosi)** *Fig. 14:* germinal cysts from testis of diapausing pupa x 650. go = cyst containing spermatogonia cyI = cyst containing spermatocyte I. *Fig. 15:* elongated cysts with bundles of spermatozoa x 650.

The chitinases released in the culture fluids by cockroach cell lines are chromatographically and electrophoretically the same as those from hemolymph. They did not present any lytic activity against *M. lysodeikticus* between pH 3 and 9 (Bernier *et al.*, 1974).

The concentration of chitinases remains relatively constant in the culture medium. The same is true in the blood for a defined stage of insect development.

The chitinase activity increases abruptly after the addition of ecdysterone *in vivo,* even in the presence of usual inhibitors of protein synthesis (Kimura, 1973). In cell culture, we were able to find this result and to determine that it was a specific effect of this hormone which begins to show up as soon as it reaches a concentration of 10^{-7} M. These facts argue in favor of a simple mobilization by ecdysone of the enzymatic reserves stored in the hemocytes. They should be compared to the hormonal variations in insect blood (Lafont *et al.*,) and to the chitinase activity at the level of the integument (Bade, 1974). By attributing a hemocyte origin to the chitinase, it becomes very easy to interprete the prevention of of the exuviation following a selective irradiation of the hematopoietic tissue (Joly, *et al.*,).

Various authors reported the direct intervention of hemocytes in the regression of certain organs such as integument (Barra, 1969) or prothoracic glands (Scharrer, 1966), or in the construction of connective tissues such as *tunica propria* of P.G. (Beaulaton, 1968) or basement membrane of integument (Wigglesworth, 1973).

Beaulaton and Wigglesworth (Beaulaton, 1968; Wigglesworth, 1973) attribute more precisely this role of hemocytes to the mucopolysaccharides that they located in the microtubular granules.

According to Hoffmann and Stoeckel (Hoffmann and Stoeckel, 1968), the same granules are involved in the process of coagulation, while Jones (Jones, 1962) and Hagopian (Hagopian, 1971) hold them responsible for the reaction of melanisation (Hoffmann, *et al.*, 1970).

Presently available hemocyte strains which have conserved their specialized activities should permit us to choose from among these hypotheses. There is no doubt that the methods of modern cytology should bring important information about the respective functions of the two categories of granules which characterize these cells.

2. *Involvement in insect spermatogenesis:*
The *P. americana* hemocyte strains release in the culture medium a factor which is capable of inducing *in vitro* the spermatogenetic differentiation of germinal cysts of several species of Saturnides (Landureau and Szollosi, A., 1974). According to preliminary result, we think that it is a protein or a substance adsorbed on a protein.

These facts confirm the experiments of Williams in collaboration with Schmidt (Schmidt and Williams, 1953) and with Kambysellis (Kambysellis and Williams, 1971) concerning the presence of a macromolecular factor (MF) in the hemolymph of Lepidoptera.

Our observation on the *in vitro* spermatogenesis of *Samia cynthia* are summarized in Table 3. When the testes are torn up, numerous cysts containing spermatogonia and spermatocytes I are collected in the culture medium. Only the latter undergo the completion of meiosis, spermiogenesis and elongation, provided that the hemo-

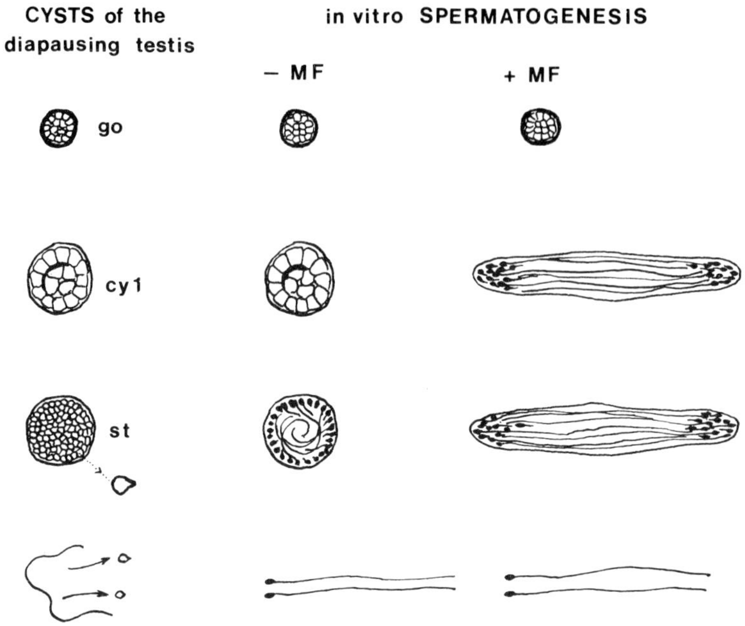

TABLE 3: *In vitro* development of germinal cysts of *Samia cynthia* (cysts containing spermatogonia (go), spermatogytes (cy 1) and spermatides (st).

cyte factor is present. Finally, the somatic sheath of cysts breaks down and the full grown spermatozoa are released outside. At this step, a strong proteolytic activity appears and, for lack of serum antiproteases, all the cultures are rapidly destroyed.

Within testis, cysts containing spermatogonia and spermatocytes may occur in different relative proportions according to the studied pupae. Very often, we found cysts with spermatids, most of them settled in earlier spermiogenesis (verification by electron microscopy), the others darker and degenerating. If the degeneration of cysts containing spermatids is not complete, these cysts are capable of developing again *in vitro*, provided that the culture medium is sufficiently rich. In the absence of MF, the developing sperm tends to form whorls while the cyst does not elongate. On some occasions, cysts were ruptured and the naked germinal cells (generally spermatids) evolved spontaneously in the culture medium.

The Saturnides used in these tests constitute an exceptionally favourable material. We have noticed that only the insects whose diapause is cyclic show an interruption of spermatogenesis. The cysts containing spermatids and spermatozoa degenerate shortly after the start of diapause, probably for lack of the hemocyte factor (Kambysellis and Williams, 1971). When the gonad no longer contains cysts with spermatids, no culture evolves in the absence of MF; the fetal bovine serum has no effect. This observation suffices to reject the hypothesis of Kambysellis (Kambysellis and Williams, 1971) concerning the existence in mammalian sera of a factor comparable to the hemocyte factor.

The hemocyte factor is absolutely necessary to get a normal evolution of the germinal cysts containing spermatocytes I. It does not seem to exert a direct effect on these cells, but rather on the somatic sheath of the cyst. It could regulate the entrance of metabolites within the cyst. Moreover, in agreement with Fowler (Fowler, 1975), we must point out the important migration of somatic cells while the cyst elongates. It is undoubtedly related to the bipolar repartition of the spermatozoa. MF is still a necessary requirement.

The production of MF is not limited to one sex nor to a determined stage in the insect development. It should be noticed that the secretion of an active factor by cells of a heterometabolous insect on the cysts of a holometabolous insect shows a possible generalization of these preliminary results.

B. Cell culture from parthenogenetic cockroach.

The facility with which the primary cultures of *P. americana* cells were established and especially the possibility of establishing strains whose caryotype remains euploid permitted the cultivation of cells from parthenogenetic cockroaches (Philippe and Landureau).

Similar research has already been undertaken on Amphibians and has recently resulted in isolation of haploid cell lines (Freed and Mezger-Freed, 1970).

1. *Chromosomic regulation during ontogenesis:*

In case of thelitoque parthenogenesis of meiotic type some varied cytological mechanisms re-establish the diploid state. This chromosomic regulation intervenes at different stages of the development. In *P. americana*, the accidental parthenogenesis is accompanied by a decrease in the viability of embryos: the percentage of hatching thus passes from 77% for a normal ootheca to 40% for an ootheca containing virgin eggs. This percentage decreases to 6% in the following generation. Various authors attribute this low viability of embryos to the persistance of haploid cells.

Our observations *in vivo* on the haploid situation of the embryo stress the extreme variability of the N/2N ratio. The rate of haploidy does not seem related to age. It varies for embryos of a determined stage and, to a lesser degree, inside an ootheca. It must be noticed however that the proportion of haploid cells is higher at the beginning of the development.

Thanks to the *in vitro* techniques we were able to demonstrate the persistance of haploid hemocytes in the adult (Philippe and Landureau). The study of the caryotype of these cells is in fact impossible in the living insect because they divide only exceptionally in the hemolymph. This somatic tissue is therefore the object of an incomplete chromosomic regulation.

In the cultures derived from older embryos and from adults, we found numerous aneuploid formulae. These abnormal formulae do not disappear in the course of ontogenesis as they do in some insects (locust) where they bring about the death of the embryo at a relatively early stage. *In vivo*, we have also found numerous aneuploid cells in embryos of stage 23.

2. *In vitro maintenance of haploid and aneuploidy:*

Cultures coming from youngest parthenogenetic embryos do not evelve even after several months *in vitro*. On the contrary, as soon as ontogenesis reaches the end of segmentation of the germ band, the corresponding primary cultures lead invariably to the establishment of new cell lines. These strains are very similar to the EPa strain, notably by their abundance of hemocytes. Our culture conditions could

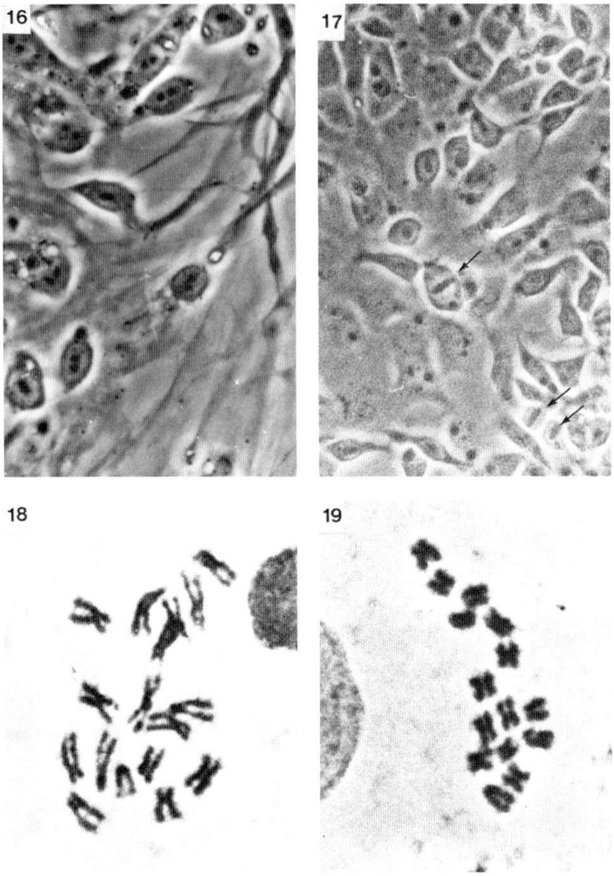

Cell cultures from parthenogenetic cockroaches. (courtesy of C. Philippe). *Fig. 16:* primary cultures from embryos of stage X, x 300. *Fig. 17:* embryonic cell strain from embryos of stage XV, x 300. *Fig. 18:* haploid caryotype: 17 chromosomes. *Fig. 19:* hypo-haploid caryotype: 15 chromosomes.

therefore select this cell category. It should be noted that the differentiation of hemocytes from the floor of the epineural sinus, takes place shortly before catatrepsis, at a time which corresponds precisely to the favourable evolution of primary cultures.

We have utilized this hypothesis to cultivate parthenogenetic hemocytes and to obtain new cell strains.

Caryological studies of all these embryonic and imaginal cell lines confirm the unperfect degree of the chromosomic regulation in the course of ontogenesis: most of the strains became diploid, often with a high percentage of tetraploid elements. Two lines coming from embryos of stages 17 and 18 contained up to 10% of haploid cells. The percentage of these decreases in older embryos and in the adult (only 1%) (Fig. 18).

Various aneuploid caryotypes were also present, especially in hemocyte strains where their proportion was equivalent to that of the diploid and tetraploid mitoses.

In controls, i.e. hemocyte lines established from normal embryos or imagos, aneuploids seem to be very rare and no haploid was ever found.

It must be noticed that we frequently observed in the lines from parthenogenetic cockroaches, besides true haploid cells, caryological formulae with reduced number of chromosomes (Fig. 19). We are attempting to clone them to ascertain their viability.

Such genetical studies are of great interest in the case of hemocytes because of the maintenance of the above mentioned activities in these cells.

IV. Insect tissue culture and developmental biology

This recent development of tissue and organ cultures was imposed by the evolution of research in insect physiology. It appeared that the schemes proposed for the exuviation mechanisms would have to be modified: the fine structure of prothoracic glands revealed that, unlike the oenocytes (Locke, 1969), they did not have the characteristics of a steroidogenous tissue. This left some doubt concerning ecdysone biosynthesis. Moreover, the irradiation experiments of Hoffmann *et al.*, (Joly *et al.*, 1973) suggested the existence of a humoral factor elaborated by the hematopoietic organ or its hemocytes. Finally, the recent observations of Wigglesworth (1973) contribute to the hypothesis of a possible participation of these hemocytes in the molting phenomenon.

A. Cultivation of endocrine glands.

The *in vitro* culture of insect endocrine glands has been the purpose of numerous studies recently. The results contributed to the localization of biosynthetic sites of insect hormones.

1. *Corpora allata and juvenile hormone:*

The cultivation of corpora allata separated from neuro-endocrine complexes in medium containing insect hemolymph has provided some evidence that this gland might be the genuine source of juvenile hormone (Rollet and Dahm, 1970). Direct proof was provided by Judy *et al.*,) using a culture medium only supplemented by bovine serum albumine.

The possibility of cultivating the complex c.c./c.a. of *P. americana* for several weeks in the synthetic medium S20 (Table 2) allowed Mueller *et al.*, (1974) to isolate and characterize the juvenile hormone released *in vitro*. The molecule is methyl (2E,6E) - 10,11 - epoxy - 3,7,11 - trimethyl-2,6 - dodecadienoate (JH III).

The absence of terpene procursors in the medium ascertains that the biosynthesis of the molecule is entirely realized de novo. The glands of the gravid females are less active than those of the larvae and of the males.

In vitro, the explanted complex becomes firmly attached to the bottom of the culture flask. Important cell migration and fiber outgrowth take place from corpora cardiaca, starting a few days after the beginning of the culture and continuing for many months (Fig. 20).

These results were utilized in experiments on parabiosis with effector tissues,

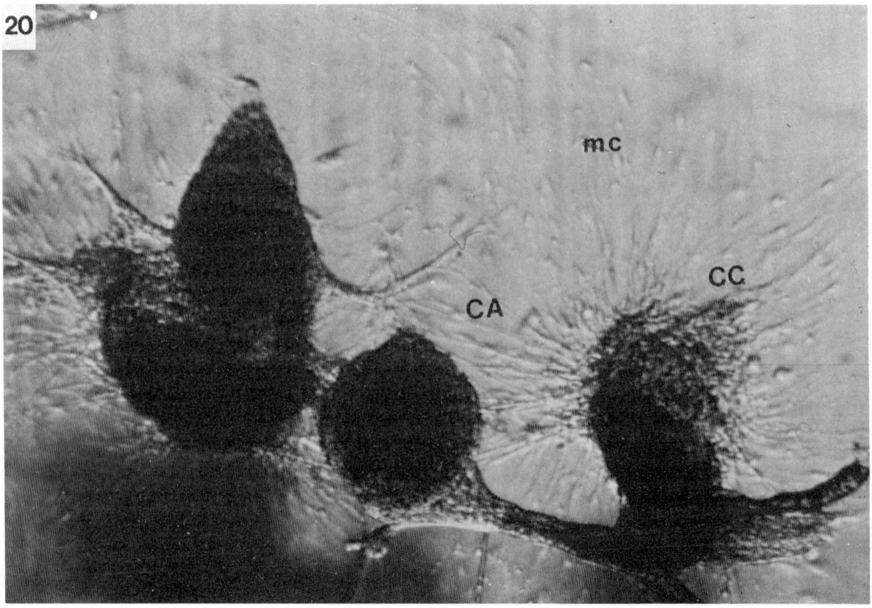

Culture of corpora cardiaca/allata from *P. americana* **(courtesy of P.J. Mueller).** *Fig. 20:* cc = corpora cardiaca, ca = corpora allata cell migration and fiber outgrowth x 100.

notably in the case of *S. gregaria* integument where the juvenile hormone induces secretory activity of the glandular cells (Casier and Delorme, 1974).

2. *Prothoracic glands:*

Very recently, the publication of Chino, King Romer, Borst *et al.*, (Borst and Engelmann, 1974; Romer *et al.*, 1974) have confirmed the secretion of α-ecdysone by the prothoracic glands of several insects. Various methods have been utilized to characterize the hormone, but incubation *in vitro* is always carried out over a short period of time, about 24 hours.

In order to detect the hormone released in the medium it is necessary to collect many glands from larvae in the middle of intermolt. In the hemolymph, the concentration of α-ecdysone is very weak. It is probable that the secretion from P.G. is rapidly converted into β-ecdysone by peripheral organs such as fat body (King, D.S., 1972) and oenocytes (Romer *et al.*, 1974).

Removed during the imaginal molt, the P.G. of *P. americana* cultured in the medium S20 degenerate rapidly as they do in the living insect. After a month, nothing subsists but the muscular axis. On the other hand, the glands taken at the beginning of the last intermolt show no sign of degeneration even after a long period of time. It is therefore probable that the *in vitro* techniques will make it possible in the future to interprete other roles of the molt gland. In effect the importance of the protein synthesis in this gland may suggest a more complex role in the regulation of apolysis, notably in the transportation of the hormone or its protection against the breakdown mechanisms (Hoffmann and Koolman, 1974).

In vitro, the prothoracic glands produce an important quantity of cAMP (Vedeckis and Gilbert, 1973). It appeared that they were as active as mammalian tissues which utilize this nucleotide as a second messenger. The brain hormone might therefore exert its effects via the cAMP.

3. *Oenocytes:*

A real competition has begun on the international level to determine the role of oenocytes whose ultrastructure is that of a steroidogenous organ.

Romer *et al.* 1974), were the first to establish that the oenocytes of *T. molitor* synthesize β-ecdysone from cholesterol or α-ecdysone. The cultures in parabiosis indicate that the hormone derives principaly from the α-ecdysone produced by the prothoracic glands.

The abundance of agranular endoplasmic reticulum in oenocytes could be related to the property of these cells to synthesize cuticular waxes (Diehl, 1973).

Using my *in vitro* techniques Caruelle (1975) and Delachambre (Delachambre and Delbecque) are studying respectively the oenocytes of locust and of *T. molitor.* The osmotic pressure of the culture medium is adjusted to 380 mosmol instead of 520 for Romer (1974). The cells survive well in long term cultures; they maintain their essential characteristics; abundance in dense bodies and in smooth endoplasmic reticulum (Fig. 21).

The cultured oenocytes store the cristalline inclusions already described by Locke (1969). These structures (Figs 22, 27), less numerous in the living insect, might contain steroids extracted by the fixation method.

B. Morphogenesis of insect integument.

The information which we possess on the integument morphogenesis of arthropods show the necessity of carrying on the study of its role by *in vitro* cultures.

In 1967, Miciarelli *et al.,* (1967) explanted fragments of nymphal abdominal sterna of *S. gregaria* in a medium supplemented by insect hemolymph. They obtained lysis of the old cuticle and a small thickening of new cuticle.

Since then, the deposition of cuticular material *in vitro* has been reported many times; in imaginal discs (Fristrom, *et al.,* 1973; Mandaron, 1971; Oberlander, *et al.,* 1973) or in leg regenerates (Marks, 1973). A series of experiments are performed to obtain further informations about the induction of cuticle deposition by molt hormones and sometimes about the antagonistic effect of juvenile hormone (Chiara and Fristrom, 1973). The results obtained to day remain preliminary and they are sometimes contradictory. Thus, the evagination of *D. melanogaster* discs was better obtained with α-ecdysone (Mandaron, 1971) or with β-ecdysone (Fristrom *et al.,* 1973). Recently, Millner and Sang (1974) have continued the techniques of these authors, but they have not been able to explain this contradiction.

1. *Cuticular secretions by cockroach integument:*

Fragments of abdominal terga are taken three days before the last molt. At this stage, the imaginal procuticle is thin and devoid of any pigmentation.

In all cases, the integument continues its cuticular secretions: the cell cytoplasm of hypodermic cells stores a large amount of glycogen and shows evident signs of secretory activity. The apex of the cells bristles with numerous microvilli.

21

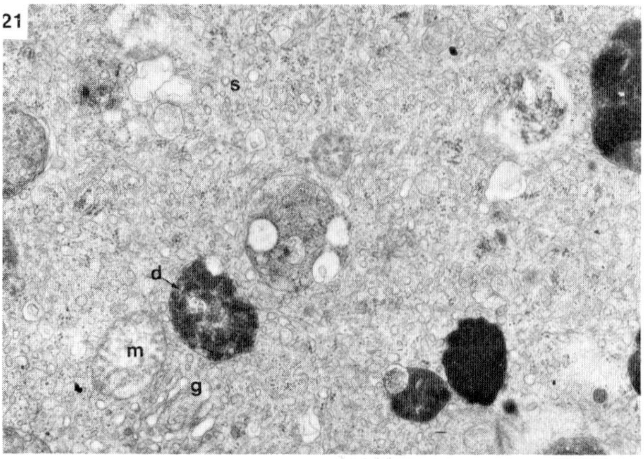

22

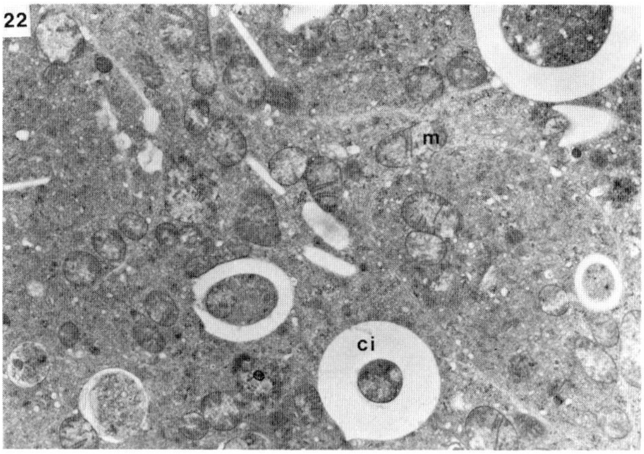

Culture of ***Tenebrio molitor*** **oenocytes (courtesy of J. Delachambre).** *Fig. 21:* cultured cells without ecdysterone x 20000. g = small Golgi, m = mitochondria, d = dense body, s = smooth endoplasmic reticulum. *Fig. 22:* cultured cells with ecdysterone (10^{-7} M.) x 10000. ci = cristalline inclusions.

On some sections, the thickness of the endocuticle reaches 120 nm after several weeks *in vitro* instead of only 50 nm in the living cockroach.

At the margin of the explant, an important accumulation of epithelial cells occurs during the first 15 days. Following this step, migrations result in several waves of neofomations (Fig. 23). In only 5 cases out of 130 cultures, the newly formed integument became as large as the initial explant after 6 months *in vitro*. On the cellular margin thus formed, a thin layer of cuticle was secreted which had lost its typical pattern (Cals, 1973) and whose pigmentation was delayed compared to that of the explant.

The study of cell populations (Fig. 24) shows that in the peripheral zone of the explant, each cuticular scale no longer corresponds to an epithelial cells while,

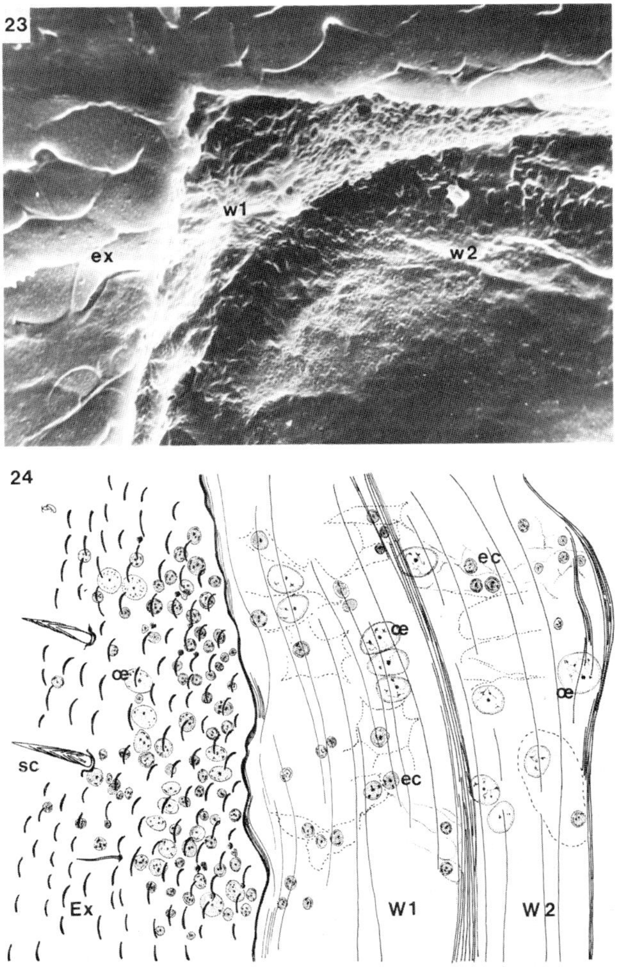

Culture of *P. americana* **integument (courtesy of P. Cals).** *Fig. 23:* Scanning electron micrograph of *in vitro* wound healing, x 5000 ex = explant, W1 and w2 = successive waves of cell migrations. *Fig. 24:* diagrammatic representation of cell migrations. ec = epithelial cells, oe = oenocytes.

at the margin, these cells congregate thickly. Besides normal epithelial cells, oenocytes and perhaps a few glandular cells participate in the neoformation; no sensory element can be found there.

Therefore the hypodermic cells of imaginal origin are capable of migrating and realizing *in vitro* a sort of wound healing. These observations agree with those of Wigglesworth who has studied how the insect repairs injuries to its integument (Wigglesworth, 1940).

The synthesis of endocuticle *in vitro* shows clearly that the epithelial cells have received the necessary information for the expression of their specialized biosynthetic activities. In the case of *P. americana* integument, the presence of numerous oenocytes in the explant could interact by secreting ecdysterone. This does not hold true for *T. molitor;* in effect, although no oenocytes exist in the larval abdominal sterna, the explants continue their cuticular secretions (Delachambre and Delbecque).

2. *Imaginal differentiation of locust integument:*

According to Strich-Halbwachs (Miciarelli, *et al.,* 1967), during each instar the hypodermic cells pass through five successive phases: (Bade, 1974) completion of the endocuticle deposit, (Barra, 1969) mitotic crisis, (Beaulaton, 1968) secretion of the new exocuticle, (Bernier, *et al.,* 1974) digestion of the old endocuticle, (Best-Belpomme and Courgeon, 1975) beginning of the formation of the new endocuticle.

As previously reported by Miciarelli *et al.,* (1967), the most favourable time for starting the cultivation of locust integument should be in the sixth day.

From the same material, Caruelle, Cassier and myself are now attempting to study *in vitro* the evolution of fragments of abdominal terga, taken from larvae of the fourth instar instead of the fifth in the above mentioned experiments.

At 30°C, the best results were obtained two days after the molt. Before this time, the epithelial cells tend to separate, especially when the culture medium has been shortly conditioned by the embryonic EPa clones.

The choice of male *S. gregaria* allows us to follow the imaginal differentiation of the larval explants. Indeed, the last molt induces the appearance of two associated cell categories: glandular cells producing the sex linked pheromone and canalicular cells releasing the secretion (Cassier and Delorme, 1974). (Figs 25, 26).

- Up to 8 days *in vitro*, the larval cells store a large amount of glycogen. The same storage occurs in the living locust, it seems to precede the molt. Within epitelial cells, the pigment granules are gradually decreasing by inclusion in autolysosomes.

- The apolysis occurs between 9 and 12 days. It is related to the immediate secretion of the new cuticle. A thin epicuticle recovers the normal lamellar endocuticle (Fig. 28). At this step, the integument lacks pigment, but shows glandular and canalicular cells, characteristic of the imago.

When adding prothoracic glands (4 P.G./2ml), locust integument evolves more rapidly *in vitro* (Figs 28,29).

High concentrations of α-ecdysone (up to 25 µg/ml) induce successive waves of apolysis and cuticular secretions. After 12 days, 3 or 4 layers of normal cuticle (epi- and endocuticles) are deposited; but, at this high hormonal level, hypodermic cells and especially glandular cells tend to be destroyed.

In the absence of juvenile hormone in the culture medium, the glandular cells are inactive. It remains therefore to reproduce *in vitro* the experiments of implantation of corpora allata carried out in locusts (Cassier, P. and Delorme, C., 1974). Some attempts are now beeing made to utilize synthetic hormones as well as the simultaneous culture of various endocrine formations.

25
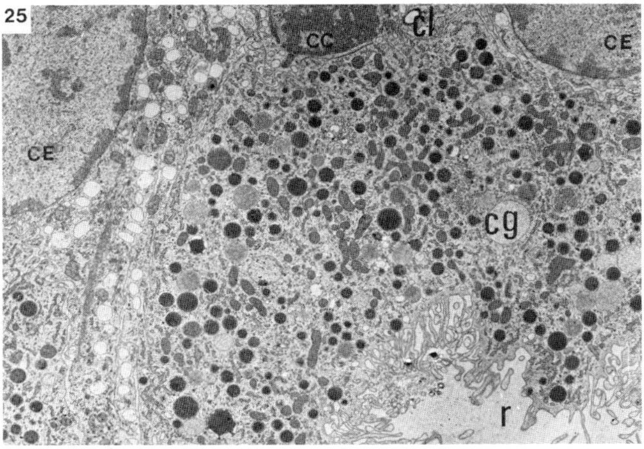

26
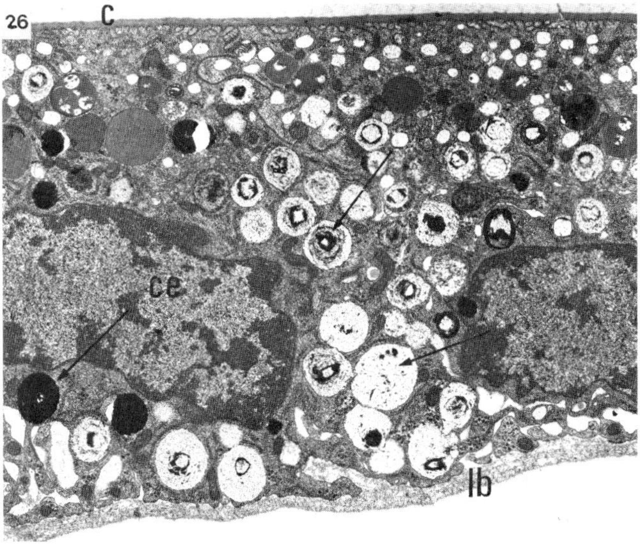

Imaginal differentiation of *S. gregaria* integument (courtesy of P. Cassier). *Fig. 25:* The 3 cell categories of ♂ imaginal integument, x 6800. ce = epithelial cells, cg = glandular cells, cc = canalicular cells, r = reservoir of glandular cell. *Fig. 26:* larval integument (fourth instar), x 9200. ce = epithelial cells rich in pigment granules (⟶), c = cuticle, lb = basement membrane.

C. Action of ecdysones on lepidoptera testis and spermiduct.

The *in vitro* response of an effector organ to a stimulation by molt hormones has been studied in the case of *Samia cynthia*. We were able, in collaboration with A. Szollosi, to show the specific effect of the α-ecdysone on the spermatogenetic differentiation of the testicle and that of the β-ecdysone on the imaginal differentiation of the spermiduct.

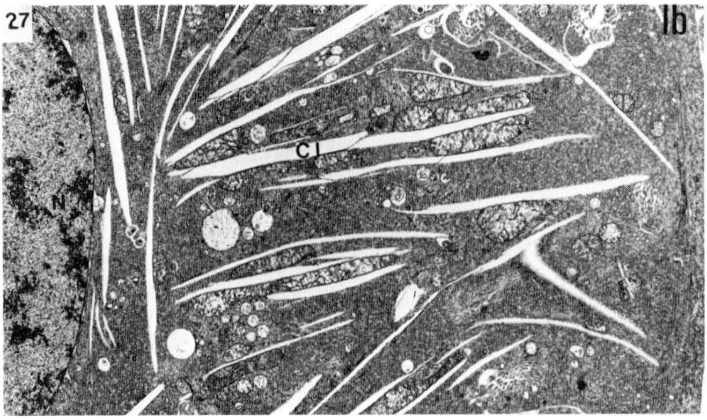

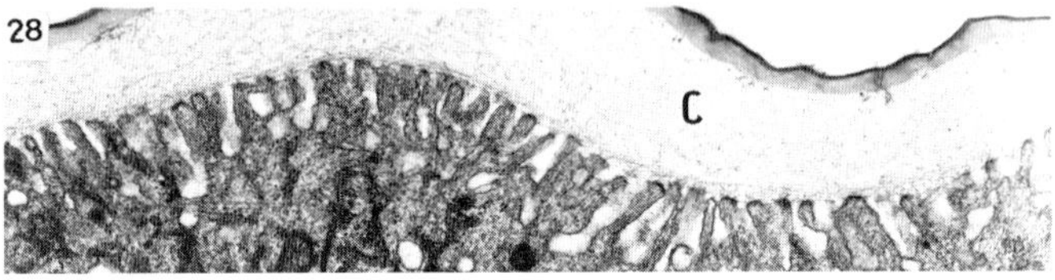

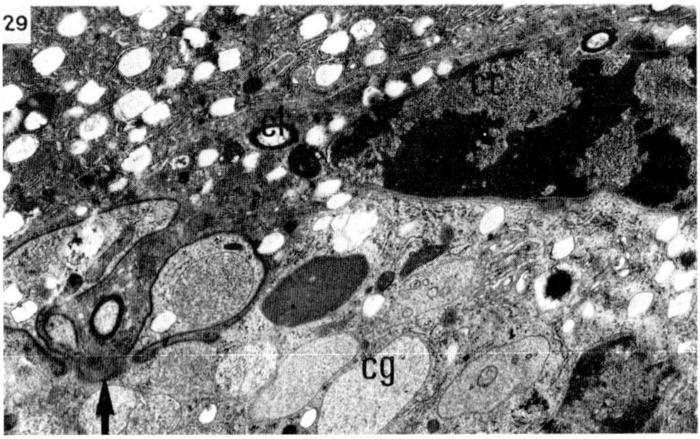

Fig. 27. Culture of S. gregaria oenocytes (12 days), x 6700. ci = cristalline inclusion. *Fig. 28.* Apical microvillosities of hypodermic cells cultivated in presence of active prothoracic glands (6 days), x 10000. *Fig. 29.* Imaginal differentiation of larval integument cultivated in presence of active prothoracic glands (4 days), x 6800.

1. *Gonad permeability to the hemocyte factor:*

As for the completion of spermiogenesis within the gonad, the only positive results obtaines to date have necessitated extra-physiological concentrations of ecdysterone (from 20 to 200 mg/l.) (Kambysellis and Williams, 1971; Yagi, *et al.*, 1969). The usual solvents of the steroid exert an activity by themselves for, by altering the testis wall, they facilitate the penetration of MF (Kambysellis and Williams, 1971). The hormone would therefore act by modifying the permeability of the organ to MF.

These data must be examined with care: the excessive dose of β-ecdysone used by Yagi *et al.*, (1969) introduces a quantity of α-ecdysone ranging from 0.1 mg/l. On the other hand, Kambysellis and Williams, (1971) cultivated testicles in pupal hemolymph, but this includes traces of α-ecdysone (Lafont, personal communication).

In our experiments, we have not been able to confirm the hypothesis of Kambysellis on the permissive role of ecdysterone. On the contrary, the addition to the culture medium of 0.1 mg/l of α-ecdysone insures normal spermatogenesis within the gonad in the presence of MF.

The experimental conditions must be precise. For example, I shall mention the recent work of Cohen and Gilbert (1974): in order to investigate the action of ecdysterone on RNA and protein synthesis in germinal cysts of *S. cynthia,* they inject the hormone into diapausing pupae, taken from the gonads after 18 or 24h. and they followed *in vitro* at 30°C the incorporation of usual labelled precursors. They were quite surprised to note that the hormone exerted an inhibitory effect which contradicts most of the present information. On this subject, I have some comments: the temperature of 30°C is lethal for *S. cynthia.* In the cyst, germinal cells are killed in less than 24h. and nothing is left but the somatic sheath. The differentiation of an organ like the spermiduct is also inhibited by this temperature. In our experiments β-ecdysone, contrary to α-ecdysone, does not act on the permeability of the testis wall, but only on the metamorphosis of the spermiduct. Finally, nothing permitted the authors to affirm that the gonads of pupae were physiologically competent and that the injected hormone was not rapidly broken down in the organism.

Thus, insect spermatogenesis seems to be regulated both by humoral and hormonal factors which are respectively efficient on the cyst and testis walls. For lack of the humoral factor during the cyclic diapause of Saturnides (Kambysellis and Williams, 1971), germinal cells can not differentiate beyond the growing phase; spermatozoa and spermatids gradually disappear within the gonad.

When the diapause is rupturing, the humoral factor is released again from hemocytes (Kambysellis and Williams, 1971; Landureau and Szollosi, 1974). At this time, ecdysterone might be produced too for we noticed the beginning of enlargment of the spermiduct (see further). Then, if we were cultivating both testes and spermiducts, they always evolved spontaneously *in vitro:* germinal cysts elongated and ducts metamorphosed. We can conclude that the induction of these *in vitro* differentiations might already start in the living pupa.

In this case, Saturnides resemble other insects. Spermatogenesis may be carried on *in vitro* for a short period of time, in the absence of any humoral or hormonal factor (Fowler, 1975), provided that the culture medium is suitable for the studied species.

It is necessary to extend these preliminary results. Significant attempts have already been made with *P. americana,* but the removal of fat body from the explant is difficult and, above all, it is necessary to avoid a rapid destruction of the hormonal factor (King, 1972). Consequently, we proposed to choose a more favourable material in order to pursue the characterization of this peculiar action of α-ecdysone.

2. *Spermiduct differentiation and myogenesis:*

The spermiducts were taken from diapausing pupae of the silkmoth *S. cynthia.* They appear as thin cellular cords, composed of very similar cells characterized by a high nucleoplasmic ratio.

The metamorphosis of the organ proceeded in two successive steps: (1) enlargment of the duct and widenning of its lumen and (2) myogenesis and contraction of the walls.

The first step was induced by a very low dose of β-ecdysone ranging from 10^{-9} to 10^{-8}M. The enlargment occured the third day following the addition of the hormone, even if the organ had been cultured for a long time before the hormonal treatment.

After seven days and in the presence of a concentration which was 10 times higher, muscular elements differentiated in the wall of the spermiduct which contracted regularly (Szollosi and Landureau, 1975) (Fig. 30). The contractions moved from proximal to distal parts, starting from a cellular mass located at the base of the organ, in the concave part of the testis. It was therefore not surprising to notice that the spermiduct never contracted when detached from the gonad.

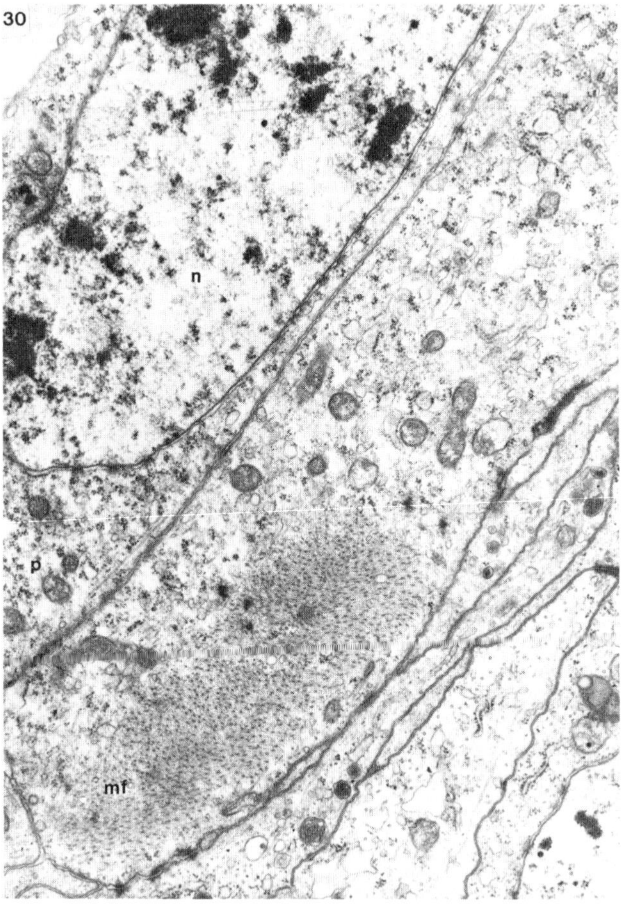

Myogenesis in *Samia cynthia* spermiduct (courtesy of A. Szollosi). *Fig. 30:* mf = thin and thick myofilaments, p = polysomes. x 40000.

If β-ecdysone was added after several weeks of culture, the enlargment of the duct intervened normally, but it did not contract either.

If we increase the concentration of ecdysterone in the culture medium, so as to approach that reached in insect hemolymph (10^{-6} to 10^{-5}M.), the metamorphosis of the spermiduct occurs in the described conditions, neither more rapid nor more important.

α-ecdysone alone is not efficient (up to 10^{-7}M). When added to ecdysterone, it seems even to counteract its action.

By electron microscopy, we characterize thin and thick myofilaments (actin and myosin) which appear in the most peripheral cells (Fig. 30) in response to the hormonal induction. They attach to the external sheath of the spermiduct and acquire the typical structure of striated insect muscle.

As for most steroid hormones, one of the effects of β-ecdysone is the stimulation of protein synthesis.

V. Conclusion

In recent years, *in vitro* cultures have brought important contributions to the study of the biology of insect development. This short review comprises the results obtained in our laboratory or in collaboration with other scientists both French and foreign.

One of the main results concerns the establishment of true hemocyte strains which maintain *in vitro* their specialized biosynthetic activities. Thanks to these cell cultures, I was able to study the role of hemocytes in insect immunity and to characterize their involvement in such different phenomena as ovulation and spermatogenesis.

Unfortunately, to my knowledge, none of the other insect cell lines established to date retains such differentiated activities. It must be recognized that they have, for the most part, an unknown tissue origin and this prevents research on specific physiological functions.

In such conditions, in order to approach the problems of insect physiology, it is necessary to have recourse to the techniques of tissue and organ culture. This experimentation permits a priori easier interpretation than experimentation with living insects: it eliminates indirect effects and makes possible the cultures in parabiosis which combine secretion and the affected tissue.

However, numerous results obtained by these *in vitro* methods are nonetheless difficult to explain; they are often contradictory and sometimes inexact. These relative failures are largely due to deficient culture conditions and to a lack of knowledge of the physiological state of the explant.

First, the composition of the culture medium is well defined only for of short duration. For a longer period of time, a protein supplementation is needed. When it consists of hemolymph (Kambysellis and Williams, 1971; Roller and Dahm, 1970) or tissue extracts (Miciarelli, *et al.*, 1967; Oberlander, *et al.*, 1973) it is capable of greatly modifying the characteristics of the medium.

On the other hand, care in selecting the cell type poses many problems; it would often impose the choice of the insect material. Thus, for integument cultures, it is difficult to exclude the presence of oenocytes in cockroach and locust explants (Caruelle, 1975), while this becomes possible in *T. molitor* (Delachambre and Delbecque).

In the absence of precise information on the *in situ* evolution of the explant it would be misleading to study *in vitro* the response to high levels of hormones. Let us note that commercial ecdysones, supply by contamination a considerable quantity of one of the two main forms. For example, the concentration of 200 mg/ml of β-ecdysone used by Yagi *et al.,* (Yagi *et al.,* 1969) introduces more α-ecdysone than we found in the blood of diapausing pupae (Lafont *et al.,*) needed in our attempts to produce the expected stimulation.

The imaginal differentiation of *S. cynthia* spermiduct by ecdysterone takes place in two steps, the second occuring by the appearance of contractile proteins in the peripheral cells of the organ wall. As for the *in vitro* action of an insect hormone on effector tissue, it was necessary to raise the level of β-ecdysone to about 10^{-7}M in order to increase the secretion of chitinase by the hemocyte strains and to induce the myogenesis of the spermiduct. Two authors arrived at the same conclusions in cell culture (Best-Belpomme, M. and Courgeon, A.M. 1975) as in organ culture (Fristrom, *et al.,* 1973).

These results however depart from almost all the other data established *in vivo* and *in vitro*. Can we deduce from such sources that the concentrations utilized by these authors are generally extra physiological? It seems that we can not. Indeed, levels equal to or lower than 10^{-5}M. are in good agreement with the maximal hormonal values found in insects. In addition, the endocrine glands do not store the products of their secretion: for ecdysones, titration from hemolymph is equivalent to that from the whole organism (Lafont *et al.*). Finally, the occurence of specific binding proteins has been well established only for juvenile hormone (Emmerich and Hartmann, 1973).

Numerous problems persist in regard to the action of insect hormones at the cellular level. The *in vitro* cultures might seen resolve them.

VI. References

Bade, M.L. (1974). *Biochem. biophys. acta 372,* 474

Barra, J.A. (1969). *C.R. acad. sci. Paris 269,* 902.

Beaulaton, J. (1968). *J. ultrastruct. res. 23,* 474.

Bernier, I., Landureau, J.C. and Jollès, P. (1974). *Comp. Biochem. Physiol. 47B,* 41.

Best-Belpomme, M. and Courgeon, A.M. (1975). *C.R. Acad. sci. Paris, 280,* 1397.

Boman, H.G., Nilsson-Faye, I., Paul, K. and Rasmuson Jr. T. (1974). *Infection and immunity 10,* 136.

Borst, D.W. and Engelmann, F. (1974). *J. exptl zool. 189,* 413.

Cals, P. (1973). *C.R. acad. sci. Paris 277,* 1021.

Caruelle, P. (1975). Thèse de 3ème cycle, Paris.

Cassier, P. and Delorme, C. (1974). *Ann. zool. ecol. an. 6,* 174.

Chiara, C.J. and Fristrom, J.W. (1973). *Develop. biol. 35,* 36.

Cohen, E. and Gilbert, L.I. (1974). *Insect biochem. 4,* 75.

Curtis, A.S.G. (1969). *Symp. Zool. Soc. London 25,* 335.

Daniel, R.S. and Brooks, M.A. (1967). *Experientia 23,* 499.
Davidson, R.L. and de la Cruz, F.F. (1974). Somatic cell hybridization. Raven Press, New York.
Delachambre, J. and Delbecque, J.P. in press.
Devauchelle, G. (1972). Thèse de doctorat, Amiens.
Diehl, P.A. (1973). *Nature 243,* 468.
Emmerich, H. and Hartmann, R. (1973). *J. insect physiol. 19,* 1663.
Freed, J.J. and Mezger-Freed, L. (1970). *Proc. natl Acad. sci. 65,* 337.
Fowler, G.L. (1975) in this book.
Fristrom, J.W., Logan, W.R. and Murphy, C. (1973). *Develop. biol. 33,* 441.
Grace, T.D.C. (1962). *Nature 195,* 788.
Hagopian, M. (1971). *J. ultrastr. res. 36,* 646.
Henry, S.M. and Block, R.J. (1960). *Contrib. Boyce Thompson Inst. 20,* 317.
Hoffmann, J.A. and Koolman, J. (1974). *J. insect physiol. 20,* 1593.
Hoffmann, J.A., Porte, A. and Joly, P. (1970). *C.R. Acad. sci. Paris 270,* 629.
Hoffmann, J.A. and Stoeckel, M.E. (1968). *C.R. Soc. biol. 162,* 2257.
Joly, L., Weins, M.J., Hoffmann, J.A. and Porte, A. (1973). *Z. Zellforsch. 137,* 387.
Jones, J.C. (1962). *Amer. zool. 2,* 209.
Judy, K.J., Schooley, D.A., Dunham, L.L., Hall, M.S., Bergot, B.J. and Siddall, J.B. *Proc. natl. Acad. sci. 70,* 1509.
Kambysellis, M.P. and Williams, C.M. (1971). *Biol. bull. 141,* 527 and 541.
Kimura, S. (1973). *J. insect physiol. 19,* 115.
King, D.S. (1972). *Am. zool. 12,* 343.
Lafont, R., Delbecque, J.P., de Hys, L., Mauchamp, B. and Pennetier, J.L. *C.R. Acad. sci. Paris, 279,* 1911.
Landureau, J.C. (1968). *Exptl cell res. 50,* 323.
Landureau, J.C. (1969). *Exptl cell res. 54,* 399.
Landureau, J.C. (1970). *C.R. Acad. sci. Paris 270,* 3288.
Landureau, J.C. and Grellet, P. (1975). *J. insect physiol. 21,* 137.
Landureau, J.C. and Jollès, P. (1969). *Exptl cell res. 54,* 391.
Landureau, J.C. and Steinbuch, M. (1969). *Experientia, 25,* 1078.
Landureau, J.C. and Steinbuch, M. (1970). *Z. für naturf. 25B,* 231.
Landureau, J.C. and Szollosi, A. (1974). *C.R. Acad. sci. Paris 278,* 3359.
Locke, M. (1969). *Tissue and cell 1,* 103.
Malke, H. (1964). *Nature 204,* 1223.
Mandaron, P. (1971). *Develop. biol. 25,* 581.
Marks, E.P. (1973). *Gen. comp. endocrinol. 21,* 472.
Miciarelli, A., Sbrenna, G. and Colombo, G. (1967). *Experientia 23,* 1.
Millner, M.J. and Sang, J.H. (1974). *Cell 3,* 141.
Mohrig, W. and Messner, B. (1968). *Biol. Zentralbl. 87,* 439 and 705.
Mueller, P.J., Masner, P., Trautmann, K.H., Suchy, M. and Wipf, H.K. (1974). *Life sciences 15,* 915.
Oberlander, H., Leach, C.E. and Tomblin, C. (1973). *J. insect physiol. 19,* 993.
Philippe, C. and Landureau, J.C. Exptl cell res. in press.
Roller, H. and Dahm, K.H. (1970). *Die Naturwiss. 9,* 454.
Romer, F., Emmerich, H. and Nowock, J. (1974). *J. insect physiol. 20,* 1975.
Sanford, K.K. and Dupree, L.T. (1964). *Ann. N.Y. Acad. sci. 112,* 823.
Scharrer, B. (1966). *Z. für zellforsch. 69,* 1.
Scharrer, B. (1972). *Z. für zellforsch. 129,* 301.
Schmidt, E.L. and Williams, C.M. (1953). *Biol. bull. 105,* 174.

Szollosi, A. and Landureau, J.C. (1975). *251 intern. Coll. C.N.R.S., Lille.*
Vedeckis, W.V. and Gilbert, L.I. (1973). *J. insect physiol. 19,* 2445.
Wallis, C., Ver, B. and Melnick, J.L. (1969). *Exptl cell res. 58,* 271.
Welsch, M. (1966). *Pathol. microbiol. 29,* 571.
Wigglesworth, V.B. (1940). *J. exptl biol. 17,* 180.
Wigglesworth, V.B. (1973). *J. insect physiol. 19,* 831.
Yagi, S., Kondo, E. and Fukaya, M. (1969). *Appl entomol. zool. 4,* 70.

Chapter 9

IN VITRO ESTABLISHED LINES OF *DROSOPHILA* CELLS AND APPLICATIONS IN PHYSIOLOGICAL GENETICS

G. Echalier

I. The resurgence of *Drosophila* research utilizing cell cultures.

Cell cultures provide a very convenient experimental material for analysing the structure and the mechanisms of functional regulation of the genome of eukaryotic cells of higher organisms. They make indeed available unlimited amounts of fairly homogeneous cellular material which might be grown under controlled conditions.

Most of the laboratories which are engaged in this research field are using mammalian cells. The reasons for such a quasi-exclusive choice are essentially technical ones: efficient techniques for growing *in vitro* of mammalian cells from various species were devised during the late fifties, that is some ten years in advance of the current development of invertebrate cell cultures. Many useful cell lines were established, especially from human and murine tissues. Culture conditions are now well standardized and a variety of ready-made media are commercially available.

In this fundamental quest it would be regrettable to disregard the rich diversity of all other zoological groups. Thus, it appears to an increasing number of scientists that insects, because of the extreme variety and originality of their phenomena of growth and metamorphosis, might well be among the best objects for studying the basic processes of cell differentiation.

This belief is the "raison d'être" of our research group and we have been trying, for the last few years, to improve the technique of *in vitro* culture of insect cells, in order to explore their capacities in physiological genetics (1).

We are working with *Drosophila melanogaster* cells, and there is no need to insist at length upon the reasons for such a choice. During three quarters of a century, an unrivalled amount of data have been collected about this small fly. *Drosophila's* genome is the best known among higher organisms. Moreover, its chromosomal set is relatively simple. At last, another relevant remark has to be made: at the biochemical level, the genetic material of *Drosophila melanogaster* (0.18 picogramms of DNA in a diploid cell) stands precisely at mid-distance between the DNA content of an *E. coli* and of a diploid mammalian cell.

II. *In vitro* cell cultures: a tool in *Drosophila* research

Let us go rapidly over some technical aspects of the culture of *Drosophila* cells in order to give a general idea of the present state of the field. The growth and use of established cell lines will be emphasized.

1. Established lines of DROSOPHILA cells

The first *in vitro* established lines of *Drosophila melanogaster* cells were derived from primary cultures that were initiated from dissociated fragments of young embryos at the beginning of 1968 (Echalier and Ohanessian 1969, 1970). Since then, some lines have been regularly subcultured.

The main interest in these lines rests on two points. They are now as easily grown as mammalian cell lines. Under standard conditions, the population doubling time of one of the most commonly used strains (line "K", Fig. 1) was estimated, from autoradiographic studies, to be about 18 hours at 25°C (Dolfini *et al.*, 1970). After many years of cultivation *in vitro*, their karyotype remains predominantly diploid, at least, in the case of two of the strains. The percentage of tetraploid cells in the population does not usually rise higher than 10-12 per cent (Dolfini, 1971, 1973).

During recent years, several other permanent cell lines have been grown from *Drosophila melanogaster* with different genotypes, in laboratories in several countries: Kakpakov *et al.*, 1969; Schneider, 1972; Mosna and Dolfini, 1972; Paradi, 1973; Richard-Molard, 1975; Echalier and Ohanessian (unpublished); Sang *et al.* (personal communication); Gehring (personal communication); Dübendorfer (personal communication)

All of these cell lines were derived from embryonic material. It would be interesting to grow cell lines from the other species of *Drosophila* which are used in developmental biology, such as *Drosophila hydei* or *Drosophila virilis*.

It must be admitted that, in spite of recent technical improvements (Echalier and Ohanessian, unpublished), the establishment of new cell lines still remains a delicate task.

---

(1) I want to apologize for quoting many unpublished results from my own laboratory. Only a few groups are, so far, devoting themselves to the culture of *Drosophila* cells, and still fewer are using them for genetic purposes.

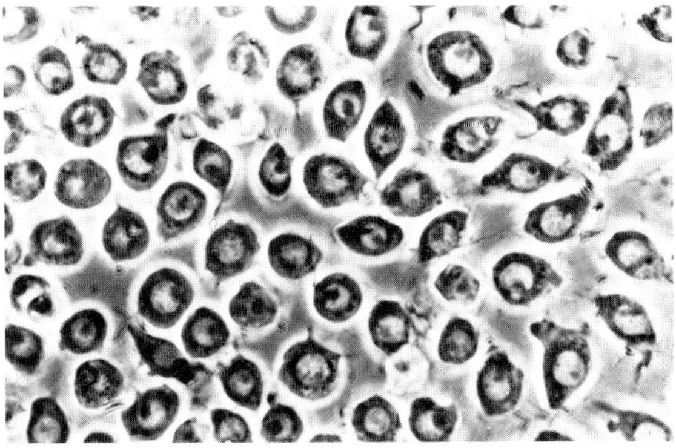

Fig. 1. Drosophila melanogaster embryonic cells of the line K (living. Phase contrast X450)

2. *Culture media*

Various media were devised for the *in vitro* culture of Drosophila tissues or cells (see Table I). Although all of them are supposed to duplicate the main physico-chemical features of the body fluid of the insect (as it can be deduced from analytical data collected from 3rd instar larvae hemolymph), they sometimes differ very much from one another (Table II).

TABLE I

Culture media for *Drosophila* tissue or cells

Authors	Imaginal Discs	Embryonic cells Primary cultures	 Established lines
KURODA and TAMURA 1956	+		
HORIKAWA and KURODA 1959	+		
HORIKAWA and FOX 1964		+	
SCHNEIDER 1964-1966	+	+	+
ECHALIER and OHANESSIAN 1965-1970		+	+
GVOSDEV and KAKPAKOV 1968		+	+
MANDARON 1970-1971	+		
ROBB 1969	+		
SHIELDS and SANG 1970		+	+

The most extensively used media are Schneider's medium (1964, 1966; commercially available) and Echalier and Ohanessian's medium D22 (a slight modification of our D20, 1970; Table III). As an example, let us describe briefly the composition of this last medium: its osmolarity is 360 milliosmoles (i.e. equivalent to a 10.5 gms per liter sodium chloride solution) and its pH is adjusted to 6.8 - to conform with the pH of the hemolymph.

TABLE II

Salt and sugar concentration of currently used culture media for *Drosophila* cells.

	Hanks (1949) Buffered Saline Solution for Vertebrates cells		*I. Schneider* (1966) (the salt composition of Mandaron's medium "M" is the same)		*Echalier* and *Ohanessian* "D22"		*Gvosdev* and *Kakpakov* "C-15"		*Shields* and *Sang* (1970)		*Shields* and *Sang* (1975) "M3" (to be published)	
	gm/l	mM	gm/l	mM	gm/l	mM	gm/l	mM	gm/l	mM	gm/l	mM
NaCl	8.	137.	2.1	35.9			4.	68.4	0.86	14.7		
KCl	0.4	5.4	1.6	21.5			1.56	20.9	3.13	42.		
$CaCl_2$	0.1[illegible]	1.3	0.6	5.4	0.8	7.3	0.5	4.5	0.89	8.	0.76	6.8
$MgSO_4.7H_2O$	0.2	0.8	3.7	15.	3.36	13.6			5.13	20.8	4.4	17.8
$MgCl_2.6H_2O$					0.9	4.5	2.5	12.3				
$Na_2HPO_4.12H_2O$	0.1[illegible]	0.3	1.76	4.9								
$NaH_2PO_4.2H_2O$					0.43	2.7	0.5	3.2	0.88	5.6	0.88	5.6
KH_2PO_4	0.06	0.4	0.45	3.3								
$NaHCO_3$	0.35	4.2	0.4	4.8			0.35	4.2				
$KHCO_3$									0.18	1.8	0.5	5.
Monosodium Glutamate. H_2O					8.	42.7			2.72	14.5	7.22	38.6
Monopotassium Glutamate. H_2O					4.98	24.5					7.88	38.8
Sodium Acetate. $3H_2O$					0.023	0.17	0.025	0.18				
Glucose	1.	5.5	2.	11.	1.8	10.	5.	27.8	4.6	25.5	11.	61.
Sucrose							5.	14.6				
Trehalose.$2H_2O$			2.	5.3								
NaOH									+	+	+	0.6
KOH					+	18.						

Comparison between ionic concentrations of hemolymph of 3rd instar larvae, currently used culture media for *Drosophila* cells and a classical saline solution for Vertebrates cells.

	Larvae Hemolymph (3rd instar)	*I. Schneider's* medium + 15% fetal calf serum(*)	*Echalier Ohanessian* "D22" + fetal calf serum	*Gvosdev Kakpakov* "C-15" + 15% fetal calf serum	*Shields Sang* (1970) + 10% fetal calf serum	*Shields* (1974) + 10% fetal calf serum	*Garcia-Bellido Nöthiger* Buffered Saline Solution (1974)	*Hanks* Buffered Saline Solution for Vertebrates cells
Na^{+}	55(50-56)	64.6	54.5	86.2	54.5	54.8	139.	41.6
K^{+}	38(36-40)	22.	39.	18.5	40.	39.9	7.6	5.8
$\frac{Na}{K}$	1.4	< 3.	< 1.4	> 4.5	1.35	1.37	18.	> 24.
Ca^{++}	8.	5.	6.9	4.2	7.5	6.5	1.8	1.3
Mg^{++}	20.	14.	16.5	10.6	19.	16.	1.	0.8
Cl^{-}	35(30-42)	73	30.7	116.	76.	22.	138.	144.6
PO_4^{---}	< 3.	(8.)	(2.7)	(3.2)	(5.6)	(5.6)	8.5	0.8
SO_4^{--}	?	13.	12.6	0.4	19.	16.2		0.8

(*) "D22" is supplemented by adding 100 ml of fetal calf serum to 1,000 ml of media "D22"

TABLE III

D22 Medium for *Drosophila melanogaster* embryonic cells

For 1 liter	mM	gm
Potassium Glutamate (I H_2O)	24.5	4.975
Sodium Glutamate (I H_2O)	42.7	7.98
Glycine	67	5.00
$MgCl_2$.$6H_2O$	4.47	0.9
$MgSO_4$. $7H_2O$	13.6	3.36
NaH_2PO_4 .$2H_2O$	2.74	0.43
$CaCl_2$ (to be dissolved apart)	7.3	0.8
Sodium Acetate, $3H_2O$	0.17	0.023
Succinic Acid		0.055
Malic Acid		0.6
Glucose		1.8
Lactalbumin Hydrolysate (Difco)		13.6
Difco Yeastolate		1.36
Vitamins B (see T. Grace 1962)		
Streptomycin		0.1
Penicillin 250,000 I.U.		
Adust to pH 6.6-6.7 with N KOH		

Likewise, the various cationic concentrations are considered, especially the characteristic Na/K ratio (1:4). Because it is well known that Cl is not the dominant anion in the body fluid of many insects, such as *Drosophila*, and that it is often replaced by organic anions, the basic saline solution of our medium D22 is essentially a mixture of sodium and potassium glutamates and glycinates - according to a suggestion by Shaw (1956).

A high content of free amino-acids is supplied by lactalbumin hydrolysate and all unknown nutritional requirements are covered by addition of yeast extract, some vitamins B and organic acids. Glucose is present in the concentration of 1.8 gm per liter.

Besides, most of the media have to be supplemented with 10 to 20% of decomplemented fetal calf serum. In the case of long established lines, this serum content may be lowered to 5% or, sometimes, 2%. We have even selected a subline from our line K, that grows in medium D22 without serum or any other protein supplementation. It is very economical when large amounts of cells are needed for biochemical investigations (see below). Unfortunately, the karyotype of this serum-independent strain was found to be rather unstable. No connection between this aneuploidy and the absence of serum was established.

Nevertheless, medium D22 cannot be considered as a chemically defined medium, because of its lactalbumin and yeast extract contents. Various simplified versions have been successfully used for short labelling experiments, but nobody has, so far, succeeded to grow *Drosophila* cells in a truly "synthetic" medium for a significantly long period.

To conclude, it must be emphasized that the extreme diversity of the formulae which are proposed for the culture of *Drosophila* tissues of cells, proves in itself that there is not yet available a perfect, polyvalent medium, which might be used for any tissue of *Drosophila* and that would promote cell multiplication as well as *in vitro* differentiation. It is quitc likely that such a "universal" medium simply does not exist, because every type of specialized cell must have its specific nutritional and environmental requirements.

3. Cloning method

It is necessary to clone frequently cells for genetic investigations, to assure homogenous cell populations. Cloning of insect cells has been successfully carried out in several laboratories in recent years (McIntosh & Rechtoris, 1974; McIntosh, 1975; Nakajima & Miyake, 1975), but the cloning of *Drosophila* cells remains difficult. The simple dilution technique, so extensively used for mammalian cells does not work here. Some years ago, nevertheless, we succeeded in isolating a few clones using the dilution method, provided that dispersed cells were covered with an agar solidified medium (Echalier, 1971). The plating efficiency was so low that this method could not be used as a standard technique.

In close connection with A. Ohanessian'group, we have recently devised a cloning method which is an adaptation of the "feeder-layer" technique contrived by Puck for mammalian cells. Single cells are helped to grow and multiply into small colonies, when they are dispersed among large populations of "feeder" cells, that is cells whose capacity to divide has been destroyed with a sufficient dose of X-rays, but which keep on metabolizing and "conditioning" the culture medium. The quantitative aspects of the technique will be described by Ohanessian and Richard-Molard in 1975 (personal communication).

The safe doses of irradiation for *Drosophila* cells vary with different strains, but were found to be much higher than for mammalian cells. Our line K can withstand at least 25,000 rads, with gamma rays.

In practice, the cells to be cloned are diluted in standard medium, and distributed into the small wells of a Falcon plastic dilution plate (on an average, one cell per well). Then, some 50,000 irradiated cells are added to each well. After 4 - 6 weeks colonies reach a sufficient size to be subcultured. To ascertain their clonal nature the process has to be repeated successively two or three times.

III. Experimental possibilities offered by *Drosophila* cell cultures

Four main approaches will be described, for the simplicity of presentation. But it should be clearly understood that they all are interrelated and represent various facets of the same endeavour: to try to understand someting about the structure and the functional modulation of the genetic material of eukaryotic cells of higher organisms.

1- Basic biochemical investigations

To grow *in vitro* large amounts of *Drosophila* cells is an easy way to overcome the limiting factor which is, for biochemical work, the small size of the *Drosophila* organism.

In order to illustrate it, we will briefly mention several types of research carried out in various laboratories with mass cultures of *Drosophila* cells:

– Thomas and collaborators, in Boston, are studying the DNA of *Drosophila melanogaster*, by analysing the fragments obtained from digestion of this DNA by several restriction enzymes (Mantueil, personal communication). By using various cell lines with definite and stable karyotypes, they hope to indentify from which chromosome come some specific fragments and perhaps to assign to them some physiological functions. Garen's group, in New Haven, has been preparing *Drosophila* RNA polymerase. They are mainly interested in the proteins that bind to DNA, especially certain minor species that require large amounts of cell material for their isolation (personal communication). They wish to compare this class of proteins in the culture cells with the corresponding proteins of embryonic stages (where they are suspected to play a decisive role in the mechanisms of cell determination).

It was thought that it might be relatively easy, on account of the simple chromosomal set of *Drosophila*, to isolate, on a large scale, /c metaphase chromosomes from cell cultures and eventually, to separate every chromosome pair. This exciting project seems to encounter technical difficulties, apparently because of an unusual stickiness of the mitotic /a spindle. However, a preparative method was described by Hanson and Hearst (1973).

Jordan, in Marseille, is studying the structure and maturation of ribosomal RNAs from *Drosophila* cultured cells. He is establishing the nucleotide sequence of 5S RNA to compare it to those of other organisms. Besides, he has discovered a new species of ribosomal RNA, "28", which seems to be not co-valently linked to 26S RNA (Jordan, 1974).

In Toulouse, Zalta has devised a technique for preparing a very pure nucleolar fraction from *Drosophila* cells (personal communication).

The above list comprises only a few significant examples and demands are now rapidly increasing for cell lines especially for our strain that requires no serum.

2- Genetics of somatic cells

Several technical obstacles must be overcome before *Drosophila* cells can become a standard tool of geneticists. The aim is not simply to repeat, with insect cells, what has already been achieved with mammalian cells. There would be no point in using somatic hybridization for mapping genes in *Drosophila,* as it is currently and successfully done with murine or human cells. But rather we should take advantage of the specific problems which can be solved by this insect material.

a) *Selection of genetic markers*

Only features which are expressed at the cellular level and under *in vitro* conditions may be used as markers. Mutations disturbing the main metabolic pathways are suitable for this purpose. Following the establishing of new *Drosophila* cell lines, workers engaged in such studies will avail themselves in the near future of the rich collection of *Drosophila* mutants:

– *Morphological features* of the cells are too unstable in culture *in vitro* to be reliable markers.

– *Chromosomal markers* are extensively used in genetics of somatic cells. When one thinks of the hard work which is required for the identification of the various chromosomal pairs of any vertebrate cell, it will be easily admitted that the simplicity of the karyotype of *Drosophila melanogaster*, with its four recognizable pairs of chromosomes is a great asset.

An extensive karyological analysis has been carried out by Faccio-Dolfini and other members of the Instituto di Genetica at Milano, Italy, on our main cell lines as well as on new ones they had recently grown. The polymorphism of chromosomal formulae in cell populations was carefully studied (Dolfini, 1971, 1973; Faccio-Dolfini, 1974b; Mosna and Dolfini, 1972). Moreover, information was collected, with autoradiographic methods, on the fine structure of mitotic chromosomes and especially on DNA replication in euchromatic and heterochromatic sections (Barigozzi *et al.*, 1967; Halfer and Barigozzi, 1972; Faccio-Dolfini, 1974a). In recent years, techniques of differential staining ("banding") of specific chromosome segments, and mostly fluorescence technique (fig. 2) permitted specialized studies on purine metabolism (Becker, 1970) and of fluorescence patterns (Zuffardi *et al.*, 1971).

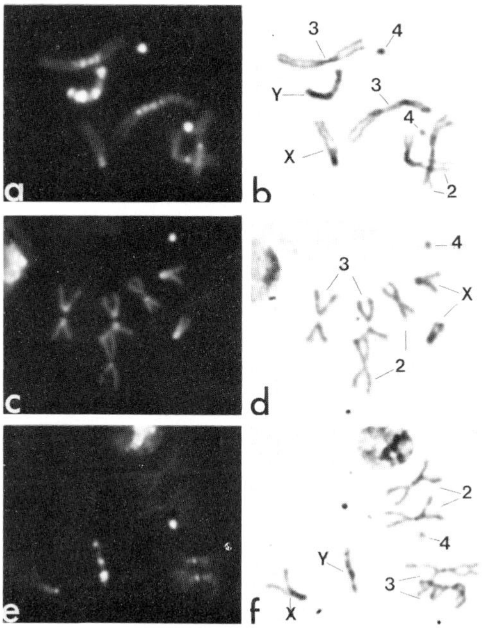

Fig. 2. Fluorescence pattern of *Drosophila* cells. Quinacrine (left) and orcein (right). (a) and (b) Embryonic male cell of the wild-type stock Varese. Bright fluorescence on the Y (four blocks) and on the chromosomes IV, weak fluorescence on the terminal portion of the X and on the centromeric regions of the III pair. (c) and (d) Established line *in vitro:* female cell of the K_c line with only one IV chromosome. (e) and (f) Established line *in vitro:* male cell of the C_a line with only one IV chromosome. (Courtesy of Dr. S. Faccio Dolfini and Dr. C. Halfer)

– *Drug resistance* and especially resistance to several purine and pyrimidine analogues provided convenient genetic markers of mammalian somatic cells, because the enzymatic deficiencies, by which most of them are accounted for, may be used in efficient "selective systems" for sorting out hybrid cells (see below the special paragraph on this question).

– *Isozymes* are excellent natural markers (see the descriptions devoted to isozymes and functional state of *in vitro* cells.)

b) Cell fusion

Genetics of somatic cells began with the initial observation of Barski *et al* (1960) that, when two recognizable strains of murine cells are grown in the same culture vessel, a few cells may accidentally fuse their cytoplasm, then their nuclei and give rise to viable hybrid cells. This event remains rare and several methods had to be devised to increase the chances of cell fusion and to sort out such hybrid cells from the parental populations.

– Spontaneous cell fusion occurs most probably in *Drosophila* cell cultures as well. Fortuitously, Dolfini (Barigozzi 1971) happened to observe karyological figures which are reasonable evidence of such an event: they were tetraploid metaphases with 3X chromosomes and IY chromosome, which can only be explained by the fusion, amid a mixed population, of a "female" cell with a "male" one. Those "androgyne" cells were only rarely observed, however.

– The treatment with UV inactivated virions of Sendai para-influenza virus - the classical method devised by Harris and Watkins (1965) to induce polykaryocytosis in vertebrate cells - has no effect on *Drosophila* cells. We were unable to find any receptor for adsorption of Sendai virus on the surface of *Drosophila* cells.

There is, nevertheless, in the litterature a brief report of cell fusion induction with Sendai virus, between insect cells and human Hela cells (Zepp *et al.*, 1971). It has been recognized since, that the insect cells were lepidopteran, and not dipteran in origin (Greene *et al.* 1972). Confirmation of this cell fusion work is lacking.

Other attempts by us, with some reputedly fusing viruses (at least for mammalian cells), such as the Herpes-type Aujeski virus or the Simian parainfluenza SV5 virus, were also unsuccessful (Echalier, 1971). A paper, by E. Suitor and Paul (1969), recording the cytopathic effects of Dengue2 virus on Singh's *Aedes* cells and the formation of large syncytia prompted us to investigate the possible fusing activity of this virus and many other Arboviruses. Most of the 18 Sero-group B Arboviruses that we tested (Hannoun and Echalier, 1971) multiplied in *Drosophila* cells for very long periods, but no cell fusion was ever observed.

– Still with the same purpose in mind, we became interested in several substances which are known to modify the cell surface. Finally, Becker, from our group, was able to establish that the phytohemagglutinin Concanavalin A not only agglutinates *Drosophila* cells – as it is extensively shown for mammalian cells – but does induce a rapid cell fusion and the formation of apparently viable synkaryons (Becker 1972).

The technique devised by Becker is rather simple: cells are maintained in a dense suspension ($3X10^6$ cells per ml) by a slowly rotating magnet and treated, during 20 minutes at room temperature, with a concentration of 100 μg of Con A (Calbiochem, Los Angeles) per ml. We knew that Con A activity is strengthened by the addition of Ca and Mn cations (10^{-4} M) to the standard medium D22. Then, the cells are seeded into culture flasks and, as soon as they adhere, washed carefully with fresh medium.

Shortly after the beginning of the treatment, cells gather in pairs or larger groups. At least for our line K cells, the efficiency is very high: almost 50% of the population may be affected by the phenomenon. Special connections are created between the partners, as observed by phase-contrast microscopy (Fig. 3a). With an electron microscope (Fig. 3b), a complex web of microvilli can be seen in this con-

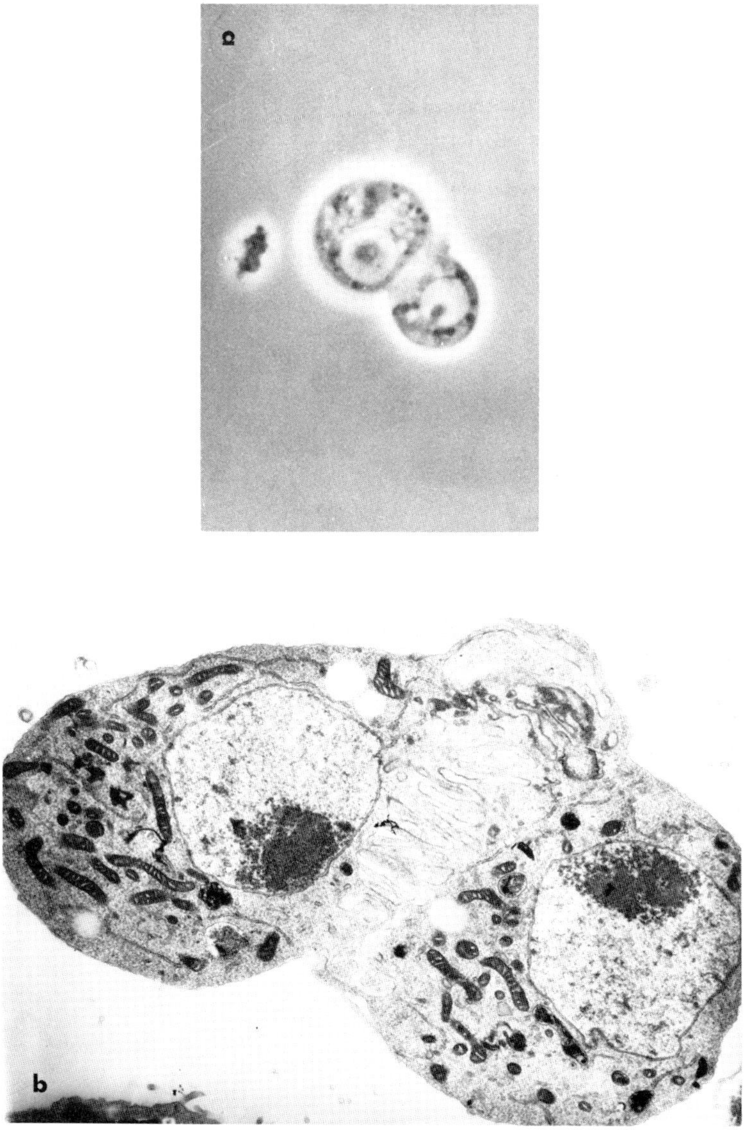

Fig. 3. Concanavalin A - induced fusion of *Drosophila* cells, one hour after the beginning of the treatment. (a) with phase contrast (X 2500) (b) with E.M. (X 8000) Observe the complex intrication of microvilli in the contact area. (Courtesy of J.L. Becker and P. Grellet. The Electron micrography is still unpublished).

tact area. They fuse during the next few hours, with the formation of di - or polykaryons, then, some 24 hours later, of synkaryons. Autoradiographic pictures were obtained of dikaryons formed between one subline whose nuclei had been labelled with tritiated thymidine and another unlabelled cell line.

Unfortunately, in the absence of an operative selective device, we have not yet been able to study the fate of such Con A induced hybrid cells. It must not be forgotten that, in mammalian cells, even with the classical Sendai method, viable hybrid cells never exceed more than 5% of the fused cells. Let us note that, among treated cells of our very stable line K, an unusually high percentage of tetraploid cells was observed during the first few days. It rapidly dropped to the normal level (about 10%), which might very well mean that diploid cells outnumbered hybrid ones. Attempts of isolating hybrids by cloning have been unsuccessful so far.

The reality and efficiency of *Drosophila* cell fusion by Concanavalin A were confirmed by Gvosdev *et al.*, in Moscow, and Gehring *et al.*, in Basel (personal communications). The rate of cell fusions seems to vary, nevertheless, from one line to another.

– More recently, two quite different types of compounds were reported to induce cell fusion in *Drosophila* cells: Lysolecithins (Barigozzi *et al.*; Lane DeCamp; personal communications) and polyethyleneglycol (Gehring *et al.*, personal communication). Nothing is known about the survival of the observed dikaryons.

So, although a decisive step may have been taken, it still remains crucial to contrive an efficacious selective method to sort out somatic hybrid cells.

c) Selective system for hybrid cells

– The universally used technique for selecting mammalian somatic hybrid cells was devised by Littlefield (1964). It consists in "crossing" a cell line which is deficient for an enzyme of the purine "salvage" pathway (hypoxanthine guanine phosphoribosyl transferase, H GpT) with another line which lacks one enzyme of the pyrimidine "salvage" pathway (thymidine kinase). In the presence of aminopterine (a folic acid analogue which blocks the neosynthesis routes to nucleotides), and provided that the culture medium is supplemented with hypoxanthine and thymidine(1), only hybrid cells, by complementation, will be able to grow, but not the parental ones. Such enzyme-deficient lines are, in practice, rather easy to isolate, because they prove to be respectively resistant to various guanine and thymidine analogues. Several variants of this selective method were described.

First attempts to adapt the system to *Drosophila* cells encountered unexpected difficulties: All our cell lines were found to be resistant to high doses of azaguanine. Finally, it was established, by direct enzymatic assay as well as by autoradiographic studies, that the enzyme HGPRT does not exist in *Drosophila* established cell lines, nor in any *Drosophila* normal tissue (Becker, 1974a). This surprising situation had been previously suspected, but not directly proved, in ovarian tissues of a fly from another genus, *Musca*. A complete survey of the interconversion routes of purine metabolism in *Drosophila* cells was necessary (Becker 1974b, 1975). It was concluded that, in *Drosophila*, the only working "salvage" pathway for purine nucleotides is through adenine phospho-ribosyl transferase (APRT is distinct from HGPRT). (see fig. 4).

Spontaneous deficient mutants were, therefore, selected against azaadenine. Their frequency, in our main lines, is about 10^{-6}. They were found in one selection step, to be resistant to relatively high doses (10 μg/ml, i.e. 10^{-6} M) and their resistance may indeed be accounted for by an absence of APRT activity (Becker 1974a).

---

(1) This selective medium is called HAT, for hypoxanthine, aminopterine and thymidine.

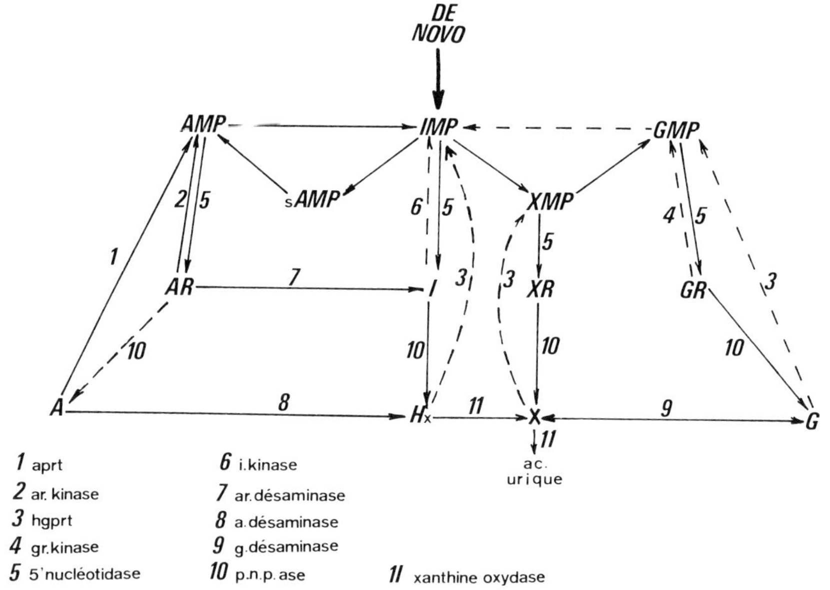

Fig. 4. Interconversion routes of purine metabolism in *Drosophila* (dotted pathways do not exist in *Drosophila* cells) (Becker 1974b).

In order to complete the selective system, TK-deficient mutants had to be looked for. Unfortunately, new difficulties arose. Those are not, as a matter of fact, specific to *Drosophila* cells and, even in mammalian cells, BUdR resistance is not always correlated with TK deficiency. In *Drosophila* cell lines, a few variants resistant to 10^{-4} M BUdR were isolated, but their measured TK activity was never found to be lower than about 30% of the normal value (which exceeds values due to the mitochondrial enzyme). Processes of BUdR penetration and utilization seem to be complex: when using labelled analogues, only traces of the radioactivity were recovered in normal cell DNA (Becker, personal communication). Besides, *Drosophila* cells were observed to be highly sensitive to FUdR, and no spontaneous resistance could be found.

– A distinct selective system, based on the use of ouabain-resistant variants, was recently suggested for murine cells by Baker *et al.* (1974). The drug is known to inhibit the plasma membrane Na/K ATPase. The fact that ouabain resistance behaves as a codominant trait might be of some use for selecting hybrids between ouabain resistant and ouabain sensitive cell lines.

The method proved disappointing in *Drosophila* cells: a few spontaneously resistant cells were isolated, but they were useless because they no longer fused with concanavalin A. Their Na/K ATPase activity was not measured and such a resistant to 10^{-4} M ouabain might very well be due to complex modifications of the plasma membrane.

– To conclude, further studies are required of selecting somatic hybrid *Drosophila* cells. However, a semi-selective method, using APRT cells, is already available, similar to that contrived by Davidson and Ephrussi (1965) for mammalian cells. It may be applied to investigations on fusion of *Drosophila* established cell lines with various freshly explanted tissue cells.

3- Analysis of functional states of in vitro cultured cells: isoenzymatic patterns

Under *in vitro* conditions, at least in liquid media, most cell lines adopt a simple morphology, without apparent differentiated structures, which is usually described as "fibroblast-like". This term, however, does not tell anything about physiological functions. It is now obvious, from experiments with mammalian cell cultures, that, even after several years of growth *in vitro* cell lines may have kept open one or a few of the specialized metabolic pathways which were specific for the differentiated tissues they were derived from.

It was thought that an analysis of the isoenzymatic patterns of our *Drosophila* cell lines could reveal some degree of functional differentiation.

It is indeed well known that, in higher organisms, the same enzymatic activity may correspond to several distinct molecular forms and that the distribution of such "isozymes" seems to be tissue specific and to vary with the developmental stages of the animal.

During the last ten years, such isozymic variations have been extensively studied in the fly *Drosophila*, by workers in many laboratories all over the world (see recent reviews: Ursprung, 1971; Fox *et al.*, 1971; Dickinson and Sullivan, 1975). A large number of enzymatic systems (about 40) have been explored, with the usual electrophoretic identification methods. In many cases, their repartition was systematically analysed in the various organs and during the successive stages, from embryo to imago. There is now available a vast synoptic table giving the specific isoenzymatic equipment of any particular tissue (for instance, of nervous ganglia at the beginning of the third larval instar). Moreover, it was possible to map the genes corresponding to some 30 of those enzymatic systems, with the usual precision of *Drosophila* genetics.

Debec (1974; personal communication), in our laboratory, studied the functional states of our cell lines, looking for two kinds of information.

– First an attempt was made to characterize significant differences in the isozymic patterns of various strains or clones, in order to use them as reliable markers.

– Secondly, an attempt was made to compare such isozymic profiles with respective patterns of the different tissues of *Drosophila*, to provide information about the states of functional differentiation, at the molcoular level, possibly retained by these established cell lines.

In a first series, 13 enzymatic systems were explored in some 20 cell strains or sublines and more recently, 12 other enzymes have been studied. The diagram (Fig. 5) sums up Debec's results concerning our line K. It can be seen, at a glance, on the genetic map of the corresponding enzymes of *Drosophila melanogaster*, which structural genes are currently expressed in the *in vitro* cells, under standard conditions. Interesting remarks may be made on the possible clustering of active or inactive genes. This pattern is typical and, with but few exceptions, similar to general profiles observed in all studied cell strains.

Three main explanations may be put forward to account for such a relative homogeneity of the observed isoenzymatic profiles:

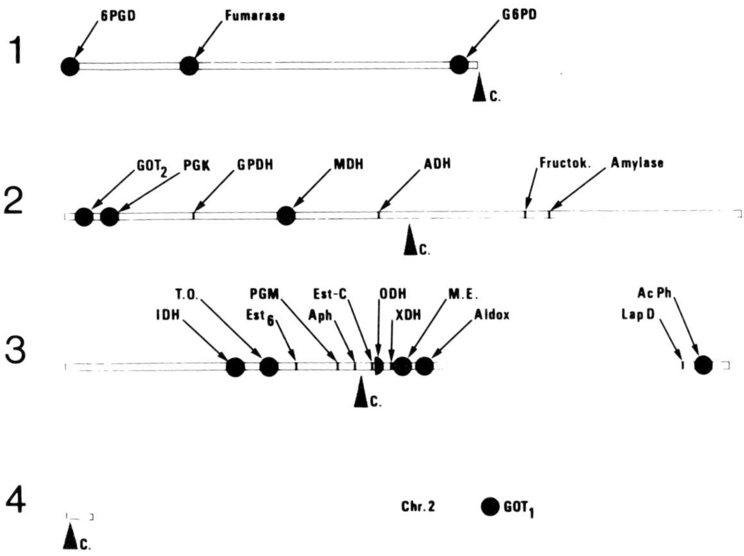

Fig. 5. Isozymic pattern of the cell line K, on the genetic map of *Drosophila melanogaster:* Dark circles correspond to loci which are active under the standard conditions of *in vitro* culture.

ADH	Alcohol Dehydrogenase	Lap	Leucine Amino Peptidases
Aldox	Aldehyde Oxydase	MDH	Malate Dehydrogenase
AcPh	Acid Phosphatase	ME	"Malic Enzyme"
Aph	Alkaline Phosphatase	ODH	Octanol Dehydrogenase
Est	Esterases	6PGD	Phospho Gluconate Dehydrogenase
Fructok.	Fructokinase	PGK	Phospho Glycerate Kinase
GOT	Glutamate Oxalate Transaminase	PGM	Phospho Glucomutase
G6PD	Glucose 6 Phosphate Dehydrogenase	TO	Tetrazolium Oxidase
GPDH	Glycerophpsphate Dehydrogenase	XDH	Xanthin Dehydrogenase
IDH	Isocitrate Dehydrogenase	(Courtesy of A. DEBEC 1975, to be published).	

– The first one is that our cell strains (mostly sublines from our two original lines C and K, derive from flies with a common genetic background. As a matter of fact, the most clear-cut difference in several enzymatic systems concerned a few lines recently grown by Richard-Molard (1975), which have a distinct genotype. It is likely that in the future when a large number of new lines with very different genotypes will become available, their isozymic profiles will differ considerably.

– The second possibility is that all *Drosophila* cell lines, derived from fragmented embryos might consist of closely related tissues: only a few cell types might be able to multiply under our present *in vitro* conditions. In regard to this interesting question isozymic patterns can perhaps provide evidence concerning the precise tissue origin of our cell lines. By comparison with the known profiles of the tissues of embryos, larvae, and imagos of *Drosophila,* it is tempting to point out some resemblances, for instance, between our line C and nervous tissues of the larva. The pattern of our line K would be more akin to the profile of imaginal disks. Supporting data, however, are still too few and we have to be extremely cautious with such assumptions.

– The third hypothesis - and the most attractive one - is to suppose that cells of any origin, under the very peculiar conditions of *in vitro* cultures, would all carry on a sort of "basic program" which allows them only to grow and divide (what B. Ephrussi calls "house keeping activities"). Even then, as it was pointed out at the beginning of this chapter, it is now well known, from results obtained with mammalian cell lines, that, beyond this common program, cells *in vitro* can keep open (or only partially open) a few specialized metabolic pathways ("luxury activities", according to Ephrussi's terminology).

In this respect, the relatively rapid and widely spread screening that can be done, in *Drosophila* cell lines, with an electrophoretic analysis of isozymic patterns may reveal some possible degree of functional differentiation. It must be understood, however, that in spite of the considerable work already accomplished, more date are needed, especially about cell lines with distinct genotypes.

When the functional states of many cell lines become better known, it might be possible to correlate some modifications of their specialized programs with specific variations of their karyotype. It will be worth while to study the biosynthesis of a particular enzyme and the resulting monosomy or polysomy of the chromosome that carries the structural gene of this enzyme. Preliminary results, in this field, were reported about X-linked G6PD and gene dosage in *Drosophila* cultured cells (Gvosdev *et al.*, 1971).

Moreover, it will be tempting to manipulate those cell lines to exceed their usual and monotonous program either by changing completely their culture conditions or by applying specific hormonal stimulations.

4- Action of insect hormones at the cellular level

Applications of invertebrate cell culture for the study of insect hormones have been discussed by Marks (1975). Our own work concerning the responses of *Drosophila* established cell lines is presented here.

Courgeon, from our group, was able to demonstrate that some cell strains, and mainly our line K, seem to respond specifically to the synthetic insect moulting hormones, α and β-ecdysones, and to several of their analogues. The effective doses may indeed be considered as very close to physiological values, as low as 0.006 µg/ml for β-ecdysone. Moreover, in order to make sure that the observed responses are not unspecific pharmacological effects of any steroid compound, many other steroids were checked, such as cholesterol and mammalian hormones, but none were found to have any discernable activity, even at much higher doses.

– It could be established that, in addition to their strikingly different threshold concentrations (β-ecdysone is 500 times more active than α-ecdysone), the two main ecdysones have rather distinct types of action: - β-ecdysone mostly induces a characteristic series of morphological modifications (Courgeon 1972a), which coincides with an arrest of multiplication: hormone treated cells of line K consisting usually of small round cells, loosely attached to the flask bottom - are seen, after 16 to 24 hours, to flatten and elongate. Then, they tend to aggregate and bebecome fibroblastic or very flattened, and the membrane shows an apparent expansion; their cytoplasm is full of inclusions. The rapidity of the phenomena is dose dependant. As for α-ecdysone, it stimulates cell multiplication (Courgeon 1972b).

These observed effects on cultured cells are in good agreement with current ideas about the distinct roles of the two ecdysones in the physiological regulation of insect growth and moulting.

– It was found that the presence of fetal calf serum in the culture medium is necessary for hormone action. This fact is reminiscent of the synergistic "Macromolecular Factor" described by Kambysellis and Williams (1971) for the *in vitro* spermatogenesis of an insect.

– Beyond such morphological variations induced by β-ecdysone in *Drosophila* cells, it should be pointed out that after hormonal treatment a few modifications of protein patterns of the cells have been revealed by immunological methods. This work, in collaboration with Robert's group (personal communication), is in a phase.

– Besides, several clones have been selected from line K, which display a wide range of receptivity to ecdysone, from the utmost sensitivity to a true resistance (i.e. no effect with concentrations of ecdysone 10,000 times exceeding the usual active doses). This trait was found to be very stable in clonal populations (Courgeon, 1975). This differential cellular response to ecdysone may give us a good tool for analysing the complex mechanisms of hormone action.

In this connection, recent preliminary results will be mentioned: Best-Belpomme and Courgeon (1975), by using labelled β-ecdysone, have just been able to characterize saturable "receptors" in a "membrane fraction" from sensitive cells, while none could be found in resistant clones.

IV. Prospects in *Drosophila* cell culture

– In recent years, slow but significant progress has been made, in setting up long term as well as primary cultures of *Drosophila* cells. A major factor, in this progress, is the recent involvement of an increasing number of research groups. One can expect that many new cell lines with interesting genotypes will soon become available. Especially mutants of *Drosophila* which are being extensively studied at the organism level: such as, for instance, "bobbed" mutants (reduction of the number of ribosomal genes) or "rudimentary" mutants (deficiency of one of the first enzymes of the pyrimidine biosynthetic pathway) and many others might thus provide a decisive advantage and incomparable possibilities in physiological genetics *Drosophila melanogaster.*

– So far, all recorded successes, at least in long term cultures, concerned embryonic cells. It would be extremely important to grow *in vitro*, for prolonged periods, tissues of larvae or imagos. None of the presently available media seems to be suitable for this purpose, even if it is claimed that they imitate the composition of the body fluid of 3rd instar larvae.

It would seem worth while to attack this problem again, without any prejudice, and to devise patiently new formulae. It is likely, indeed, that differentiated tissue cells are living, *in vitro*, in a micro-environment which might differ considerably from the environment provided by circulating hemolymph. Moreover, much immagination would have to be devoted to modify, perhaps in a radical way, the standard conditions of the culture in liquid media - in order to preserve the supposedly important cell to cell connections of differentiated tissues.

Let us note that the above remarks correspond to a new general trend, in vertebrate cell culture as well. Ten years ago (and during the fifties for mammalian cells), methods were devised to favor cell divisions and to grow large cell populations. At that time, monolayers of fibroblast cells were mostly used as substrate for viral multiplication. The point of view changed entirely when the main current of biological research shifted to physiological genetics of eukaryotic cells of higher organisms. Now, less emphasis is put on cell multiplication and greater care is taken in trying to grow the cells under conditions which would allow them to maintain, at least partly, their differentiated structures and metabolic programs.

– Undoubtedly, among the various *Drosophila* tissues to be grown and propagated *in vitro*, imaginal disc cells have received the greatest attention.

Imaginal disc cells indeed constitute one of the most convenient materials for analysing the complex mechanisms of cell determination. This can be deduced from the remarkable series of investigations carried out *in vitro* by Hadorn and his coworkers in Zurich (Ursprung, 1972).

It must be recalled that imaginal discs are small groups of cells which are kept apart during the whole larval period of the insect and, at the time of the metamorphosis, will permit the building of an entirely new organism, the imago. The determination of those cells, that is the selection of the program they will have to accomplish, goes back to early embryonic life, even though the expression of this differentiated program is postponed for several days and will take place only under the hormonal stimulations of metamorphosis. The basic stability of this determined state has been clearly established.

There are, in the litterature, many reports on more or less successful development of imaginal discs *in vitro.* Among the most recent reports, let us mention the partial success obtained by Schneider (1964, 1966), in her liquid medium, with eye-antennal discs and, mostly, the outstanding results of Mandaron (1970, 1971): in a synthetic culture medium and under the stimulation of ecdysone, leg or wing discs from late 3rd instar larvae can evaginate and differentiate completely in a few days. The various segments of the appendage, with their specific cuticular structures, are perfectly recognizable. It must be pointed out, however, that one is dealing with intact imaginal discs and not with individual cells; there is no or very little, cell multiplication. Therefore despite the attractiveness of this experimental model for an analysis of late stages of differentiation, a biochemical approach remains difficult and restricted, even though it offers the possibility of isolating relatively large amounts of discs, according to the method of Fristrom and Mitchell (1965).

A promising step was taken when Schneider (1972) observed the growth of floating vesicles in primary cultures from late embryos. They continued to swell and bud during a few passages and, when finally transplanted into larvae, they gave rise, at the time of their host metamorphosis, to identifiable cuticle and bristles. This means that they were disc cells. Unfortunately, the established lines which were derived from the same cultures apparently differed from this imaginal material.

Quite recently, Dubendorfer, in collaboration with the group of Sang (personal communication) established that similar vesicles, developed in primary cyltures from dissociated early embryos, can differentiate and secrete cuticular structures, not only when transplanted *in vivo,* but also *in vitro* when treated with ecdysone. Unfortunately, such vesicles, in the present conditions of culture, do not survive longer than a few weeks.

A new frontier will open when veritable permanent lines of such determined imaginal disc cells became available.

In conclusion, in recent years *Drosophila* cell cultivation has given rise to a spectacular revival of interest in developmental biology. It is quite probable that almost all predictable experimental approaches employing *Drosophila* material *in vitro* will yield significant and perhaps decisive results.

V. References

Baker, R.M., Brunette, D.M., Mankovitz, R., Thompson, L.H. Whitmore, G.F., Siminovitch, L. and Till, J.E. (1974). *Cell, 1,* 1.

Barigozzi, C. (1971). In: Current Topics in Microbiology and Immunology, *55,* 209. Springer-Verlag, Berlin, Heidelberg, New York.

Barigozzi, C., Dolfini, S., Fraccaro, M., Halfer, C., Rezzonico Raimondi, G., Tiepolo, L. (1967). *Atti Asso. Genet. Ital. 12,* 291.

Barski, G. Sorieul, S. and Cornefer, F. (1960). *C.R. Acad. Sci. Paris, 251,* 1825.

Becker, J.L. (1970). *C.R. Acad. Sci. Paris, 271,* 2131.

Becker, J.L. (1972). C.R. *Acad. Sci. Paris, 275,* 2969.

Becker, J.L. (1974a). *Biochimie, 56,* 779.

Becker, J.L. (1974b). *Biochimie, 56,* 1249.

Becker, J.L. (1975). *C.R. Acad. Sci. Paris* (in press).

Best-Belpomme, M. and Corgeon, A.M. (1975). *C.R. Acad. Sci. Paris, 280,* 1397.

Courgeon, A.M. (1972a). *Exp. Cell Res., 74,* 327.

Courgeon, A.M. (1972b). *Nature, New Biol., 238,* (86) 250.

Courgeon, A.M. (1975). *Exp. Cell Res.* (in press).

Davidson, R. and Ephrussi, B. (1965). *Nature, Lond., 205,* 1170.

Debec, A. (1974). *W. Roux'Arch., 174,* 1.

Dickinson, W.J. and Sulivan, D.T. (1973). In: Results and Problems in Cell Differentiation: vol. 6, Springer-Verlag, Berlin, Heidelberg, New York.

Dolfini, S. (1971). *Chromosoma (Berl.), 33,* 196.

Dolfini, S. (1973). *Proc. III Intern. Collq. Invertebrate Tissue Culture,* 143.

Dolfini, S., Courgeon, A.M. and Tiepolo, L. (1970). *Experientia, 26,* 1020.

Dubendorfer, A., Shields, G. and Sang, J.H. (1975). *(personal communication)*

Echalier, G. (1971). In: Current Topics In Microbiology and Immunology, *55,* 220. Springer-Verlag, Berlin, Heidelberg, New York.

Echalier, G. and Ohanessian, A. (1969). *C.R. Acad. Sci. Paris, 268,* 1771.

Echalier, G. and Ohanessian, A. (1970). *In Vitro, 6,* (3) 162.

Echalier, G., Ohanessian, A. and Brun, G. (1965). *C.R. Acad. Sci. Paris, 261,* 3211.

Faccio Dolfini, S. (1974a). *Chromosoma (Berl.) 50,* 383.

Faccio Dolfini, S. (1974b). *Chromosoma (Berl.) 47,* 253.

Fox, D.J., Abacherli, E. and Ursprung, H. (1971). *Experientia 27,* 218.

Fristrom, J.W., Logan, W.R. and Murphy, C. (1973). *Developmental Biol. 33,* 441.

Fristrom, J.W. and Mitcell, H.K. (1965). *J. Cell Biol. 27,* 445.

Grace, T.D.C. (1962). *Nature, Lond. 195,* 788.

Greene, A.E., Charney, J., and Nichols, W.W. (1972). *In Vitro, 7,* 313.

Gvosdev, V.A., Birstein, V.Y., Kakpakov, V.T. and Polukarova, G.L. (1971). *Ontogenes 2,* 304.

Gvosdev, V.A. and Kakpakov, V.T. (1968). *Genetika, 4,* 129.

Halfer, C. and Barigozzi, C. (1972). *Chromosomes Today, 2,* 181.

Hanks, J.H. and Wallace, R.E. (1949). *Proc. Soc. Exp. Biol. Med. 71,* 196.

Hannoun, C. and Echalier, G. (1971). In: Current Topics in Microbiology and Immunology, *55,* 227. Springer-Verlag, Berlin, Heidelberg, New York.

Hanson, C.V. and Hearst, J.E. (1973). *Cold Spring Harbor Symposia.*

Harris, H. and Watkins, J.F. (1965). *Nature, Lond. 205,* 640.

Horikawa, M. and Fox, A.S. (1964). *Science, 145,* 1437.

Horikawa, M. and Kuroda, Y. (1959). *Nature, 184,* 2017.

Horikawa, M., Ling, L.N., and Fox, A.S. (1966). *Nature, 210,* 183.

Jordan, B.R. (1974). *FEBS Letters, 44,* 39.

Kakpakov, V.T., Gvosdev, B.A., Platova, L.G. and Polukarova, L.G. (1969). *Genetika, 5,* 67.

Kakpakov, V.T. and Polukarova, L.G. (1971). *Ontogenes, 2,* 3.

Kambysellis, M.P. and Williams, C.M. (1971). *Biol. Bull. Woods Hole 141,* 527.

Kuroda, Y. and Tamura (1956). *Med. J. Osaka Univ. 7,* 137.

Littlefield, J.W. (1964). *Science, N.Y., 145,* 709.

McIntosh, A.H. (1976). In: Invertebrate Tissue Culture in Research. (K. Maramorosch, ed.), Academic Press, New York.

McIntosh, A.H. and Rechtoris, C. (1974). *In Vitro, 10,* 1.

Mandaron, P. (1970). *Dev. Biol. 88,* 298.

Mandaron, P. (1971). *Dev. Biol. 25,* 581.

Marks, E.P. (1976). In: Invertebrate Tissue Culture: Application in Medicine, Biology and Agriculture'' (E. Kurstak and K. Maramorosch, eds.) Academic Press, New York.

Mosna, G. and Dolfini, S. (1972). *Chromosoma (Berl.) 38,* 1.

Nakajima, S. and Miyake, T. (1976). In: Invertebrate Tissue Culture: Applications in Medicine, Biology and Agriculture, (E. Kurstak and K. Maramorosch, eds.) Academic Press,New York.

Ohanessian, A. and Richard-Molard, C. (1975). *(Personal communication).*

Paradi, E. (1973). *Biologiai Kozl. 21,* 11.

Richard-Molard, C. (1975). *Arch. Virol. 47,* 139.

Robb, J.A. (1969). *J. Cell. Biol. 41,* 876.

Schneider, I. (1964). *J. Exp. Zool. 156,* 91.

Schneider, I. (1966). *J. Embryol. Exp. Morph. 15,* 271.

Schneider, I. (1972). *J. Embryol. Exp. Morph. 27,* 353.

Shaw, E.I. (1956). *Exp. Cell Res. 11,* 580.

Shields, G. and Sang, J.H. (1970). *J. Embryol. Exp. Morph. 23,* 53.

Shields, G., Dubendorfer, A. and Sang, J.H. (1975). *J. Embryol. Exp. Morph. 88,* 159.

Suitor, E.C. and Paul, F.J. (1969). *Virology, 38,* 482.

Ursprung, H. (1971). *Naturwiss. 58,* 383.

Ursprung, H. et al. (1972). In: Results and Problems in Cell Differentiation. Springer-Verlag, Berlin, Heidelberg, New York.

Zepp, H.D., Conover, J.H., Hirschhorn, K., and Hodes, H.L. (1971). *Nature, New Biol. 229,* 119.

Zuffardi, O., Tiepolo, L., Dolfini, S., Barigozzi, C. and Fraccaro, M. (1971). *Chromosoma (Berl.) 34,* 274.

Chapter 10

METAMORPHOSIS OF IMAGINAL DISC TISSUE GROWN *IN VITRO* FROM DISSOCIATED EMBRYOS OF *DROSOPHILA*

A. DÜBENDORFER

I. Introduction

In early embryogenesis of the fruitfly *Drosophila melanogaster* a number of cell groups are set apart and do not participate in the development of the larval organs but form the imaginal primordia from which, during metamorphosis, the cuticle and most of the internal organs of the adult fly arise (review by Nöthiger, 1972). At a very early stage of embryogenesis (blastoderm formation) these groups of cells are determined to develop into specific parts of the fly (Chan and Gehring, 1971; Illmensee and Mahowald, 1974). As these primordia grow during embryogenesis and larval development they are gradually subdivided into various compartments (Garcia-Bellido *et al.*, 1973; Garcia-Bellido, 1975). This process of compartmentalization results in the establishment of the specific patterns observed after metamorphosis. During the second larval instar the imaginal discs become competent to react to the metamorphotic hormones by differentiation of imaginal structures (Bodestein, 1939, 1943; Mindek, 1972, Mindek and Nöthiger, 1973; Gateff and Schneiderman, 1975). The molecular bases of the processes of embryonic determination, of the specification of patterns within an anlage (Bryant, 1974), and of the acquisition of competence for metamorphosis, respectively, are not understood. There is some evidence that pattern formation and acquisition of competence for metamorphosis are independent processes: Hadorn (1966) has shown that in long term cultures *in vivo* the capacity to form a normal pattern can be lost although the cells remain competent to metamorphose (anormotypic cultures). Similarly, imaginal disc cells from embryos may differentiate *in vitro*, however no recognizable patterns are formed.

In the course of imaginal disc development a number of *morphological* changes have been described (Auerbach, 1936; Poodry and Schneiderman, 1970; Sprey, 1970; Mandaron, 1974; van Ruiten and Sprey, 1974): The epithelium of a first larval instar disc is smooth and consists of cuboidal cells. During the second and third instar, the cells elongate and form a columnar epithelium that becomes folded, each disc acquiring its characteristic pattern of folds. At the onset of metamorphosis, the cells return to cuboidal shape again and finally flatten to form a squamous epithelium. It is at this stage that the cells differentiate into the characteristic cuticular structures, such as bristle organs, trichomes (hairlike protuberances) and sensillae.

When the cells of early embryos of *Drosophila* are dissociated and dispersed in a culture chamber a number of larval cell types (e.g. fat body cells) can be observed to differentiate *in vitro* (Shields *et al.*, 1975; for literature see Shields and Sang, 1970). Imaginal disc cells divide in such cultures and differentiate into adult cuticle following the induction for metamorphosis *in vivo* by transplantation into metamorphosing larvae (Shields *et al.*, 1975) and *in vitro* after treatment with ecdysones (Dübendorfer *et al.*, 1974). Embryonic imaginal disc cells can proliferate *in vitro* giving rise to either bloated, monolayered vesicles of flattened cells, or unorganized masses of densely packed rounded cells, respectively (Fig. 1). We have shown that vesicular imaginal disc tissue always differentiates into non-recognizable "simple patterns" irrespective of whether the vesicles metamorphose *in vitro* or along with a host larva of the late third (= last) instar. Recognizable patterns such as wing, leg, eye, genitalia, or labium, can only be obtained if the tissues are allowed to remain *in vivo* for a minimum culture period of 48 h prior to metamorphosis (Dübendorfer *et al.*, 1974, 1975).

This paper describes an analysis of the developmental capacity of isolated imaginal disc vesicles and of isolated masses of packed cells under different experimental conditions. *In situ*, flattened cells in vesicular arrangement are observed only at the onset of metamorphosis, whereas packed cuboidal cells are found in first instar discs. Therefore, it may be assumed that the masses of packed cells grown *in vitro* correspond to an earlier developmental stage, whereas the cells in vesicular arrangement might represent a more advanced stage of development.

II. Materials and methods

For each culture, several hundreds of embryos were collected from wild type *Drosophila melanogaster* ("Oregon" and "Sevelen") and incubated for 7 ± ½ h at 25°C. Mechanical dissociation, setting up of the cultures, and preparation of the culture medium were performed following the description of Shields *et al.* (1975). The medium was changed every four days.

Vesicular and dense imaginal disc tissue was isolated from 3-5 weeks old cultures with electrolytically pointed tungsten needles and transplanted into the body of host larvae (for metamorphosis *in vivo)* or adult female flies (for culture *in vivo)* as described by Ursprung (1967). α-ecdysone (a generous gift from Roche-Maag AG, Basle) was dissolved directly in the medium and applied to the cultures at a concentration of 10 μg/ml (1.1 x 10^{-5} M).

III. Results and conclusions

1. *Metamorphosis of imaginal disc cells IN VITRO*

Vesicles as well as masses of densely packed imaginal disc cells are only observed in cultures containing well-developed fat body cells. Upon treatment with α-ecdysone, the first obvious change in the cultures is the lysis of the fat body cells which release their large globular cytoplasmic inclusions into the medium (Fig. 2). In some of the vesicles, cytoplasmic granulation, an indication of trichome formation, can be observed by the third day following hormonal treatment. Vesicles may also form from solid tissue under the influence of the hormone. The first cuticular differentiation in the cells of the vesicles is observed one or two days later when trichomes arise from the granulated cells (Fig. 3 A). By the 6th day after application of α-ecdysone bristle organs develop from two cells each, one cell forming the socket and the other the bristle shaft (Fig. 3 B). One vesicle may eventually contain hundreds of trichomes alone, bristles alone, or bristles and trichomes, respectively. The absence of any specialized structures such as sensillae, claws, or eye facets, confirms earlier findings that vesicles grown from embryonic imaginal disc cells and induced to metamorphose *in vitro* can only differentiate into "simple patterns" (Dübendorfer *et al.*, 1975).

Under the same experimental conditions, tissues of densely packed imaginal disc cells, unlike the vesicles, never yielded any of the aforementioned cuticular structures. This indicates that only the flattened cells of the vesicles, and not those of the dense tissues, are competent to differentiate *in vitro*.

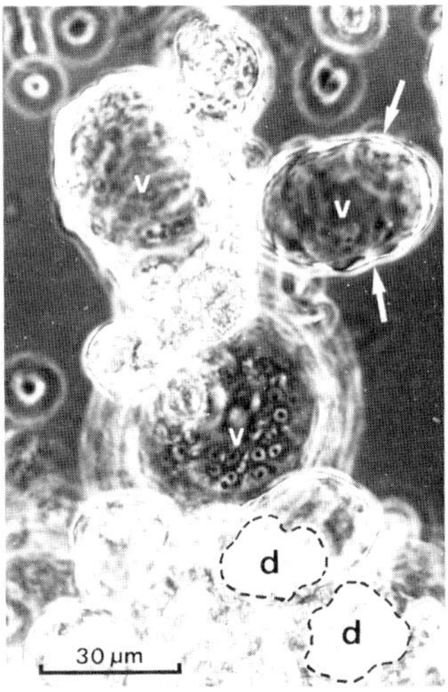

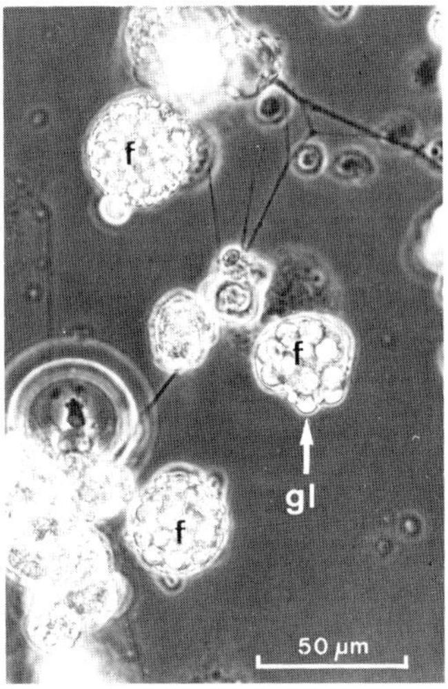

Fig. 1. Vesicles (v) and densely packed imaginal disc cells (d, encircled) grown *in vitro* for 15 days. The cells of the vesicles are extremely flattened (arrows). Phase contrast phtomicrograph by G. Shields.

Fig. 2. Fat body cells (f) in the second week of culture *in vitro*, 14 h after α-ecdysone was applied. The cell in the center has just started to release some of its globular inclusions (gl). Phase contrast.

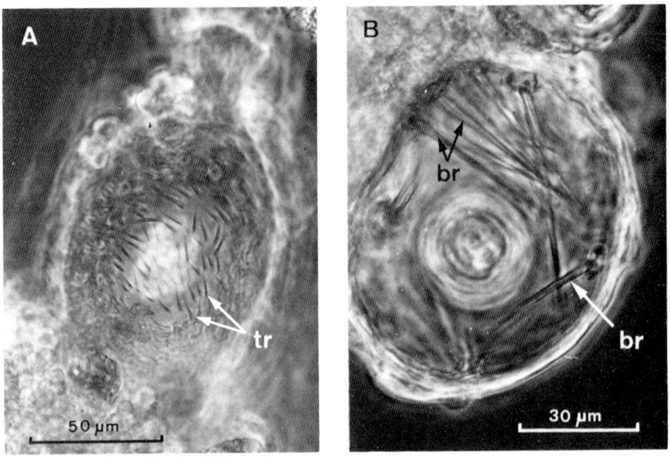

Fig. 3. Differentiation *in vitro* of cuticular imaginal structures by the cells of vesicles after 4 weeks of culture. α-ecdysone was applied on the 13th and 22nd day. (A) Formation of trichomes (tr); (B) Differentiation of bristle organs (br). Phase contrast.

2. *The differentiative capacity after metamorphosis IN VIVO.*

Single vesicles or isolated dense tissue, when transplanted into late third instar larvae 10-20 h before puparium formation, can be recovered from the hatching host fly and the differentiated cuticular structures can be analyzed. The results of this transplantation series are given in Table 1 (A): Vesicles as well as dense tissues differentiated into "simple patterns", i.e. cuticular sheets of varying size, covered with trichomes and/or bristles of a pattern not unambiguously identifiable as a specific part of an adult fly (Fig. 4). Only one single implant, derived from a piece of dense tissue, yielded, adjacent to an area of "simple pattern", a small part identifiable beyond doubt as belonging to the adult proboscis. Apparently, the solid tissues only become competent to differentiate after implantation into a living larva.

In a second series, vesicles and dense tissue were cultured in adult female flies for 9 days, after which the implants were recovered and their gross morphology examined for the presence of folds (see below). Subsequent transplantation into mature larvae resulted in some 55% of these implants metamorphosing into identifiable patterns, irrespective of whether they originated from vesicular or densely packed cells (Table 1, B).

Of the total of 37 implants recovered after culture *in vivo*, 17 (10 from vesicles and 7 from dense tissue) were clearly organized into a folding pattern. The remaining 20 implants (14 from vesicles and 6 from dense tissue) demonstrated no apparent folding. After metamorphosis *in vivo* 14 of the 17 folded implants yielded identifiable patterns (80%) and 3 exhibited the "simple patterns". Of the 20 unorganized im-

TABLE 1

Number of transplanted samples of vesicular and dense imaginal disc tissue from primary embryonic cultures, and types of patterns formed after metamorphosis without (A) or with additional culture IN VIVO (B).

Type of tissue	A Metamorphosis 20 h after transplantation		B Metamorphosis after 9 days of culture in adult flies	
	Identifiable pattern	"Simple pattern"	Identifiable pattern	"Simple pattern"
vesicular	0	40	13	11
dense tissue	1	21	7	6

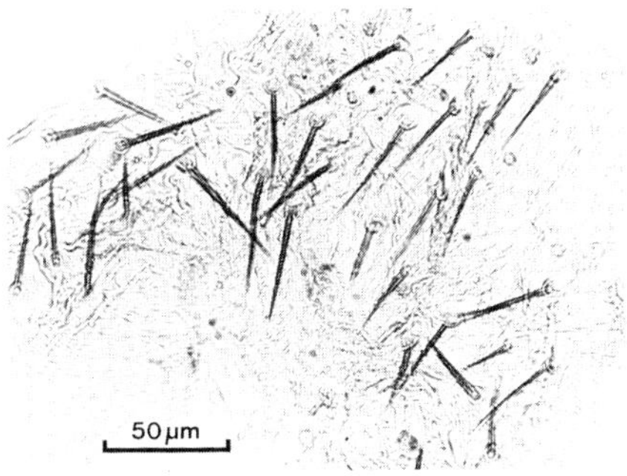

Fig. 4. "Simple pattern" derived from a vesicle of imaginal disc cells grown *in vitro* for 38 days and brought to metamorphosis *in vivo* by transplantation into a mature larva.

plants, only 6 metamorphosed into recognizable patterns (30%), and 14 were of the "simple pattern" type.

The identifiable patterns obtained from this series corresponded to the adult antenna, wing, thorax, haltere, leg, abdomen, and genital apparatus, respectively. Some of these implants are shown in Fig. 5.

The following conclusions are drawn from these results: (i) Vesicles as well as solid imaginal disc tissue from embryonic primary cultures can acquire an organisation into folds after additional culture *in vivo*, (ii) under these conditions identifiable adult cuticular structures are formed from both types of tissues (vesicular and dense), and (iii) the occurrence of folds in the cultured tissue is positively correlated with the development of normal, identifiable patterns.

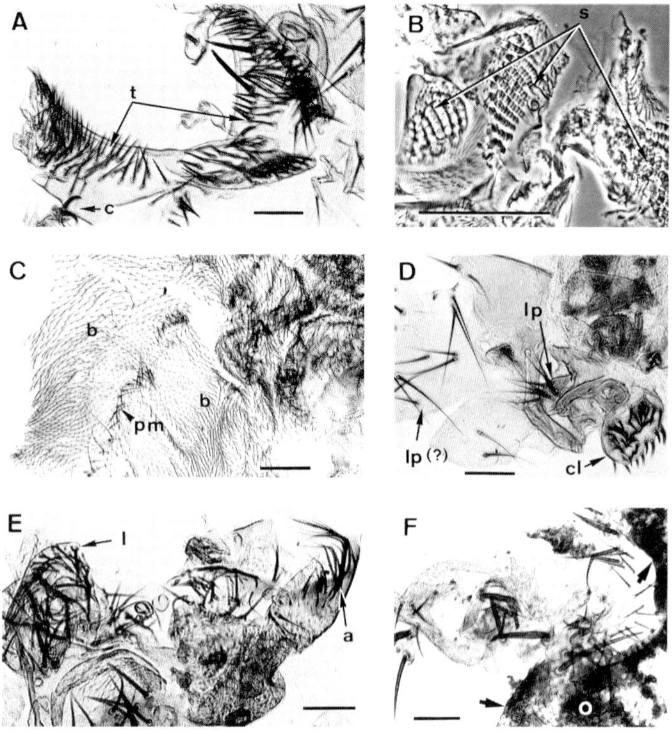

Fig. 5. Recognizable patterns differentiated by imaginal disc tissues from primary embryonic cultures after additional culturing *in vivo* for 9 days and subsequent metamorphosis *in vivo*. (A) leg structures including transversal rows (t) and claws (c); (B) groups of sensillae (s) of the haltere; (C) posterior margin (pm) and blade (b) of a wing; (D) lateral plate (lp) and clasper (c) of the male genital apparatus; (E) structures of the antenna, including the arista (a), and of the leg (1); (F) parts of the head and eye with ommatidia (o) and eye pigment (arrows). The bars in the figures represent 75 µm

IV. Discussion

1. The significance of tissue vesicularisation and cell flattening

Vesicle formation appears to be a general characteristic of imaginal disc tissue cultured *in vitro*. Schneider (1972) observed large vesicles growing out from fragments of advanced (ready to hatch) *Drosophila*-embryos. The tissue was demonstrated to be imaginal disc by transplantation. An attempt to induce metamorphosis *in vitro* resulted in the secretion of membranous structures and small setae, however differentiation of bristle organs did not occur (Kambysellis and Schneider, 1975). It is interesting that other insect embryos behave similarly *in vitro:* Sohi (1968) found vesicles growing out from embryonic fragments of *Choristoneura fumiferana (Lepidoptera)* in a bilaterally symmetrical fashion, assuming that there was a pair of vesicles in each segment. This would suggest the imaginal disc character of the vesicle cells, however direct evidence by induction of metamorphosis was not presented.

Metamorphosis of imaginal discs never occurs without vesicularisation of the epithelium and flattening of the cells, either *in vivo* (Auerbach, 1936; Poodry and Schneiderman, 1970; Sprey, 1970) or *in vitro* (Mandaron, 1974). Investigations on the mechanisms of evagination suggested that cell flattening is actively promoting disc evagination (Fristrom *et al.*, 1970; Poodry and Schneiderman, 1971), but that other mechanisms, such as cellular displacement, may also be involved (Fristrom and Fristrom, 1975). The present report shows that cell flattening and tissue vesicularisation are two of the factors involved not only in disc evagination but also in metamorphosis itself, i.e. in the acquisition of competence to differentiate. In previous reports (loc. cit.), imaginal discs were called "competent for metamorphosis" when they had reached a physiological state which allows differentiation to occur after transplantation into a mature larva. From this viewpoint, vesicles as well as dense tissues from cultures *in vitro* may be considered competent to metamorphose, since both can respond with imaginal differentiation if transplanted into mature larval hosts. *In vitro*, however, induction of metamorphosis reveals the potency of the cells to differentiate upon hormonal treatment, devoid of the complex physiological environment with which tissues are supplied when transplanted into a metamorphosing host. This capacity is exhibited only by the flattened cells of vesicular tissues.

Vesicles capable of differentiation *in vitro* can grow out from dense tissues after treatment with α-ecdysone. Thus it appears that the cells of solid tissue can also respond to the hormonal stimulus. This response may be visualized in two phases: (i) vesicularisation and cell flattening, and (ii) cellular differentiation into cuticular structures of the imago. Vesicule formation *in vitro* without treatment with α-ecdysone could therefore indicate that low levels of the hormone exist in the primary embryonic cultures. The identification of the hormone in the culture medium, and the search for hormone producing cells in the primary cultures, are subjects of further investigations.

2. Hormone action

In this study metamorphosis *in vitro* was induced with α-ecdysone. This does not mean that α-ecdysone is the active hormone leading to differentiation of the cells of imaginal disc vesicles. In fact we have shown earlier (Dübendorfer *et al.*, 1975), that β-ecdysone is some 100 times more active and that it usually acts within shorter time after application than α-ecdysone. This might indicate that β-ecdysone is also the active hormone *in vitro*, which is in agreement with the suggestions of other investigators of the action of ecdysones on imaginal disc cells (King, 1972; Chihara *et al.*, 1974). α-ecdysone, in contrast to β-ecdysone, allows cell divisions and morphogenetic changes (vesicularisation) to occur between the time of application and the onset of imaginal differentiation. This has also been found in cultured wing discs of *Galleria mellonella* (Oberlander, 1972). The dramatic changes of the fat body cells *in vitro* after treatment with α-ecdysone, could be interpreted in terms of a functional engagement of the fat body cells in the action of the hormone, possibly by converting it to β-ecdysone, or by the production of a cofactor. A similar effect of the fat body tissue has been found previously in cultured imaginal discs of Lepidopterans (Richman and Oberlander, 1971; Dutowski and Oberlander, 1973; Benson *et al.*, 1974).

3. Pattern formation

Imaginal discs of second instar larvae, just after the acquisition of competence for metamorphosis, often give rise to incomplete patterns, which, however, are always clearly recognizable as a region of the adult fly (cf. Mindek and Nöthiger;

1973). Therefore, it can be concluded, that in normal development the organization of patterns occurs prior to the acquisition of the competence to metamorphose. Embryonic imaginal disc tissues grown *in vitro*, on the other hand, become competent to metamorphose even if the pattern has not previously been specified. They differentiate into "simple patterns", although it must be assumed that the cells, when taken into culture, were disc-specifically determined (Illmensee and Mahowald, 1974).

My results indicate that pattern specification is correlated with a folding of imaginal disc tissue. Evidence that pattern formation and differentiation do not occur in imaginal disc tissue unless it is organized into folds has been presented by Gateff *et al.*, (1974): Eye-antennal discs cultured *in vivo* for 50 transfer generations lost the folded organisation concomitant with the capacity to differentiate. One of these long term cultures remained in this state for four years. However, under appropriate experimental conditions the folded organisation of the tissue, and simultaneously the differentiative capacity, could be restored.

The results discussed indicate that the morphology of imaginal disc cells is correlated with a particular stage of disc maturation. Patterns may only become specified when the tissue is in a defined morphological state. That is, the cells must form a columnar epithelium organized into folds. This type of cellular organisation would guarantee a maximum area of cell-to-cell contact and a minimum distance between different regions of the blastema. Such an organisation could facilitate intercellular communication and tissue interaction.

Acknowledgments

I wish to thank Professor J.H. Sang of the University of Sussex, in whose laboratory I have carried out some of the presented work when I had a fellowship of the European Exchange Programme, and his co-worker, Mr. G. Shields, who introduced me into the technique of *in vitro* culture. Helpful discussions with my colleagues in Zürich, especially Drs. R. Nöthiger and G.T. Baker, and the skilful technical assistance of Miss J. Nägeli, are gratefully appreciated.

The study was partially supported by grants from the Science Research Council of Great Britain and from the Swiss National Science Foundation 3.081.73.

V. References

Auerbach, C. (1936). *Trans Roy. Soc. Edinburgh 58*, 787.

Benson, J., Oberlander, H., Koreeda, M., and Nakanishi, K. (1974). *Wilhelm Roux' Archiv 175*, 327.

Bodenstein, D. (1939). *J. exp. Zool. 82*, 1.

Bodenstein, D. (1943). *Biol. Bull., Woods Hole 84*, 34.

Borst, D.W., Bollenbacher, W.E., O'Connor, J.D., King, D.S., and Fristrom, J.W. (1974) *Develop. Biol. 39*, 308.

Bryant, P.J. (1974). In: Current Topics in Developmental Biology (ed. Ruddle), *Vol. 8*, 41-80, Academic Press, New York.

Chan, L.N. and Gehring, W. (1971). *Proc. Nat. Acad. Sci. USA 68*, 2217.

Chihara, C.J., Petri, W.H., Fristrom, J.W., and King, D.S. (1972). *J. Insect Physiol. 18*, 1115.

Dübendorfer, A., Shields, G., and Sang, J.H. (1974). *Heredity 33*, 138.

Dübendorfer, A., Shields, G., and Sang, J.H. (1975). *J. Embryol. exp. Morph. 33*, 487.

Dutowski, A.B., and Oberlander, H. (1973). *J. Insect Physiol. 19,* 2155.

Fristrom, D., and Fristrom, J.W. (1975). *Develop. Biol. 43,* 1.

Fristrom, J.W., Raikow, R., Petri, W., and Stewart, D. (1970). In: Problems in Biology: RNA in Development (ed. E.W. Hanley), 381-401, Salt Lake City, University of Utah Press.

Garcia-Bellido, A. (1975). In: Cell Patterning (Ciba Foundation Symposium 29), 161-182, Elsevier, New York.

Garcia-Bellido, A., Ripoll, P., and Morata, G. (1973). *Nature N.B. 245,* 251.

Gateff, E., Akai, H., and Schneiderman, H.A. (1974). *Wilhelm Roux' Archiv 176,* 89.

Gateff, E.A., and Schneiderman, H.A. (1975). *Wilhelm Roux' Archiv 176,* 171.

Hadorn, E. (1966). *Develop. Biol. 13,* 424.

Illmensee, K., and Mahowald, A.P. (1974). *Proc. Nat. Acad. Sci. USA 71,* 1016.

Kambysellis, M.P., and Schneider, I. (1975). *Develop. Biol. 44,* 198.

King, D.S. (1972). *Am. Zoologist 12,* 343.

Mandaron, P. (1974). *Wilhelm Roux' Archiv 175,* 49.

Mindek, G. (1972). *Wilhelm Roux' Archiv 169,* 353.

Mindek, G., and Nöthiger, R. (1973). *J. Insect Physiol. 19,* 1711.

Nöthiger, R. (1972). In: The Biology of Imaginal Discs (ed. H. Ursprung and R. Nöthiger), 1-34, Springer Verlag, New York.

Oberlander, H. (1972). *J. Insect Physiol. 18,* 223.

Poodry, C.A., and Schneiderman, H.A. (1970). *Wilhelm Roux' Archiv 166,* 1.

Poodry, C.A., and Schneiderman, H.A. (1971). *Wilhelm Roux' Archiv 168,* 1.

Richman, K., and Oberlander, H. (1971). *J. Insect Physiol. 17,* 269.

Schneider, I. (1972). *J. Embryol. exp. Morph. 27,* 353.

Shields, G., Dübendorfer, A., and Sang, J.H. (1975). *J. Embryol. exp. Morph. 33,* 159.

Shields, G., and Sang, J.H. (1970). *J. Embryol. exp. Morph. 23,* 53.

Sohi, S.S. (1968). *Can. J. Zool. 46,* 11.

Sprey, Th.E. (1970). *Neth. J. Zool. 20,* 253.

Ursprung, H. (1967). In: Methods in Developmental Biology (eds. F. Wilt and N. Wessels), 485-492, Crowell, New York.

van Ruiten, Th.M., and Sprey, Th.E. (1974). *Z. Zellforsch. 147,* 373.

Chapter 11

SINGLE-CYST *IN VITRO* SPERMATOGENESIS IN *DROSOPHILA HYDEI*

G. Fowler and R. Johannisson

I. Introduction

The dynamics of spermatogenesis in Drosophila (particularly *D. melanogaster)* has been extensively studied and well described at the cytological level with both electron (Tokuyasu *et al.*, 1972a,b; Stanley *et al.*, 1972; Tates, 1971) and light microscopy (Cooper, 1950; Peacock and Miklos, 1973; Gould-Somero and Holland, 1974). The process of spermatogenesis is virtually the same in all species of Drosophila: the spermatogonia in the apical end of the hollow, sac-like testis (Fig. 1) divide mitotically and ultimately give rise to groups of primary spermatocytes encased in a cyst formed by two cyst cells. The spermatocytes on the cyst then undergo synchronous meiotic division forming a group of immature spermatids connected intracellularly in a syncytium. Following their differentiation, these spermatids ultimately are individualized into active sperm. Differences in this process of spermatogenesis which occasionally are observed between some Drosophila species primarily reflect the initial number of primary spermatocytes present in each cyst. For example, in *D. melanogaster*, sixteen primary spermatocytes differentiate to form sixty-four mature sperm per bundle. In *D. hydei*, on the other hand, only one-half this number do so (Fig. 2).

A system of chromosome associated loops, analogous to the lampbrush loops found in developing amphibian oocytes (Callan, 1963) has been described in the nuclei of the primary spermatocytes during their growth phase. Even though such structures appear in virtually all species of Drosophila, especially distinct ones have been described in the spermatocytes of *D. hydei* (Meyer *et al.*, 1961; Hess and Meyer, 1968). That these loops structures are absent in X/O males and are found in duplicate in X/Y/Y males strongly suggests that they arise exclusively associated with the Y chromosome (Hess and Meyer, 1968). During the first meiotic prophase, the Y chromosome goes through a series of structural modifications, probably in part mediated by the X chromosome (Lifshytz, in press), resulting in the formation of distinct loop-like structures in the nucleus of the primary spermatocytes. The locations of these intranuclear structures (loops) have been mapped on the Y chromosome of *D. hydei* and have been labeled according to their distinctive individual morphology as "threads", "pseudonucleolus", "tubular ribbons" and "clubs" (Figs. 2, 3 and 4).

Although the exact function of these structures is not yet completely known, there is ample genetic and cytologic evidence to support the hypothesis that they represent localized sites of RNA synthesis and RNP storage. Indeed, the presence of these loop structures on a relatively silent, somatically non-essential chromosome just prior to the differentiation and morphogenesis of sperm has led to the hypothesis that the products of the Y loops play a crucial role in directing the ensuing processes of spermiogenesis (Hess and Meyer, 1963, 1968; Hennig, 1968, Hennig *et al.*, 1974). As proof of the validity of this hypothesis, it has been genetically shown that the absence of any one of these structures in the nucleus of the primary spermatocyte results in an aberrancy in the process of sperm differentiation leading to male sterility.

Most of the work which has been done on the lampbrush system in *D. hydei* has revolved around the morphological and biochemical characterization of the loop structures. Beyond this point, however, not much is presently known about this system. There are, for example, two basic questions which remain to be unequivocally answered: (1) Under what conditions is the DNA matrix of the Y chromosome induced, at a particular time in development, to enter into the complicated lampbrush configurations? (2) What is the precise nature and mode of action of the gene products (i.e., the lampbrush loops)of the Y chromosome? Specific answers to either of these questions would be of considerable importance with respect to getting at some of the complex questions regarding the genetic control of differentiation in higher organisms.

The lampbrush system in *D. hydei* is particularly well suited to studies of this kind. There are several reasons why this is so: First, the Y chromosome structures have already been well characterized and due to their distinct morphology they can be microscopically studied without difficulty. Second, genetic mutations of the various loop structures exist and, therefore, the necessary genetic test of the cytological situation are possible to perform. Finally, the primary spermatocytes of *D. hydei* can be studied *in vitro*.

This last point is a particularly significant one since, free of the whole organism, the critical experiments can be done in a much more definitive way than when carried out under *in vivo* conditions (as has been the case in the majority of the experiments up to the present time). Even though whole-testis culture has already shown itself to be a useful means of studying certain aspects of Drosophila spermatogenesis (Fowler, 1973; Gould-Somero and Holland, 1974), an *in vitro* study of *individual cysts* of primary spermatocytes is a considerable improvement in the technique and constitutes a firm basis for further experimentation in studies of the

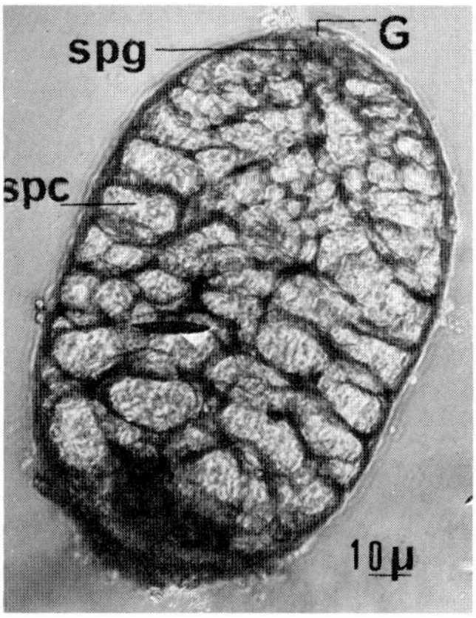

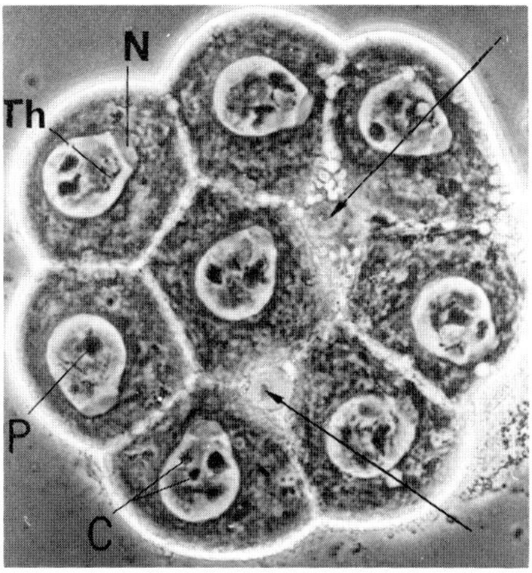

Fig. 1. The *in vivo* testis or wild-type (+) *D. hydei* at 40 hours after pupation. Cellular contents at this time of development are spermatogonia (spg) in the "germinal zone" (G) and bundles of primary spermatocytes (spc).

Fig. 2. Cyst of 8 primary spermatocytes in the "lampbrush phase" of Prophase I of meiosis. C, clubs; P, pseudonucleolus; Th, threads; N, nucleolus. The primary spermatocytes are surrounded by and individually contained within two cyst cells each of which possess a large nucleus (arrows). Phase contrast photograph of a squash preparation, x1000.

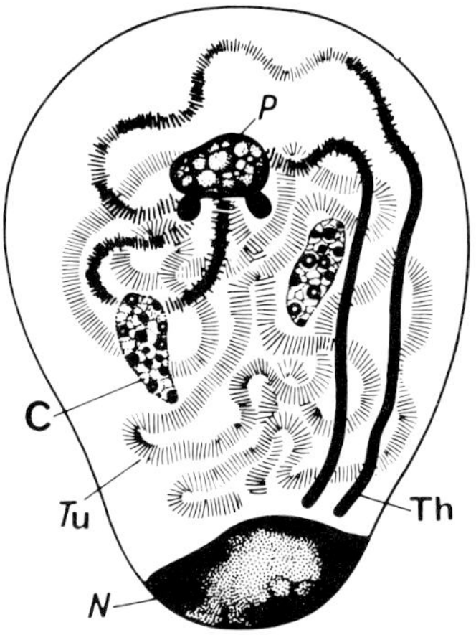

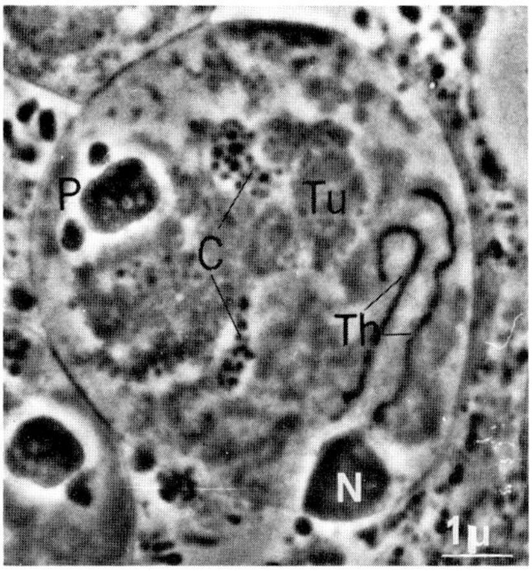

Fig. 3. Schematic representation of the loops in the nucleus of the primary spermatocyte. C, clubs; Tu, tubular ribbons; P, pseudonucleolus; Th, threads; N, nucleolus. Courtesy of O. Hess.

Fig. 4. Full grown (mature) spermatocyte nucleus. Phase contrast photograph of a living cell. C, clubs; Tu, tubular ribbons; P, pseudonucleolus; Th, threads; N, nucleolus. Courtesy of O. Hess.

genetic control of development in Drosophila. The results of the experiments which lead to this conclusion are the subject of the present paper.

II. Materials and methods

Cultures of single cysts of primary spermatocytes were obtained in the following manner: Intact testes were aseptically removed from wild-type (+) pupae of *D. hydei* about 40 hours after the formation of the puparium (Fig. 1). At this stage of development, the testis contains only spermatogonia and primary spermatocytes in various stages of maturation. After washing the testes several times in sterile culture medium (Mandaron, 1971), individual cysts of primary spermatocytes were liberated from each testis by pipetting the entire testis through a smaller-diameter pipette, a technique used to physically rupture the fragile sheath of the testis thereby liberating the cellular contents. Cysts in the desired stage were then selected and transferred to a sterile culture chamber (Cruickshank *et al.*, 1959) containing approximately 0.3 ml. of the Mandaron medium (1971) supplemented with 10% fetal calf serum, and then incubated at room temperature (about 25°C) until the end of the differentiation process (about 30 hours). At various times during the process each cyst was observed and photographed with an inverted microscope under high resolution phase contrast microscopy.

For electron microscopy, the fixation of individual cysts was performed in 5% glutaraldehyde in 0.1M sodium cacodylate buffer (pH 7.2) at room temperature for 1 hour. Appropriate stages of cysts were transferred with a drawn-out pipette to a drop of agar solution (3% agar in a Drosophila Ringer-buffer mixture slightly above melting point). After hardening, the agar drop was washed with three changes of Ringer for 30 minutes each. Specimens were post-fixed with 2% osmium tetroxide in cacodylate buffer for 2 hours. Subsequently, the agar drops were trimmed to small blocks and washed in Ringer for 1 hour. Specimens were dehydrated with alcohol, embedded through propylene oxide in araldite and allowed to polymerize for 1 day at 30°C and another 2 days at 60°C in a vacuum. Sections prepared with a Reichart microtome were stained with 10% uranyl acetate in methanol for 8 minutes and lead citrate for 4 minutes. Observations were made with a Zeiss EM 9 S2. Half-thin sections were stained with a mixture of methylene blue and azur B. These techniques are presently being modified and will be discussed more fully in a subsequent publication (Johannisson, in preparation).

III. Results

As can be seen in Fig. 5, spermatogenesis of a single cyst of primary spermatocytes, beginning with a late stage of Prophase I of meiosis (a) and ending with a stretched-out bundle of coiled spermatids (n), requires approximately 30 hours under the present *in vitro* conditions (Fowler and Uhlmann, in press). The actual time of differentiation is measured from the onset of Metaphase I (b) since the exact state of meiotic prophase (i.e., the extent of the lampbrush loop unfolding) of the explant is impossible to determine from living preparations. Once the spindle apparatus has been formed and the characteristic degenerative changes in the lampbrush structures of the spermatocyte nuclei can be noted, both meiotic divisions (c) and (d) are completed within 6 hours. About the same amount of time is required for the formation (e) and subsequent differentiation (f, g) of the nebenkern derivatives. The formation of a "cytoplasmic cap" on one side of the differentiating cyst (at approximately 13 hours after the beginning of the division of the

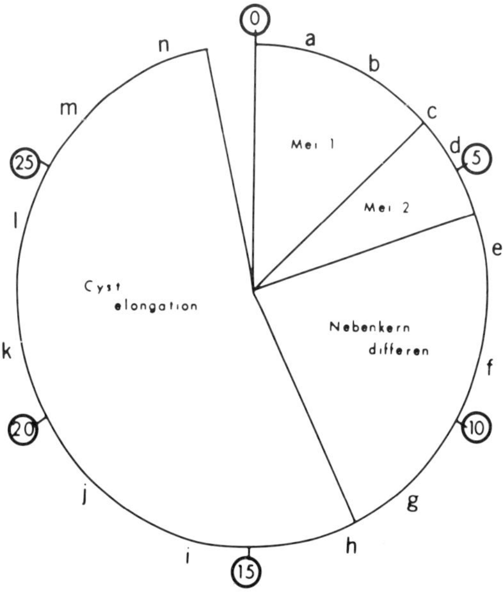

Fig. 5. In vitro differentiation (in hours) of a single cyst of 7 primary spermatocytes removed from the testis of *D. hydei*. *a*, late lampbrush stage of Prophase I; *b*, Metaphase I; *c*, Prophase II; *d*, Anaphase II; *e-g*, differentiation of the nebenkern derivatives; *h-n*, elongation of the cyst. Phase contrast photographs of a living cyst. *a-i*, X500; *j-n*, X112.

spermatocyte nuclei) signals the beginning of cyst elongation and the distal growth of the nebenkern derivatives. Further elongation of both the cyst and differentiating spermatids continues (i-m) during the next 15 hours until, at about 28 hours after the onset of division, the spermatids within the cyst begin to coil at the distal end (n). Beyond the coiling stage, no differentiation of either the cyst or the spermatids is observed if, indeed, it is occurring. With continued time in the culture chamber the cyst begins to show signs of degeneration and finally bursts.

IV. Discussion

In spite of the fact that further development of the cyst does take place, the degree of its differentiation (at least morphologically) at 30 hours *in vitro* is roughly equivalent to that seen in cysts of intact testes which have developed *in vivo* for the same period of time. It is difficult to determine exactly what stage of differentiation the spermatids attain after this period of times in culture. It must be, however, a rather advanced one since according to observations in the testis of *D. melanogaster* (Tokuyasu *et al.*, 1972b) spermatid coiling precedes the release of motile mature sperm. There is no *a priori* reason to believe that the situation in *D. hydei* would be different.

The individual steps by which the coiling stage of spermatid differentiation (Fig. 6) occurs *in vitro* are, furthermore, closely parallel to those described *in vivo* by Hess and Meyer (1968), an observation which attests to the fact that the *in vitro*

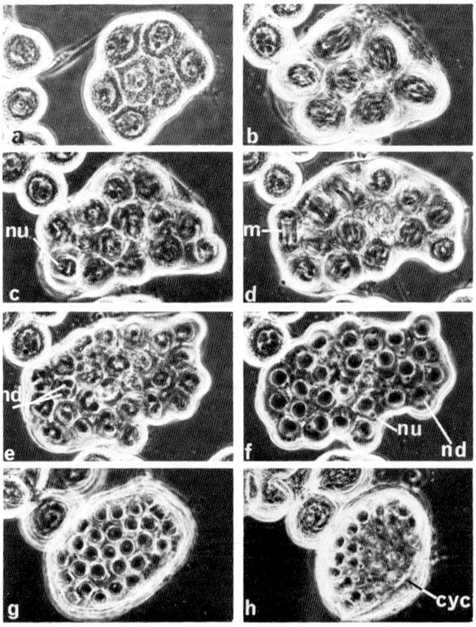

a. The cyst of primary spermatocytes at the time of explantation (late lampbrush stage of Prophase I). The lampbrush structures of the mature spermatocytes which are completely unfolded at this time appear as in Fig. 2.

b. Metaphase I (time of differentiation, t, = 0 hours). Spermatocytes dividing in synchrony are characterized at this stage by a condensation of the lampbrush loop matrix, a loss of the integrity of the nuclear membrane with a concomitant formation of the spindle apparatus.

c. End of Meiosis I. 14 secondary spermatocytes in Prophase II (the Interkinesis period of meiosis in *D. hydei* is too short to be seen) each containing a small nucleus (nu) lying within a fully developed spindle apparatus. (t = 3 1/2 hours).

d. Anaphase II. Asynchronous division of the secondary spermatocytes some with coalescing mitochondria (m) arranged on the spindle apparatus.

e. End of Meiosis II. Mitochondria have coalesced to form the early nebenkern derivatives (nd) which surround the small rounded nucleus of the 28 young spermatids. (t = 6 hours).

f. Somewhat older spermatids (t = 7 1/2 hours) at which time both the nebenkern derivates (nd) and the nucleus (nu) have grown in size.

g. Spermatids at t = 11 hours. The nebenkern derivatives (See Fig. f above) are fully differentiated and are assuming a single plane or orientation within the cyst. Note the more regular morphology of the cyst from that in previous stages.

h. Elongation begins by the formation of a "cytoplasmic cap" (cyc) on one side of the differentiating cyst. (t = 13 hours).

conditions are probably similar enough to the *in vivo* situation so that the lampbrush loops are able to maintain their normal functional capacity for directing at least these stages of spermatogenesis. Similar results have been obtained in whole testis culture of *D. hydei* (Fowler, 1973). The late stage of spermatid differentiation will remain morphologically the same in both the *in vivo* and *in vitro* cysts until

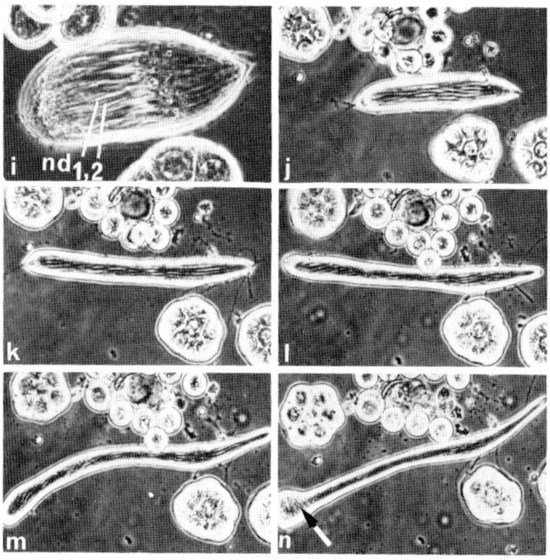

i. Continued elongation of both the cyst and the nebenkern derivatives ($nd_{1,2}$). (t = 14 1/2 hours).

j-n. (t = 15-30 hours). Continued elongation of the cyst and nebenkern derivatives until the stage is reached (t = 28 hours) when the spermatids begin to coil at the distal end (arrow).

the time of eclosion (4-5 days after the formation of the puparium). At this time the *in vivo* cysts will begin the process of spermiogenesis leading ultimately to motile sperm.

Why cysts *in vitro* fail to do this is not clear. There are, perhaps, several reasons to account for this: (1) There may be some developmental factor (hormone) in the *in vivo* system at the time of eclosion which functions to initiate spermiogenesis and which is absent in the *in vitro* culture. If this *is* the case, the hormonal factor required for the process is probably *not* ecdysone. This conclusion can be drawn from the extensive series of implantation experiments carried out by Garcia-Bellido (1964a,b), following those of Stern and Hadorn (1938), which together argue convincingly against the environmental (hormonal)-dependent differentiation of spermatids in the testes of Drosophila. For example, the work of Garcia-Bellido (1964a,b) clearly indicates that various titers of ecdysone present in larval and adult hosts of *D. melanogaster* have virtually no effect on the completion of meiosis and spermatid differentiation in implanted 48-hour-old larval testes. Preliminary experiments in our laboratory in which ecdysones were added directly to the testes in culture, support this conclusion.

It is also possible that (2) Mandaron medium (1971) is not satisfactory for the normal maintenance of the lampbrush loops *in vitro* for periods longer than 30 hours. Whole testes of *D. hydei* have been cultured, however, in this medium for periods as long as ten days. During this time *in vitro*, normal patterns of differentiation appear to be taking place and in the absence of any obvious degeneration of the primary spermatocytes. In these organ cultures, differentiation continued until the entire testis was filled with the same late stage of spermatids as seen in single-cyst culture. These observations, then, would tend to argue against the medium as a limiting factor to the further differentiation of 30-hour-old single cysts *in vitro*.

The possibility that the metabolites present in the medium at 30 hours were no longer capable of supporting continued differentiation of the cyst could also be ruled out. A series of experiments were carried out whereby medium was continuously perfused through the chamber. In no case was any effect on cyst development noted, either in the degree of differentiation or in the speed with which the advanced stage was reached (Hildebrandt, unpublished).

Among several other possibilites, one which is occasionally alluded to in the literature of Drosophila spermatogenesis is particularly interesting; specifically, (3) the cessation of spermatogenesis in culture is a reflection of either the death or the loss of the metabolic activities of the two cyst cells (Figs. 2 and 7) which surround and contain the differentiating primary spermatocytes at the time of explantation and which can be seen *in vitro* to assume specific positions (roles?) during the elongation phase of spermatid differentiation.

Such a possibility obviously assumes that the cyst cells may be playing an active role during spermatogenesis, particularly at the time of eclosion when the pro-

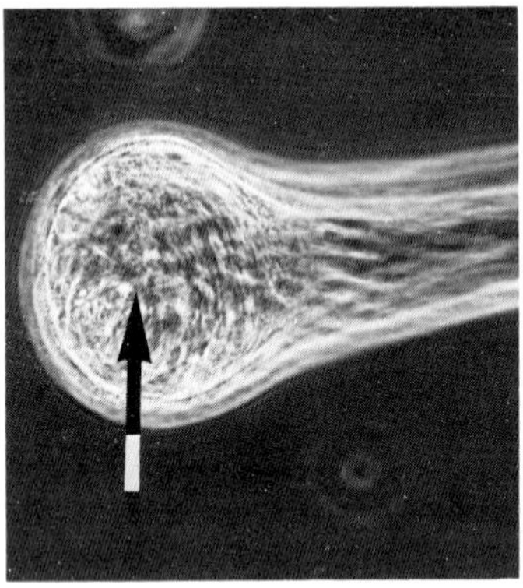

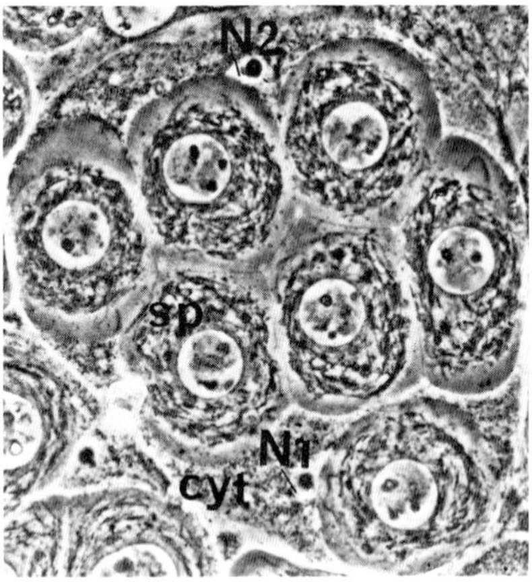

Fig. 6. Enlarged view of coiled spermatids (arrow) in the distal portion of the most advanced stage of cyst differentiation (See Fig. 5, n). X500.

Fig. 7. Single cyst of 7 primary spermatocytes just prior to Metaphase I. Note the loss of mature lampbrush loop morphology and early formation of the spindle apparatus (sp). The cytoplasm (cyt) of the two cyst cells (designated by their nuclei, N_1,N_2) completely surrounds the differentiating gametes at this stage of development. Phase contrast photograph of a squash preparation. X1000.

cess of spermiogenesis begins. Even though this hypothesis has not yet been investigated and, therefore, there is no evidence to either support or refute it, there are a number of observations in Drosophila which tend to indirectly support it as a possibility.

Specifically, as early as 1938, Stern and Hadorn suggested the "nutritive cell" (now considered to be at least one of the cyst cells of each bundle of differentiating gametes and into which the sperm heads are inserted during their development) as having a specific local influence on the attached spermatids and one which may be determining their developmental fate. More recently, Tokuyasu *et al.*, (1972a,b) suggests, on the basis of fine structural studies of the dynamics of spermatogenesis in *D. melanogaster*, that the cyst cells probably play the major role in the process by which sperm become "individualized" into discrete units by the loss of excess cytoplasm.

As lending support to the possible role which the cyst cells may play in spermatogenesis in *D. hydei*, the following observations have been made: (1) Individual spermatocytes free of the cyst cells will not differentiate in culture media which do, on the other hand, support whole testis and intact cyst growth. Likewise, (2) individual cysts which remain intact and viable but whose cyst cells are rendered immobile (e.g., by sticking to the bottom of the culture chamber) will not undergo any differentiation. Finally (3), it has been observed (Hennig, *et al.*, *in press*) that spermatogenesis in X/O males proceeds considerably beyond meiosis and, in many instances, spermatids of advanced stages can be observed. On the basis of this finding, then, it is clear that the lampbrush loops of the Y chromosome in *D. hydei* are not the only factors which contribute to the processes of spermatogenesis. Even though such evidence is indirect, taken together with the other observations, it does give some additional support to the premise that the somatic component of the cyst i.e., the cyst cells may be playing a decisive role in controlling the immediate environment and the developmental fate of the differentiating gamete.

Most of the developmental studies on sperm development in *D. hydei* have revolved around the elucidation of the nature and function of the lampbrush loops in the primary spermatocytes. Consequently, very little is known about either the structure or the biochemical nature of the cyst cells.

With regard to their morphology, only a few investigations have characterized the cyst cells at the fine structural level (e.g., Bairati, 1967; Tokuyasu *et al.*, 1972a,b) and then primarily as only part of a more general study of Drosophila spermatogenesis. By using the technique of *in vitro* culture, whereby individual cysts can be separated from the rest of the cellular components of the testis and prepared for electron microscopy, these very interesting cells can now be accurately described.

For example, as seen in Fig. 8a, the cyst cells in the early stages of meiosis (Prophase I) are very flat cells and closely invest the individual primary spermatocyte. In all stages of development (e.g., the spermatid stage, Fig. 10), the cyst cells are essentially devoid of cell organelles except for (1) mitochondria, which in the light microscope are seen to be present in large numbers and which are undoubtedly responsible for the considerable kinetic activity typical of the cyst during its differentiation (Fig. 5), some (2) endoplasmic reticulum and (3) bundles of microtubules which are generally oriented parallel to the cell surface.

At the fine structural level, it is also possible to clearly define the boundary between the two cyst cells, the septate desmosomes (Figs. 8a and 8b). Membrane boundaries of this type are typical of Drosophila and have been described in *D. melanogaster* by Tokuyasu *et al.*, 1972a,b).

By far, the most conspicuous feature of the cyst cells is the large nucleus which each possess (Figs. 7 and 9). The nucleoli of the cyst cells are also distinctive, each composed of granular elements which are separated into a compact core and an outer, less dense area (Fig. 9). Adjacent to the outer area on one side of the nucleolus is a structure consisting of large granules, the "chromocentrum" (Meyer, 1963).

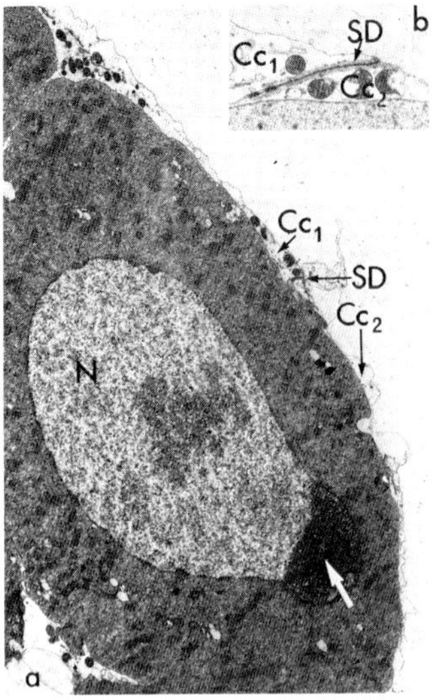

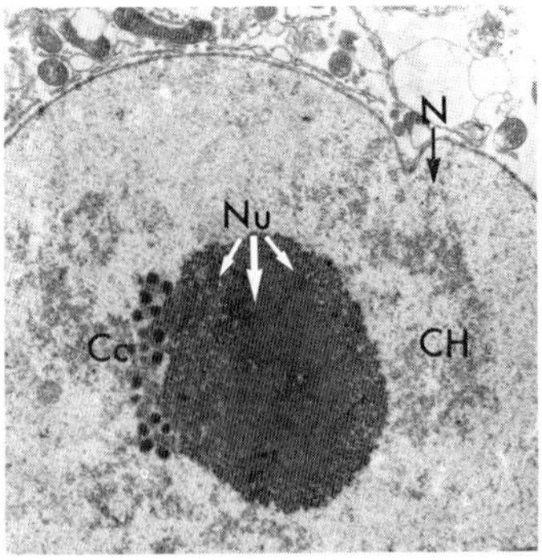

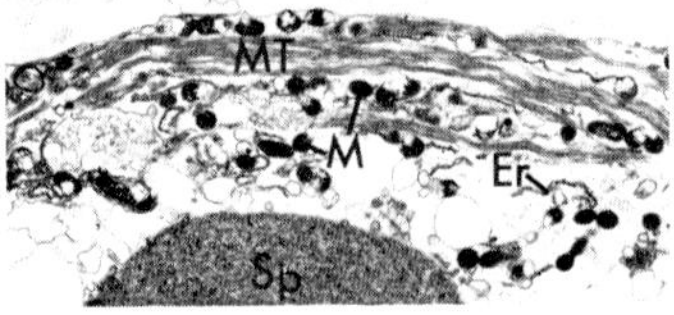

Fig. 8. a. Section through a spermatocyte cyst. The spermatocyte contains a pear-shaped nucleus (N) with a nucleolus (arrow) directly attached to the nuclear envelope. Adjacent to the spermatocyte the two flat cyst cells can be seen (Cc_1, Cc_2), separated by septate desmosomes (SD). X5300. b. Enlarged view of the septate desmosome (SD). The cyst cells (Cc_1, Cc_2) overlap in the contact area. X16,000.

Fig. 9. Section through a nucleus (N) of a cyst cell. Nucleolus (Nu) consisting of a compact inner core (large arrow) and an outer area of less density (small arrows). A "chromocentrum" (Cc) is associated with the nucleolus. The nucleolus is surrounded by chromatin (CH). X13,000

Fig. 10. Section through a spermatid (Sp) cyst before elongation. Bundles of microtubuli (MT) are oriented parallel to the surfaces of the cyst cell. Endoplasmic reticulum (Er) and mitochondria (M) are located between these bundles. X14,000.

A biochemical characterization of the cyst cells is much more difficult than a morphological one. For example, one of the important questions at the biochemical level is whether these cells are actively transcribing RNA throughout the process of spermatid differentiation (a period of spermatogenesis when the rest of the genetic complement is quiescent), or not? If so, is the transcribed mRNA somehow being received and being used by the differentiating gamete? At first glance, this question seems to have been answered in *D. melanogaster* (Gould-Somero and Holland, 1974). These authors report, on the basis of inhibitor and autoradiographic studies in whole-testis culture, that RNA synthesized by cells other than the primary spermatocytes is not needed for spermatid differentiation (thereby confirming the earlier work of Olivieri and Olivieri, 1965, and Hennig, 1967). The evidence for this statement is, however, not conclusive in so far as Gould-Somero and Holland (1974) do not specifically demonstrate that RNA synthesis is, in fact, even *occurring* in the cyst cells. Without direct visualization of the cyst cells throughout the entire process of spermatogenesis, any determination of biochemical synthetic activity must be indirect. This is especially true in experiments dealing with the intact whole testis, either in culture or *in vivo*. The present system of following individual cysts of primary spermatocytes through their differentiation *in vitro* should prove to be a more definitive technique for answering this question, as well as other of a biochemical nature.

All of the features of the cyst cells, both morphological and biochemical ones, and their relationship to the developmental morphogenesis of sperm in Drosophila, are presently under investigation in our laboratory. Some of the results of this work will be published in a forthcoming paper devoted entirely to the role of the cyst cells in spermatogenesis (Rieder, in preparation). Until the second component of the spermatic cyst (the cyst cells) are unambiguously characterized and defined, the extent to which the primary spermatocytes control their own destiny will remain unknown, as will the true significance of the lampbrush loops and the role they play in the process of spermatogenesis.

V. Conclusions

The results of the series of experiments reported here indicate that not only can (1) the late spermatid stage be reached by isolated cysts *in vitro* within a period of time similar to that of cysts *in vivo* (i.e., about 30 hours), but also that under culture conditions (2) it becomes possible to follow,the dynamics of spermatogenesis at the light microscope level with a precision that is not possible in the usual cytological preparation.

With regard to the two basic questions remaining to be answered about the lampbrush loops and the role they play in the genetic control of spermatogenesis in Drosophila (see Introduction), the development and expansion of the *in vitro* techniques discussed here will, hopefully, represent the system in which the answers to these problems can be found.

Considerable progress has already been made toward a better understanding of the processes of spermatogenesis at the fine-structural level in Drosophila. This information, combined with that from studies such as those now possible *in vitro* will, perhaps, lead ultimately to a keystone in ellucidating a general mechanism for the genetic regulation of sperm development in Drosophila and of the genetic control of fertility in higher organisms.

Acknowledgments:

The authors are indebted to A. Graf (University of Duesseldorf) and H. Howard (University of Oregon) for working so diligently to find solutions to the various photographic challenges, to the Alexander-von-Humboldt and the Richard Merton Foundations for their financial support of the project and to Prof. O. Hess for providing the space and the equipment necessary to the research and for taking a personal interest in the work during the three-year tenure of one of us (G.F.) in his laboratory.

VI. References

Bairati, A. (1967). *Z. Zellforsch. 76,* 56.

Callan, H.G. (1963). *Intern. Rev. Cytol. 14,* 1.

Cooper, K.W. (1950). In: Biology of Drosophila (Demerec, M., ed.), 1-61, Wiley, New York.

Cruickshank, C.N.D., Cooper, J.R. and Conran, M.B. (1959). *Exper. Cell Res. 16,* 695.

Fowler, G.L. (1973). *Cell Diff. 2,* 33.

Fowler, G.L. and Uhlmann, J. D.I.S. *51,* in press.

Garcia-Bellido, A. (1964a). *Wilhelm Roux' Arch. Entwickl.-Mech. Org. 155,* 594.

Garcia-Bellico, A. (1964b). *Wilhelm Roux' Arch. Entwickl.-Mech. Org. 155,* 611.

Gould-Somero, M. and Holland, L. (1974). *Wilhelm Roux' Arch. 174,* 133.

Hennig, W. (1967). *Chromosoma (Berl.) 22,* 294.

Hennig, W. (1968). *J. Mol. Biol. 38,* 227.

Hennig, W., Meyer, G.F., Hennig, I. and Leoncini, O. (1974). *Cold Spring Harbor Symp. Quant. Biol. 38,* 673.

Hennig, W., Hennig, I. and Leoncini, O. D.I.S. *51,* in press.

Hess, O. and Meyer, G.F. (1963). *J. Cell Biol. 16,* 527.

Hess, O. and Meyer, G.F. (1968). *Adv. in Genet. 14,* 171.

Lifshytz, E. in press.

Mandaron, P. (1971). *Dev. Biol. 25,* 581.

Meyer, G.F., Hess, O. and Beermann, W. (1961). *Chromosoma (Berl.) 12,* 676.

Meyer, G.F. (1963). *Chromosoma 14,* 207.

Olivieri, G., Olivieri, A. (1965). *Mutation Res. 2,* 366.

Peacock, W.J. and Miclos, G.L.G. (1973). *D.I.S. 50,* 41.

Stanley, H., Bowman, J., Romrell, L., Reed, S. and Wilkinson, R. (1972). *J. Ultrastruct. Res. 41,* 433.

Stern, C. and Hadorn, E. (1938). *Am. Naturalist 72,* 42.

Tates, A.D. (1971). Cytodifferentiation during spermatogenesis in *Drosophila melanogaster.* 162pp. 's-Gravenhage: Drukkerij. J.H. Pasmans.

Tokuyasu, K., Peacock, W.J. and Hardy, R.W. (1972a). *Z. Zellforsch. 124,* 479.

Tokuyasu, K., Peacock, W.J. and Hardy, R.W. (1972b). *Z. Zellforsch. 127,* 492.

Chapter 12

ROLE OF A MACROMOLECULAR FACTOR IN THE SPERMATOGENESIS OF SILKMOTHS

I. Kiss and C.M. Williams

I. Introduction

The time is 1915; the place, Yale University. There some eight years earlier, Prof. Ross Harrison (1907) had "inovated" (to use the term recommended by Jane Oppenheimer, 1967) the method of tissue culture. What Harrison did was to explant part of the nerve tube from frog embryos at stages before the neuroblasts had formed axonal outgrowths. The fragments were placed into clotted frog lymph on cover slips inverted over depression slides - the now familiar hanging-drop technique. In this manner Harrison was able to observe under the microscope the lively outgrowth of delicate axonal fibers from the neuroblasts. For this discovery which resolved the long-standing controversy concerning the embryonic origins of nerve fibers Harrison in 1917 was voted the Nobel Prize by the Nobel Committee. Strange to say, Harrison never received the Prize: because of the outbreak of World War I, the award was cancelled (Nicholas, 1961).

Back in those days in New Haven it so happened that a 37 year old German visitor, Richard Goldschmidt, was working in Harrison's laboratory. Goldschmidt's career also got entangled with World War I. As a German national he was actually thrown into prison and then interned for the war's duration (Goldschmidt, 1960). Despite all these vicissitudes, Goldschmidt was destined to become one of the foremost biologists of his day. It was in Harrison's laboratory some 60 years ago that Goldschmidt undertook the first invertebrate tissue culture.

His experimental subject, believe it or not, was the Cecropia silkworm *(Hyalophora cecropia)*. Goldschmidt had collected some cocoons in and about New Haven. From the cocoons he removed the diapausing pupae and from certain of the male pupae he removed the testes.

II. Results

Figure 1A shows a stained section of a diapausing testis. Internally it is divided into four chambers containing large numbers of spermatogonia and spermatocysts. Each cyst is comprised of a hollow ball of up to 64 primary spermatocytes, the

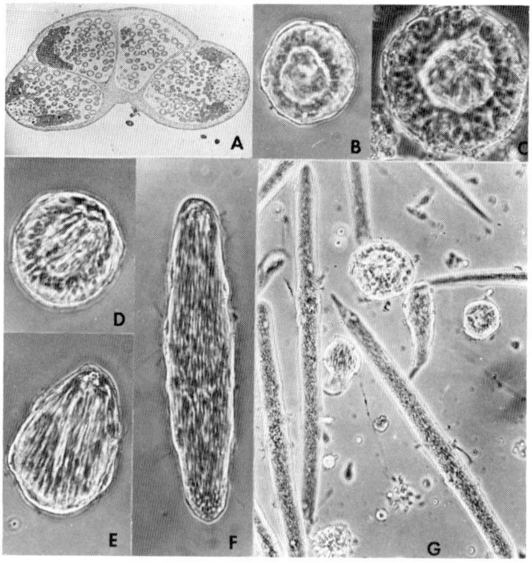

Fig. 1A. Is a fixed and stained section of a testis of a diapausing saturniid pupa; the spermatogonia and spermatocysts occupy the four chambers formed by the inflections of the inner layer of the testicular walls (40X). (All the other figures are phase contrast photographs of living cultures at several different magnifications).

Fig. 1B. Is a typical spermatocyst from a diapausing testis ; the primary spermatocytes surround the central lumen and are enveloped in a thin layer of follicle cells (470X).

Fig. 1C (660X) *and 1D* (470X). Are cysts after 24 hours of culture in MF-containing medium; meiosis has been completed and the axial filaments are oriented to the center of the lumen.

Fig. 1E. The spermatids and the cyst as a whole have begun to elongate (470X).

Fig. 1F. The cyst is now more than twice as long as wide (490X).

Fig. 1G. Shows the typical appearance of cultures after seven days in MF-containing medium; the very elongate cysts contain bundles of 256 fully developed sperm (160X). This plate is reproduced from Kambysellis and Williams (1971a).

whole being enveloped by a simple squamous layer of follicle cells (Fig. 1D). The germinal cells are the precursors of the future spermatozoa which in the insect will differentiate 6-8 months later after the overwintering pupa terminates diapause and initiates adult development.

Goldschmidt tore open the testes and prepared the first invertebrate tissue culture by suspending the spermatogonia and spermatocysts in hanging-drops of pupal hemolymph. In some of these preparations he witnessed the remarkable happenings summarized in Fig. 1B-G. On the second day of culture, all of the primary spermatocytes in many cysts initiated the two meiotic divisions and swiftly differentiated into spermatids and spermatozoa. Simultaneously, as shown in Figs. 1F and G, the cysts as a whole showed remarkable growth to form elongate objects each containing a bundle of 256 spermatozoa. Meanwhile those germinal cells which were spermatogonia underwent mitotic multiplication to form cysts containing primary spermatocytes which then proceeded to differentiate as just described. Though Goldschmidt's own work was preoccupied with the early cytological events, the *in vitro* spermatogenesis of silkmoths remains to this day the most spectacular morphogenesis observed in any tissue culture system.

Goldschmidt noted that in certain cultures no trace of development took place. The germinal cells just sat there and finally died after a week or two. To account for this baffling result, he inferred that the testes or the hemolymph of Cecropia pupae must undergo some critical change to permit *in vitro* spermatogenesis. This was a most prophetic suggestion when one recalls that seven years were to elapse before the first insect hormone - the so-called "brain hormone" - was discovered by Kopec (1922). And over 30 years were to elapse before the endocrine basis of pupal diapause was worked out (Williams, 1946, 1947, 1952).

Goldschmidt's three papers (1915, 1916, 1917) on *in vitro* spermatogenesis seem to have attracted little attention. Nor did Goldschmidt, himself, continue any further studies of this sort after returning to Germany. It was only in 1948 that a systematic study of *in vitro* spermatogenesis of silkmoths was initiated by a Harvard graduate student, the late Edmond Schmidt (Schmidt and Williams, 1949, 1953).

Schmidt found the reason for Goldschmidt's capricious results. It turns out that the decision as to whether the spermatocysts develop or fail to develop depends on the status of the pupa that donates the hemolymph used as the culture medium. The medium must contain a certain serum factor prerequisite for meiosis and spermiogenesis. The necessary factor is not present in the blood of diapausing Cecropia pupa but reappears in the hemolymph after the overwintering insect is subjected to prolonged chilling to terminate the pupal diapause. The substance in question proved to be a heat-labile, trypsin-sensitive, non-dialyzable, non-species-specific macromolecule which has provisionnaly been termed the "macromolecular factor" (MF).

In vitro spermatogenesis has continued to be studied at the Harvard laboratory since Schmidt's untimely death in 1956. Substantial advances were scored a few years ago by a postdoctoral associate, Dr. Michael Kambysellis (Kambysellis and Williams, 1971a and b). He found that as long as the spermatocysts are removed from the testes and cultured in direct contact with MF, the presence or absence of ecdysone is inconsequential. Kambysellis also discovered the startling findings summarized in Table I - namely, that substantial MF-like activity is present in mammalian sera. Subsequently, Istvan Kiss and C.M. Williams followed up this lead by screening various commercially available fractions of human serum. We now report for the first time that MF-like activity is concentrated in the β-globulin Cohn fraction III (Calbiochem).

In further previously unpublished studies we found that MF was necessary not only to induce but also to sustain *in vitro* spermatogenesis. And in quantitative bioassays of serially diluted hemolymph of the tobacco hornworm (*Manduca sexta*), substantial MF titers were encountered in larval blood. The titer doubles during pupation. It doubles again during the first week of adult development and is still high in the blood of freshly emerged male or female moths. In confirmation of the earlier finding of Schmidt and Williams (1953), MF activity could not be detected in the hemolymph of diapausing *Manduca* pupae derived from larvae reared under short-day conditions.

In previously described experiments (Kambysellis and Williams, 1971b), Dr. Kambysellis undertook the culture of intact testes. After seven days he tore open the testes to see what had happened to the spermatocysts inside. A strikingly different result was observed. Now, as shown in Table II, spermatogenesis required the presence not only of MF but also of ecdysone. When ecdysone and MF were presented sequentially, it is important to note in TABLE II that spermatogenesis took place only when ecdysone came first.

These findings strongly argue that ecdysone plays a permissive role in spermatogenesis - that its sole function is to augment the penetrability of the testis walls so that MF can get in. And, evidently, it is then MF that turns on meiosis and spermiogenesis at the cellular level.

TABLE I.

Flask cultures of cysts in Grace's medium supplemented with mammalian sera or Cynthia plasma

Sera	Number of cultures	% of cysts developing as a function of % serum or plasma				
		20	40	60	80	100
Calf serum	20	6 ± 1	34 ± 9	21 ± 4	0	0
Fetal calf serum	15	10 ± 3	40 ± 8	20 ± 4	0	0
Newborn calf serum	50	16 ± 2	15 ± 5	61 ± 8	0	0
Active Cynthia plasma	30	0	30 ± 3	65 ± 7	80 ± 8	80 ± 7

TABLE II.

*Spermatogenesis in intact testes of Cynthia pupae cultured with or without addition of MF and α-ecdysone**

Preincubation for 1 hour with:	Culture for 7 days with:	Number of testis cultures	Number of testes showing spermatogenesis	Developing cysts in responding testis (%)
—	MF	30	0	0
—	Ecdysone	21	0	0
—	Ecdysone + MF	30	30	40 - 80
Ecdysone	MF	12	12	20 - 50
MF	Ecdysone	12	0	0
Ecdysone + MF	No MF, no ecdysone	12	0	0

* α-Ecdysone, when added, was at a concentration of 1.6 μg per 100 μl medium.

III. Conclusion

What are the implications of these findings? Is MF solely concerned with spermatogenesis? Then why is it present throughout larval life when spermatogenesis is not going on? And why is its titer as high in females as in males? What is it doing in females? Is it just possible that ecdysone often plays a permissive role as in the case of spermatogenesis? Perhaps ecdysone's primary function is to turn on the synthesis of enzymes or other agents that augment the penetration of MF or kindred macromolecular factors which then undertake the necessary gene-switching inside of target cells. In that event we may begin to comprehend why the culture of so many insect cells and tissues requires the presence of insect hemolymph or vertebrate serum containing MF-like activities. And since MF is apparently synthesized by hemocytes (Kambysellis and Williams, 1971a), such cells could in principle contribute MF to most *in vitro* systems that one is likely to assemble.

Thus, in retrospect, we learn that Richard Goldschmidt was on the right track when he prepared the Mark I insect tissue culture utilizing the germinal cysts of the Cecropia silkworm.

Acknowledgments

Supported in part by the Rockefeller Foundation and NSF grant GB-26539.

IV. References

Goldschmidt, R. (1915). *Proc. Nat. Acad. Sci. USA 1,* 220.

Goldschmidt, R. (1916). *Biol. Zentrbl. 36,* 160.

Goldschmidt, R. (1917). *Archiv. Zellforschung 14,* 421.

Goldschmidt, R. (1960): *In and Out of the Ivory Tower,* Univ. of Washington Press.

Harrison, R.G. (1907). *Anat. Rec. 1,* 116.

Kambysellis, M.P. and Williams, C.M. (1971a). *Biol. Bull. 141,* 527.

Kambysellis, M.P. and Williams, C.M. (1971b). *Biol. Bull. 141,* 541.

Kopec, S. (1922). *Biol. Bull. 42,* 322.

Nicholas, J.S. (1961). *Biogr. Mems. Nat. Acad. Sci. USA 35,* 132.

Oppenheimer, J.M. (1967). Ross Harrison's contributions to experimental embryology. In: *Essays in the History of Embryology and Biology.* M.I.T. Press. pp. 92-116.

Schmidt, E.L. and Williams, C.M. (1969). *Anat. Rec. 105,* 487.

Schmidt, E.L. and Williams, C.M. (1953). *Biol. Bull. 105,* 174.

Williams, C.M. (1946). *Biol. Bull. 90,* 234.

Williams, C.M. (1947). *Biol. Bull. 93,* 89.

Williams, C.M. (1952). *Biol. Bull. 103,* 120.

Chapter 13

INSECT SPERMATOGENESIS *IN VITRO*

A.M. Leloup

I. Introduction

Since Goldschmidt's first publication sixty years ago, numerous experiments on *in vitro* spermatogenesis in Insects have been published. We were given the opportunity to review them some years ago (Demal & Leloup, 1972). Since that time, other reviews have been published (Fowler, 1973; Kambysellis & Williams, 1971 a & b, 1972; Kuroda, 1972, 1974; Levine, 1972; Gould-Somero & Holland, 1974). In most of the cases, they deal with the transformation of spermatocytes into spermatozoa. In this publication, I would like to stress the point that this is only a part of the phenomenon, to remind us that the evolution of spermatogenesis has already been followed from the stage of primary spermatogonia onwards, and to present first results concerning the last steps of spermatogenesis occuring inside the testis: the phenomena of entrapment and spiralisation. The influence of ecdysterone on these different phenomena has also been checked.

II. Material and methods

Experimental animals - Experiments presented here were performed on *Calliphora erythrocephala* (Diptera) reared under axenic conditions at 25^oC as previously described (Demal, 1961; Meynadier, 1971). Pupae, 17 days old individuals, and prepupae, 9 days old, were used.

Preparation and evaluation of the cultures - Whole testes were used as in previous experiments (Demal, 1971; Demal & Leloup, 1973; Leloup, 1964, 1969). They were cultured in liquid media, using the hanging-drop technique. Gross morphological examination was performed every day in order to check the persistence of contractions and/or, the maintenance of tissue transparency. In some cases, for prepupal testes, microphotographs were taken *in vitro*. The cultures were kept for at least three days and rarely not longer than this in the present series of experiments as the phenomena we were interested in, did not require longer incubation. The testes were then fixed in Bouin's fluid, embedded in paraffin, sectioned serially at 5 or 7 μ, stained with Heidenhain's hematoxylin -eosin, and finally studied histologically. Reexamination of histological sections of cultures made previously in M_1, M_1H, M_2H (Leloup, 1964) and M_1H_{L3} (Leloup, 1964, 1969) was also performed.

Culture media - The basic medium is the M_1 (Leloup, 1964). It was used as such or supplemented respectively with 10% fetal calf serum Difco (M_1S), egg "extract" q.v. (M_1O_{100}) and 5 or 2.5 µg/ml ecdysterone (M_1E, M_1SE and $M_1O_{100}E$). M_1O_{100} was prepared as follow: *Calliphora* eggs less than 12 hours old were washed and surface sterilized as for axenic rearing (Demal, 1961). They were then rinsed twice with the basic medium; 100 of them were transfered into a 1 ml tissue grinder Kontes with 1 ml of M_1 and carefully ground. Drops of the mixture were used as such without any subsequent filtration or centrifugation. In no case was there any antibiotic used.

III. Results

The two testes of each pupa or prepupa were systematically cultured in two different media in order to get direct comparisons of the value of the media. In a certain number of cases, one of them was fixed just after its dissection to serve as control. In such controls, we find that the stage of differentiation reached at the time of explantation was similar to that presented in a previous publication (Leloup, 1964). For more information on the structure and evolution of the testes in *Calliphora*, refer to Leloup (1974).

In vitro the pupal testes showed continuous contractions. The presence of a developing pigmentation at their surface made any other observation *in vitro* difficult but in many cases it was possible to distinguish the appearance of a white mass inside the posterior part, which proved in histological sections to correspond to the accumulation of bundles of spermatozoa. We did not check any emission of spermatozoa into the medium. The prepupal testes behaved exactly the same way as previously described (Leloup, 1969) for M_1H_{L3}: they showed an important increase in volume and then burst.

Survival - Examining the histological sections of the pupal testes, we can see that if the general survival of the organs is good in most of the cases (Fig. 1), this is not always true for different structures. In 1964 and 1969, we pointed out that cysts of young spermatogonia were more fragile than the others and that many of them did not survive very long in the basic medium. The situation was improved in media supplemented with hemolymph. This is also the case with M_1S and M_1O_{100}. Most of the germ cells at other stages seem in good condition. The terminal epithelium did not survive very well and very long in the basic medium and formed fibroblasts in the media supplemented with hemoplymph. It was much better preserved in M_1O_{100} and M_1S. The addition of ecydsterone did not seem to improve any of these results.

Gametogenesis - The different stages to be considered here are: formation of new cysts, evolution of cysts of spermatogonia, spermatocytes and spermatids, spermiogenesis, entrapment and spiralisation. The formation of new cysts is the result of mitosis of primary spermatogonia located in the socalled "germinal reserve" or "germinative zone" situated at the anterior tip of the testis. Such mitosis are never very numerous in controls. Sometimes we even do not find any. We previously reported (Leloup, 1964) the occurence of some of them (Fig. 2). It is also the case here for the different media. Here again ecdysterone does not seem to influence the phenomenon (Fig. 3). The evolution of cysts can be checked in two ways: recording cysts with germ cells in division -mitosis and meiosis- and comparison of cultured testes and controls. As previously, we call attention to the occurence of many cysts in division (Fig. 4 & 5). We can even observe cysts of young spermatogonia dividing.

Microphotographies of histological sections through pupal testes of Calliphora *cultured for 72 hours. The media are indicated between brackets.*

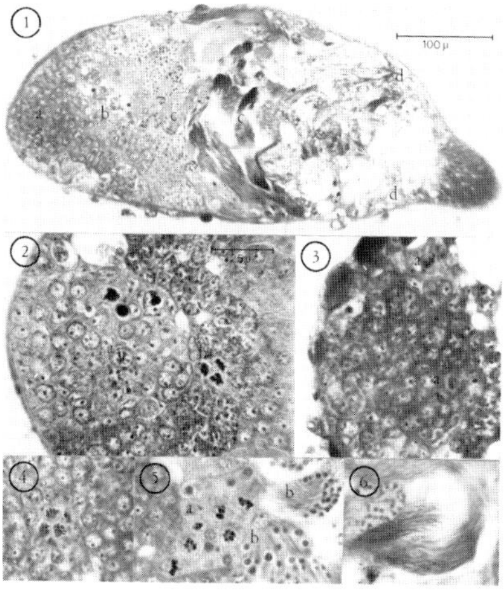

Fig. 1. (M_1O_{100}) Axial section showing (a) the zone of primary spermatogonia, (b) the zone of cell growth and multiplication, (c) the zone of spermiogenesis, (d) the terminal epithelium.

Fig. 2. (M_1H) *and 3* ($M_1O_{100}E$) Show (a) mitosis inside the germinative zone and (b) young cysts of spermatogonia in mitosis.

Fig. 4. (M_1O_{100}) Cyst of spermatogonia in mitosis.

Fig. 5. (M_1O_{100}) Shows (a) the transformation of spermatocytes II into spermatids and (b) spermatids in elongation.

Fig. 6. (M_1O_{100}) Transformation of spermatids into spermatozoa.

Comparison with controls shows a considerable increase of cysts at more advanced stages, especially of those in spermiogenesis and quantitatively more in M_1O_{100} (Fig. 1) than in the other media after 3 days. Qualitative evolution was also proved by culturing prepupal testes under certain conditions (Leloup, 1969). Explantats after 4 days in M_1O_{100} give similar results: we find spermatocytes, young spermatids and even spermatids in elongation. The addition of ecdysterone to any of the media does not improve the results.

The phenomenon of entrapment and spiralisation were described *in vivo* by Tokuyasi *et al.* (1972) at the ultrastructural level in *Drosophila* testes. In histological sections of *Calliphora* testes at the end of the pupal stage we have found pictures which correspond to them (Fig. 7). In most of the controls, entrapment has not yet started and in none of them was there any spiralisation. In testes cultured for three days in M_1S and M_1O_{100} we regularly find pictures of entrapment and sometimes of spiralisation (Fig. 8 to 10). In some testes cultured in M_1 we find traces of entrapment in the very few cells of the terminal epithelium still persisting (Fig. 11). Here again, the presence of ecdysterone dœs not seem to influence the phenomena.

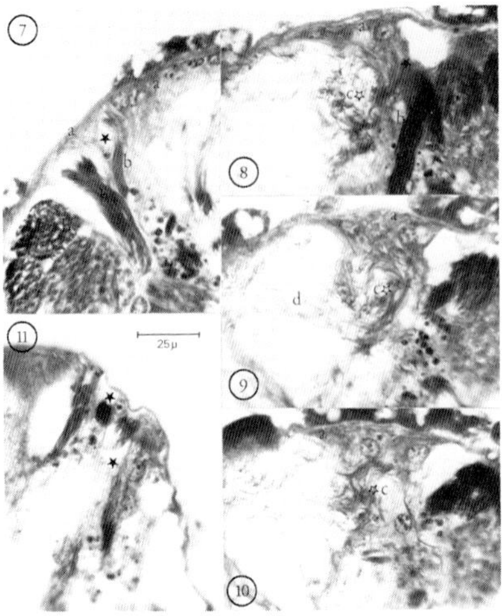

Fig. 7. (Control) Entrapment (★). See (a) Cells of the terminal epithelium and (b) bundles of fully developed spermatozoa.

Fig. 8 to 10 ($M_1O_{100}E$) Entrapment (★) and spiralisation (☆). Three successives sections showing (a) cells of the terminal epithelium, (b) bundles of fully developed spermatozoa and their entrapment, (c) spiralisation and (d) free spermatozoa.

Fig. 11. (M_1) Traces of entrapment (★) by two persisting cells of the terminal epithelium.

IV. Discussion

A first point to consider here is the action of ecdysone (s.l.) on spermatogenesis. Once again it appears that in our experiments spermatogenesis does not seem influenced by it. We have already proposed (Demal & Leloup, 1973) two different situations as far as insect spermatogenesis in concerned: diapausing, and non-diapausing insects. Whether Kroeger's hypothesis -ecdysone acting by altering membrane permeability- or Karlson's -selective gene derepression- (Doane, 1973) the results of the first group, it seems that in no case has this hormone, when added to a medium in pure form, been reported to have any positive influence in experiments belonging to the second group. This does not imply that a hormonal control of spermatogenesis does not exist in insects. In fact, most of the media used may not be a priori considered as devoid of hormones as it is well known that some of their fractions contain substances which are recognized as hormones even if chemically they do not correspond to any of those already identified in insects. We have especially in mind hormones belonging to the steroid group. Without any doubt certain of them are present in fractions such as fetal calf serum, horse serum and chick embryo extract, and we lack information as to whether some of them may affect the results as far as insect spermatogenesis is concerned. Media without such fractions and used for studying insect spermatogenesis *in vitro* are rare. Very often they give worse results, apparently because of lack of certain substances. Whether these are "permissibility factors" (Leloup, 1969) or a "macromolecular factor" (Williams & Kambyssellis, 1969) can't be decided at the present time. We think it is of a particular interest to note their presence in *Calliphora* eggs.

Another point to be stressed is that most of the results concerning *in vitro* spermatogenesis in insects are restricted to the transformation of spermatocytes into spermatozoa. In the present report, our main aim was not to elucidate factors involved in the control of spermatogenesis but to prove the feasibility of getting *in vitro* all the stages which occur inside the testis. Now we may start again trying to "dissect" them out.

A last point to deal with briefly is the behaviour of the somatic structures of the testis mainly in order to state and to regret the lack of interest shown up to now, and to stress their importance as revealed by recent progress in their study at the ultrastructural level *in vivo*. Particular reference is to be made to *Drosophila* (Peacock *et al.*, 1971; Peacock & Miklos, 1973; Tokuyasu, 1974 a & b; Tokuyasu *et al.*, 1974 a & b).

The normal advancement of certain phases of spermatogenesis depends strongly on their survival and the maintenance of their normal structure or even more on the pursuing of their differentiation *in vitro*. We think we have succeded obtaining this in *Calliphora* at the pupal stage.

The author wishes to express her thanks to Professor Demal for reading the manuscript and to Professor De Wilde for kindly supplying the ecdysterone.

V. References

Demal, J. and Leloup, A.M. (1972). In: Invertebrate Tissue Culture (C. Vago, ed.) Vol. 2, 3-39.

Doane, W.W. (1973). In: Developmental systems: Insects (Counce S.J. and Waddington C.H., eds) Vol. 2, 291-497.

Fowler, G.L. (1973). *Cell differ.*, *2*, 33.

Gould-Somero, M. and Holland, L. (1974). *W. ROUX' Archiv*, *174*, 133-148.

Kambysellis, M.P. and Williams, C.M. (1971a). *Biol. Bull.*, *141*, 527.

Kambysellis, M.P. and Williams, C.M. (1971b). *Biol. Bull.*, *141*, 541.

Kambysellis, M.P. and Williams, C.M. (1972). *Science*, *175*, 769.

Kuroda, Y. (1972). *Ann. Rep. Nat. Inst. Gen.*, *22*, 27.

Leloup, A.M. (1964). *Bull. Soc. Zool. France*, *89*, 70.

Leloup, A.M. (1969). *Ann. Endocr.*, *30*, 852.

Levine, L. (1972). *Biol. Reprod.*, *7*, 211.

Meynadier, G. (1971). In: Invertebrate Tissue Culture. (C. Vago, ed.) Vol. 1, 141-167.

Peacock, W.J. and Miklos, G.L.G. (1973). In: Advances in Genetics, Vol. 17, 361-409. Academic Press, New York.

Peacock, W.J., Tokuyasu, K.T. and Hardy, R.W. (1971). Proc. Int. Symp. "The genetics of the Spermatozoon". (Beatly, R.A. and Gluecksohn-Waelsch, S., eds), 247-268.

Takeda, N. (1972a). *J. Insect Physiol.*, *18*, 571.

Takeda, N. (1972b). *Appl. Ent. Zool.*, *7*, 37.

Tokuyasu, K.T. (1974a). *Exp. Cell Res.* *84*, 239.

Tokuyasu, K.T. (1974b). *J. Ultr. Res.* *48*, 284.

Tokuyasu, K.T., Peacock, W.J. and Hardy, R.W. (1972a). *Zeit. Zellforsch*, *124*, 479.

Tokuyasu, K.T., Peacock, W.J. and Hardy, R.W. (1974b). *Zeit. Zellforsch*, *127*, 492.

Williams, C.M. and Kambysellis, M.P. (1969). *Proc. Nat. Acad. Sci.* *63*, 231.

Chapter 14

JUVENILE HORMONE-INDUCED BIOSYNTHESIS OF VITELLOGENIN IN ORGAN CULTURES OF *LEUCOPHAEA MADERA* FAT BODIES

J. Koeppe and J. Ofengand

I. Introduction

Although there has been extensive research in recent years on insect juvenile hormone (JH), there is a large gap in our understanding of the mode of action of this hormone at the molecular level (Doane, 1973; Gilbert, 1974; Willis, 1974). This is in part due to a lack of knowledge about hormone-hormone and hormone-tissue interaction in insects, but is mostly due to the fact that the most obvious effect of JH in insects, that of maintaining or reestablishing juvenile morphological characteristics, do not lend themselves readily to analysis at the molecular level. For this reason, we have chosen to study the mechanism of action of JH in a system which, although devoid of morphogenetic changes, is amenable to biochemical analysis.

The system we chose is the juvenile-hormone induced biosynthesis of vitellogenin, the major yolk protein of oocytes. We selected this system for study since previous work in a number of insects had shown that hormone-induced synthesis of vitellogenin was localized to the insect fat body (Doane, 1973) and *Leucophaea maderae* was selected as the organism since some characterization of the vitellogenin product had been done (Dejmal and Brookes, 1972) and fat body organ culture as well as induction by externally supplied JH had already been demonstrated (Brookes, 1969; Engelmann, 1971).

The reproductive cycle of *Leucophaea* is diagrammed in fig. 1. This schematic representation illustrates that unmated adult females have a low titer of JH, do not synthesize vitellogenin, and have undeveloped oocytes. Upon mating, JH is produced which in turn stimulates the fat body to synthesize vitellogenin and secrete it into the hemolymph. After retraction of the egg case back into the abdomen, the level of JH drops, and vitellogenin synthesis ceases.

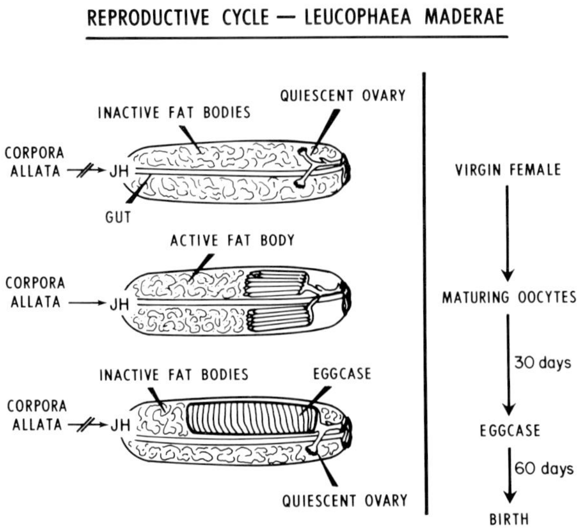

Fig. 1. Schematic representation of the reproductive cycle of *Leucophaea. maderae*. JH, juvenile hormone.

II. Synthesis of vitellogenin by fat bodies in organ culture

Vitellogenin synthesis and secretion *in vivo* can be duplicated in organ culture. Fig. 2 shows the rate of protein synthesis in fat body tissue taken from a female with maturing oocytes. The tissue was incubated as indicated, and samples assayed for total protein synthesis and for synthesis of protein precipitable by vitellogenin antibody. The antibody was developed against vitellogenin purified from oocytes by salt fractionation (Brookes, 1969; Koeppe and Ofengand, 1975). As expected for a protein destined for export, the amount of vitellogenin in the fat body increased with time and reached a low steady-state plateau value by 3 hours, while the vitellogenin in the medium showed an initial lag and then a linear increase for at least 8 hours. After 5 hours of incubation, 45% of the total protein made was vitellogenin. Of the total secreted protein 75% was vitellogenin and the secreted vitellogenin was approximately 80% of the total vitellogenin synthesized. In view of these numbers it seems clear that this system is well suited to a study of hormonal induction of protein synthesis outside of the intact animal.

III. Juvenile hormone stimulation *in vivo* of inactive fat bodies

The effect of endogenous synthesis of JH can be duplicated by injection of JH into virgin females. The plan of the experiment is illustrated in Fig. 3. Isolated abdomens of virgin adult females were prepared, and in one set the ovaries were also removed to test the possibility that an ovarian hormone might be secondarily involved. Both groups were then given a single dose of the Hoffman-LaRoche mixed isomer preparation of JH I in olive oil and at various times thereafter fat bodies were removed and assayed for the ability to synthesize vitellogenin during a 5 hour incubation. It is clear from the results (fig. 4) that injection of JH induces a more

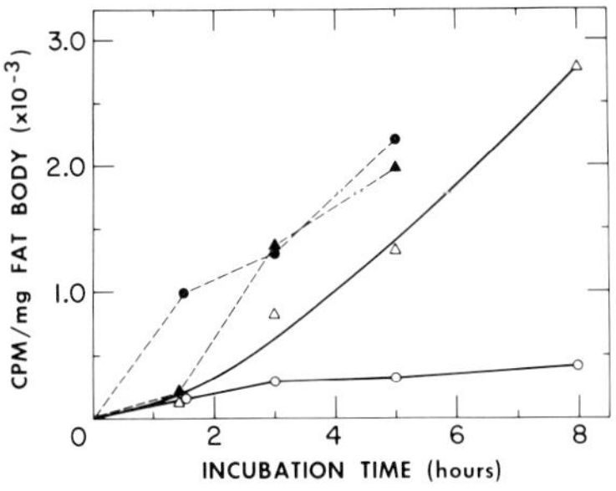

Fig. 2. Kinetics of incorporation of [^{3}H]-leucine into total protein and into vitellogenin during organ culture of isolated fat bodies taken from an active female with 4 mm oocytes. The tissue was incubated in Marks M-20 medium containing 7.5% fetal calf serum, 15 mg/l. penicillin G, 25 mg/l. streptomycin, and 0.4 mM [^{3}H]-leucine (72 mCi/mmole) at 30° with gentle shaking for the times indicated. Vitellogenin in the medium and in the tissue was assayed by immunoprecipitation. Total protein was assayed as hot TCA-precipitable material. Medium vitellogenin (△____△), tissue vitellogenin (0____0); total protein in medium (▲____▲); total protein in tissue (●____●).

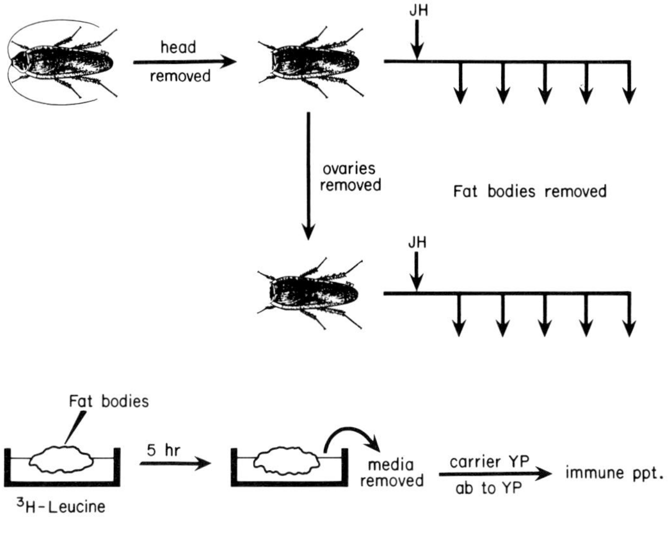

Fig. 3. Diagram of the experimental plan for determining the kinetics of juvenile hormone induction of vitellogenin synthesis. JH, juvenile hormone; YP, yolk protein (vitellogenin). Note that only protein secreted into the medium was examined.

than 50-fold stimulation in vitellogenin synthesis compared to the sham injected control. The induction was somewhat dose-dependent as 20 μg of JH produced a lesser effect. The kinetics of induction were similar to other systems (Fallon *et al.*, 1974; Clemens, 1974) in that a lag was detected (18 hours) followed by a rapid rise to a maximum (72-96 hours) and then a slower decline. Since *in vivo* vitellogenin synthesis and oocyte maturation continues for 30 days, the decline in syn-

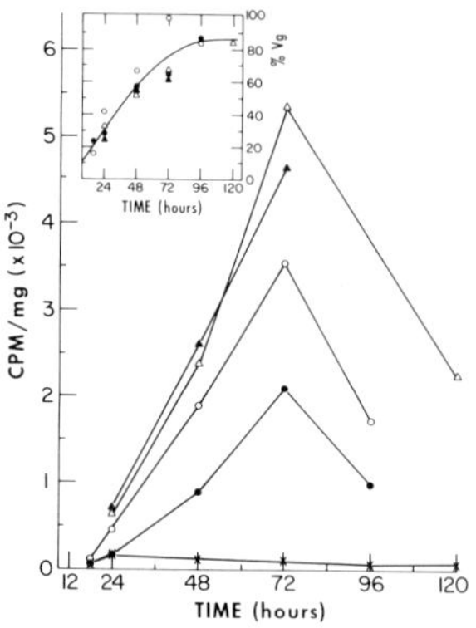

Fig. 4. Vitellogenin synthesis *in vitro* by fat bodies from headless virgin females with and without ovaries after an *in vivo* injection of juvenile hormone. Virgin females were decapitated and, as indicated, ovaries removed from the terminal abdominal segment by dissection. After a two-day resting period, the animals were given a single injection of juvenile hormone dissolved in olive oil, and at the indicated times thereafter, were sacrificed and the fat bodies removed. The capacity of the fat bodies to synthesize vitellogenin was assayed by immunoprecipitation after a 5 hr incubation with [^3H] leucine in organ culture. The results are expressed as radioactive vitellogenin per mg of fat body in the culture dish versus time of exposure to juvenile hormone. The insert shows the fraction of total protein made after juvenile hormone stimulation which is vitellogenin. Total protein was determined as hot TCA precipitable material. Δ, females with ovaries (100 μg hormone); ▲, ovariectomized females (100 μg hormone), 0, females (20 μg hormone); ●, ovariectomized females (20 μg hormone); x, sham (olive oil) females with ovaries.

thesis of vitellogenin by 5 days is probably due to catabolism of the active form of JH in this single dose experiment. Ovariectomy had no effect when 100 μg of JH was given, and only a small effect at the lower dose. Thus neither brain, corpora allata or associated organs, or ovaries are needed for the JH response.

The inset of fig. 4 shows that the massive increase in vitellogenin occurs by specifically increasing the synthesis of vitelloginin which at the peak of stimulation (72 hours) accounts for 75-80% of the total protein secreted. Non-vitellogenin protein synthesis is also stimulated by JH with a sharp ten-fold rise between 18 and 48 hours which then stays approximately constant to 72 hours when it also declines (data not shown but calculable from fig. 4). The inset also shows that the *fraction* of protein synthesized which was vitellogenin was independent of the dose of JH, although the *amount* of vitellogenin produced by a given amount of tissue was dose-dependent. The fractional synthesis of vitellogenin increased with time following JH induction but did not decrease once the maximal level was reached even when the total synthetic capacity had decreased by half. Induction of isolated abdomens of *Leucophaea* by injection of JH has also been reported by Brookes (1969) and Engelmann (1971), but the induction kinetics have not been previously described nor the lack of effect of ovariectomy. In addition, previous workers did not use a quantitative measure of the true rate of vitellogenin synthesis and secretion.

IV. Sedimentation properties of the organ culture secretion product

In an earlier study (Brookes, 1969) it was shown that all of the vitellogenin secreted from fat bodies in culture sedimented at 14S in sucrose gradients. However, since in those experiments the vitellogenin had been first purified from the medium by salt fractionation, we re-examined this point using our quantitative antibody precipitation procedure to analyze unprocessed incubation medium.

Fig. 5 shows the sedimentation profile of added carrier vitellogenin (---) compared with the product secreted by the fat body in organ culture which was analyzed by antibody precipitation (·-). For comparison the medium was also purified according to Brookes (1969) and sedimented (———). *Leucophaea* oocyte vitellogenin exists in two forms (Dejmal and Brookes, 1972). In young oocytes, the 14S (559,000 MW) monomer is found predominately, while mature oocytes contain mostly the trimer 28S form (1,590,000 MW). It is clear that only 14S material is produced in organ culture. There is nothing large, nor any smaller size molecules capable of reacting with antibody. The 20% non-vitellogenin protein sediments in a broad peak between fractions 1-5 (data not shown).

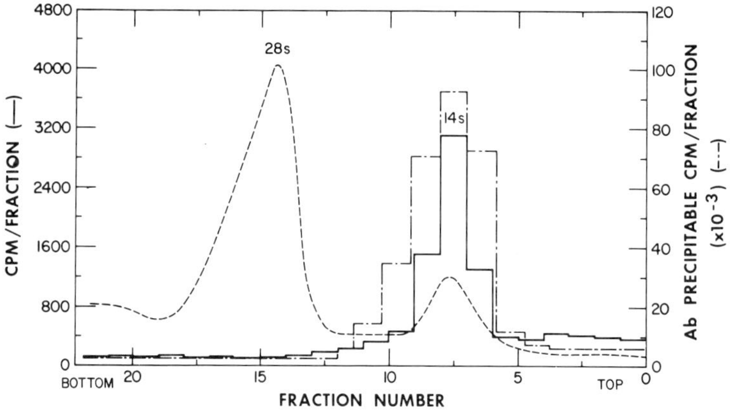

Fig. 5. Sucrose gradient profiles of vitellogenin labelled in organ culture. The gradients, linear from 0.4M NaCl, 10% sucrose to 0.8M NaCl, 30% sucorse were run at 52,000 x g for 20 hours at 4°C. Unprocessed medium was obtained from 5 hour organ cultures of fat bodies from active females with 4 mm oocytes, and layered directly on the gradient with 20 A_{260} units of unlabeled vitellogenin added as an internal marker. Antibody-precipitable radioactivity was determined on an aliquot, and fractions were analyzed by the same method. The recovery of antibody-precipitable radioactivity from the gradient was 81%, and 80% of the total hot TCA-precipitable material was found in the 14S peak. The remainder was in fractions 1-5, peaking at fraction 4 (____ · ____ · ____), antibody-precipitable radioactivity from unprocessed incubation medium; (__ __ __ __), carrier vitellogenin added before centrifugation (A_{280}). The solid line (____) is medium purified by salt precipitation (Koeppe and Ofengand, 1975) in the presence of carrier vitellogenin, and assayed by direct counting.

V. Subunit structure of oocyte vitellogenin

One assumes that a protein of 559,000 daltons must be made as subunits and indeed there was a previous reference to the existence of subunits in *Leucophaea* vitellogenin (Dejmal and Brookes, 1972), but no data were given. In order to compare the subunit structure of oocyte vitellogenin with the organ culture secretion product, it was first necessary to characterize the oocyte protein.

Oocyte vitellogenin was labeled *in vivo* by injection of $[^3H]$ or $[^{14}C]$ leucine for seven days prior to isolation from developing oocytes in order to be able to use both radioactivity and the intensity of Coomassie Blue staining for quantitation of the amount of protein present. The 28S and 14S forms were purified by salt fractionation and sucrose gradient sedimentation (Koeppe and Ofengand, 1975), and analyzed by SDS gel electrophoresis (fig. 6). The top panel shows the subunit analysis of the native 14S component (n14S), the middle panel, 28S, and in the bottom panel a mixture of $[^3H]$n14S plus $[^{14}C]$28S is shown in order to confirm the correspondence among the various peptides. Clearly, 28S is made of 3 dissimilar polypeptide subunits, while n14S contains in addition to the same three size classes an additional band, D. Thus, the conversion of n14S to 28S does not simply involve the trimerization of n14S as originally suggested by Dejmal and Brookes (1972). Moreover the occurence of an additional band in n14S was surprising since 14S has been shown to be the precursor to the 28S component (Brookes and Dejmal, 1968). A possible explanation of this apparent paradox can be derived from a consideration of the stoichiometry and molecular weights of the peptides.

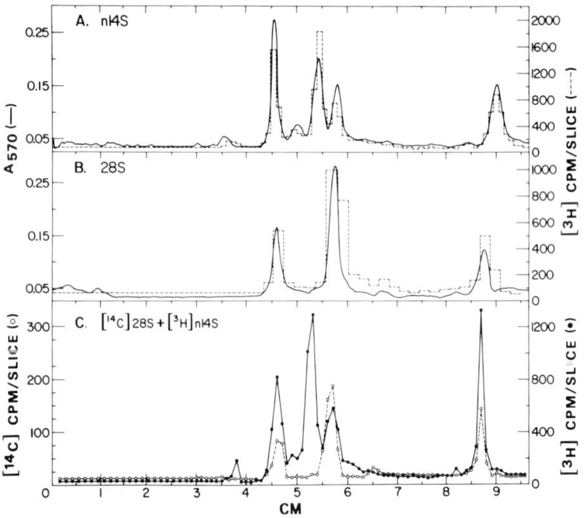

Fig. 6. SDS gel analysis of $[^3H]$n14S and 28S vitellogenin. n14S and 28S vitellogenin were labeled, purified, dissociated, electrophoresed for 4 hrs, and analyzed (Koeppe and Ofengand, 1975). The running gel begins at 0 cm. No isotope was found in the stacking gel. *Top:* SDS gel of n14S vitellogenin from both 2 mm and 5 mm oocytes which were indentical. Middle: SDS gel of 28S vitellogenin obtained from 2 mm and 5 mm oocytes. Bottom: $[^{14}C]$ 28S and $[^3H]$n14S were dissociated, mixed and co-electrophoresed. The peptides, from left to right, are denoted A, D, B, C.

The molecular weights were established by co-electrophoresis with a suitable set of reference proteins. Since vitellogenin in *Leucophaea* is known to be a glycoprotein containing approximately 8% neutral sugar (Dejmal and Brookes, 1972) the molecular weights were measured in gels of varying porosity. Under these conditions the molecular weight of glycoproteins decreases asymptotically to a plateau value at higher gel concentrations, which is empirically found to yield the true molecular weight (Segrest and Jackson, 1972). The molecular weights obtained were peptide A, 118,000; B, 87,000; C, 57,000; D, 96,000.

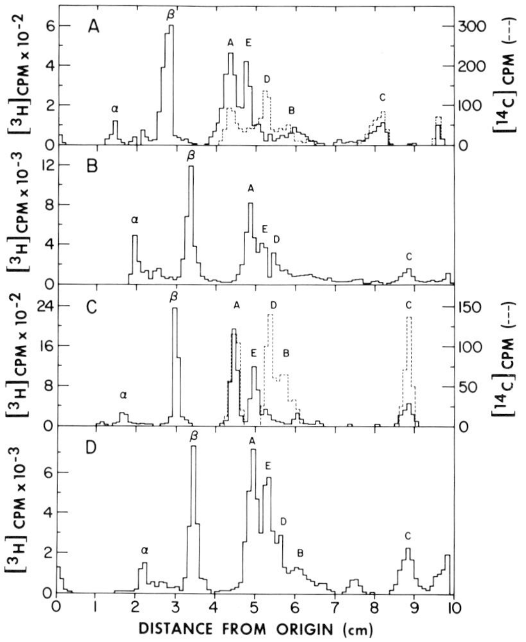

Fig. 7. SDS gel analysis of vitellogenin synthesized in organ culture. Organ culture synthesis of $[^3H]$ leucine labeled vitellogenin was carried out with fat bodies from active females (4 mm oocytes) for a period of 5 hrs. A. The entire medium from an incubation which yielded 1980 cpm in vitellogenin/mg of fat body by the anti-serum assay was layered on a sucrose gradient and fractionated as in Fig. 5. The 14S peak was collected and used directly. All solutions except the organ culture medium contained 10^{-5} M phenylmethyl sulfonyl fluoride (PMSF). The recovery of radioactivity from the gel was 94%. B. Antibody precipitation was used to purify vitellogenin from a second incubation as in part A which yielded 2031 cpm in vitellogenin/mg of fat body by the antibody assay. PMSF was absent from the solutions in this experiments. The recovery of radioactivity was 95%. C. Vitellogenin was prepared from the medium of a third incubation as in Part B, but in this case 10^{-5} M PMSF was present in all solutions, including the incubation medium. This resulted in a decreased synthesis of vitellogenin to 962 cpm/mg fat body. Radioactivity recovery from the gel. was 119%. D. Vitellogenin was prepared from the fat bodies used in the incubation of Part B by the antibody method. The recovery of radioactivity from the gel was 76%. As indicated, dissociated $[^{14}C]$n 14S was added as an internal peptide marker to the preparations. Electrophoresis was in 7.5% gels. Slicing and counting was as described in Fig. 6.

The stoichiometry of each of the peptides in n14S and 28S vitellogenin was determined from the mass ratio under each peak of fig. 7 divided by the corresponding molecular weight. Analyzes were made on immature (2 mm) oocytes where the proportion of n14S is large and on mature (5 mm) oocytes where most of the vitellogenin is as 28S. No differences were detected nor did inclusion of an iodoacetamide step in the dissociation reaction to block SH groups have any effect. Both the radioactivity data and the quantitative staining analyses were used and were in good agreement. Based on these molar ratios, a stoichiometry of $A_1B_1D_2C_2$ was found for n14S, and $A_1B_3C_2$ for 28S. These stoichiometries correspond to a minimum molecular weight of 549,000 for n14S, and 530,000 for 28S after correction for the 6.8% lipid content found in vitellogenin from mature oocytes (Dejmal and Brookes, 1972). It is assumed that the same percent lipid is also associated with the n14S component. These results are in excellent agreement with the reported values of 559,000 for 14S and 1,590,000 for 28S (Dejmal and Brookes, 1972), assuming the true 28S stoichiometry to be $A_3B_9C_6$ (MW 1,590,000).

These stoichiometries suggest that in the conversion of n14S to n28S, peptide D is converted to a smaller size which fortuitosly has the same molecular weight as B. Thus the true stoichiometry of 28S is suggested to be $A_1B_1C_2B_2^D$. We are currently trying to resolve this point by peptide mapping and immunological studies on the n14S and 28S peptides. The splitting of 9,000 daltons of proteins from D probably occurs in the oocyte itself rather than in the follicular space since in the early stage of oocyte growth, about 50% of the total isolable vitellogenin is in the 14S unprocessed form.

VI. Subunit composition of vitellogenin synthesized in organ culture.

Having established the polypeptide composition of oocyte vitellogenin, we next examined the nature of the 14S products excreted into the medium by JH-induced fat bodies in organ culture (fig. 7). Three different samples labeled with [^{3}H]-leucine were prepared. The first (fig. 7A) was obtained by sucrose gradient fractionation of the incubation medium, as in fig. 5. An aliquot of the 14S peak was taken directly from the gradient for dissociation and electrophoretic analysis. In this case, all solutions, except the incubation medium itself contained 10^{-5} M PMSF, a serine protease inhibitor. The second sample (fig. 7B) was obtained by large-scale immunoprecipitation of vitellogenin from the medium with antibody. After washing, the immune precipitate was dissociated and electrophoresed. PMSF was absent from all solutions. The third sample (fig. 7C) was also obtained by immunoprecipitation as described above, but in this case 10^{-5} M PMSF was present in all solutions, including the original incubation mixture. PMSF was added to suppress proteolytic activity during the incubation, but it also decreased the level of vitellogenin synthesis 2-fold (10^{-4} M PMSF inhibited 70%). [^{14}C] labeled n14S was included in two of the runs to serve as internal markers for peptides A,B,C, and D.

The two most striking features of these analyses were first that despite the different methods of preparation and different degrees of precaution against proteases the patterns of all three runs were very similar. The second and more important point is that the peptide pattern was very different from n14S and showed a preponderance of two larger peptides, labeled α and β. In addition, a new peptide, E, appeared between A and D. The molecular weights for β and E, obtained by electrophoresis with reference proteins, were 179,000 and 108,000, respectively. The value for α

was estimated to be 260,000 daltons from fig. 7 using the known values for β and A for calibration of this region of the gel and assuming a linear relationship between log MW and distance migrated. The actual value could be even greater. The size of α and β in comparison to the largest peptide found in n14S indicates that they are composed of at least two smaller peptides.

In order to see if some processing of vitellogenin occurred during secretion from the fat body cells, a similar analysis was performed on fat body extracts (fig. 7D). This preparation came from the same organ culture as panel B, and it is clear that essentially the same amount and size of peptides were found. It should be noted, however, that protease inhibitors were not employed in this experiment. Although it is possible that larger peptides were originally formed, they would have had to be rapidly broken down to the same size and to the same extent during the course of tissue disruption as those which were secreted during the incubation in order to account for the essential similarity between panels B and D. This experiment does not, however, rule out processing of vitellogenin precursors *inside* the fat body cells.

VII. Conclusions

As shown in this work, the juvenile-hormone induced biosynthesis of vitellogenin by adult female fat body is a good model system for the study at the molecular level of the way JH regulates protein synthesis. Hormone induction can be initiated at will, protein synthesis goes on outside of the insect in organ culture, and large amounts of a readily characterizable product are produced. JH probably acts directly on the fat body, although until *in vitro* induction has been achieved, this point remains in question. Which fat body cell type is active in producing vitellogenin is also not yet clear.

Vitellogenin is synthesized as one or more precursor proteins, which still exist in a precursor state upon secretion into the hemolymph. Presumably they are processed in the hemolymph or during uptake by the oocytes to the mature n14S form by specific proteolytic cleavages. The conversion of n14S to 28S appears to involve a specific proteolytic processing step by which peptide D is converted to a smaller size. In this regard it is worth noting that 14S and 28S species are stable when separated from each other and do not form an equilibrium mixture. Trimerization thus appears to be an active process triggered by a specific proteolytic cleavage.

Synthesis of the peptide subunits of vitellogenin as precursor proteins may be the way that coordinate synthesis of these peptides is controlled. The stoichiometry suggests that A + B are made as a precursor of at least 205,000 daltons, and C + D as a precursor of at least 154,000 daltons. It is tempting to identify these hypothetical precursors with bands α and β, respectively, but experimental evidence for such a conclusion is not yet at hand.

Recently, *Xenopus* vitellogenin was shown to be synthesized as a single precursor protein which was subsequently cleaved into lipovitellin and phosvitin (Bergink and Wallace, 1974; Clemens *et al.*, 1975). It may be that this feature of vitellogenin biosynthesis will turn out to be a general phenomenon not only in insects but throughout Nature.

VIII. References

Bergink, E.W. and Wallace, R.A. (1974). *J. Biol. Chem. 249,* 2897.

Brookes, V.J. (1969). *Develop. Biol. 20,* 459.

Brookes, V.J. and Dejmal, R.K. (1968). *Science 160,* 999.

Clemens, M.J. (1974). In: Progress in Biophysics and Molecular Biology (eds. Butler, A.J.V. and Noble, D.) *28,* 69.

Clemens, M.J., Lofthouse, R. and Tata, J.R. (1975). *J. Biol. Chem. 250,* 2213.

Dejmal, R.K. and Brookes, V.J. (1972). *J. Biol. Chem. 247,* 869.

Doane, W. (1973). In: Developmental System: Insects (ed., S.J. Counce, and Waddington, C.H.) *2,* 291-497. Academic Press, New York.

Engelmann, F. (1971). *Arch. Biochem. Biophys. 145,* 439.

Engelmann, F. (1969). *Science 165,* 407.

Fallon, A.M., Hagedorn, H.H., Wyatt, G.R. and Laufer, H. (1974). *J. Insect Physiol. 20,* 1815.

Gilbert, L.I. (1974). *Recent Progress in Hormone Research 30,* 347.

Koeppe, J. and Ofengand, J. (1975). *Arch. Biochem. Biophys.* (submitted).

Segrest, J.P. and Jackson, R.L. (1972). *Methods Enzymol. 28,* 54.

Willis, J.H. (1974). *Ann. Rev. Entomol. 19,* 97.

Chapter 15

JUVENILE HORMONE-INDUCED VITELLOGENIN SYNTHESIS IN LOCUST FAT BODY *IN VITRO*

G.R. Wyatt, T.T. Chen and P. Couble

I. Introduction

Cell and organ cultures have recently become increasingly valuable in the study of hormone action, for cellular structure which may be necessary for the expression of hormonal effects is maintained intact, while the environment can be fully controlled and cellular products harvested. Among the hormones of insects, the juvenile hormone (JH) is of exceptional interest because of the profound effect that it exerts upon the complex developmental processes of metamorphosis (Willis, 1974). In addition, as an open-chain terpene derivative, it belongs to a chemical group not represented among other known animal hormones (Wyatt, 1972). As yet, however, very little is known about the molecular mode of action of JH, although conflicting speculation and proposals have been put forward (Williams and Kafatos, 1971; Ilan *et al.*, 1970).

II. The role of juvenile hormone

The well known role of JH is to prevent metamorphosis in immature insects, thus determining the nature of the molt induced by ecdysone. The analysis of this process at the biochemical level is made difficult by the interaction of JH with ecdysone and by the lack (as yet) of effects clearly definable in biochemical terms, such as specific protein products. After falling to a low titre during the pupal stage, however, JH returns in many adult insects to take on a second role of regulating reproductive maturation, paradoxical though this may seem for a substance whose earlier function was to maintain immaturity! JH exercises this second role in many, though not all, insects, and may be required for several activities of both the male and female reproductive system (Doane, 1973). An effect which is particularly amenable to biochemical analysis is the induction of synthesis of yolk protein precursors, or vitellogenins, in the fat body. These proteins are released into the hemolymph and then taken up by the growing oocytes for deposition in the yolk, quite analogously to the vitellogenins produced in the liver of amphibia and birds as precursors of lipovitellin and phosvitin (Wallace and Bergink, 1974). In appropriate insect species, this system provides excellent material for analysis of JH action.

We hope it may serve as a model for understanding of JH action in metamorphosis, although we recognize that findings with respect to one role of JH cannot necessarily be applied directly to another.

This action of JH in vitellogenin synthesis has been investigated in the cockroach, *Leucophaea maderae* (Engelmann, 1974), and studies have been initiated with the Monarch butterfly (Pan and Wyatt, 1971). We felt, however, that experiments at the subcellular level might be complicated by the abundant intracellular bacterial symbiotes which are harboured in the fat body of cockroaches, and Monarchs are difficult to maintain in the laboratory. Seeking an advantageous species for the project, we have turned to the locust, *Locusta migratoria*. The insect is conveniently large, free of symbiotes and easy to raise in the laboratory throughout the year. It was known that removal of the corpora allata (the source of JH) form locusts prevented the maturation of the eggs and lowered amino acid incorporation in the fat body (Hill, 1965; Minks, 1967), but effects of JH on specific proteins were not clearly established (Benz *et al.*, 1970; Engelmann *et al.*, 1971).

III. Material and results

We identified the locust vitellogenin as a component in acrylamide gel electrophoresis of extracts of eggs and hemolymph of mature females, but not of males or immature females (Fig. 1). This protein was purified from hemolymph and egg extract by procedures involving two steps of DEAE-cellulose chromatography. Antisera were prepared against egg extract and rendered specific for the vitellogenin by exhaustive absorption with hemolymph from male locusts. Antigenic identity of the specific protein from eggs with the corresponding component of mature female hemolymph was demonstrated by continuity of the precipitation zone in the Ouchterlony double diffusion method, as well as immunoelectrophoresis. In egg extracts, the typical vitellogenin is accompanied by aggregation forms and also by nonspecific proteins that occur also in males. The *Locusta* vitellogenin has a molecular weight of 550,000, contains about 10% of lipid and 12% of polysaccharide and has been analyzed with respect to amino acid composition and subunit structure (Chen and Wyatt, 1976). In these properties, it is similar to other insect vitellogenins that have been characterized (Dejmal and Brookes, 1972; Engelmann and Friedel, 1974; Pan and Wallace, 1974; Kunkel and Pan, 1976).

Using the quantitative immunodiffusion assay of Oudin, we showed that vitellogenin is absent from the hemolymph of newly emerged female locusts, and first appears (in colonies including both sexes under crowded conditions) on about the eighth day of adult life. Its concentration then rises rapidly for several days to a plateau level as production is balanced by uptake into the ovaries. In females allatectomized at emergence, this antigen never appeared and no eggs matured. The effects of allatectomy were reversed by application of pure synthetic juvenile hormone.

Biochemical analysis of JH action in this system required an assay of specific protein synthesis in the fat body, and for this we turned to an organ culture technique (Fig. 2). Fat body collected under sterile conditions was incubated with ^{3}H-leucine in a synthetic medium prepared for locust tissues (Table 1). The medium was collected and, after addition of carrier vitellogenin, the protein was precipitated with specific antiserum, while total protein was precipitated from another sample with trichloroacetic acid. After appropriate washing of the two precipitates, counting for radioactivity permitted calculation of the total incorporation as well as

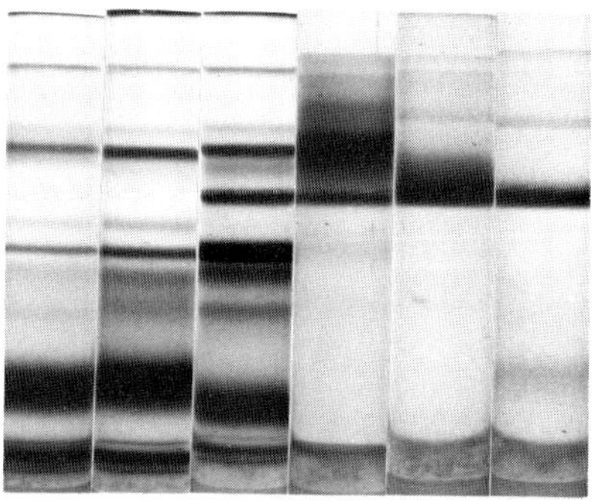

Fig. 1. Acrylamide gel electrophoresis of locust hemolymph and egg proteins. Hemolymph samples were allowed to clot at room temperature for 30 min and the supernatant after centrifugation was used. Egg proteins were extracted in 0.05 M Tris-HCl (pH 8.1) containing 0.4 M NaCl and 0.1 M phenylmethylsulfonyl fluoride, and vitellogenin was purified by DEAE cellulose chromatography (Chen and Wyatt, 1976). Electrophoresis: 5% acrylamide, Tris-glycine buffer, pH 8.3 (0.025 M Tris, 0.19 M glycine), M. male hemolymph; IF, immature adult female hemolymph; RF, reproductive adult female hemolymph; E, egg yolk extract; EV, vitellogenin purified from egg; HV, vitellogenin purified from hemolymph.

the percentage represented by the specific product. The procedure is essentially similar to that developed by Hagedorn (1974) for assaying vitellogenin synthesis in mosquito fat body [the controlling hormone in this instance is apparently ecdysone (Hagedorn *et al.*, 1975)].

It was important to know some of the characteristics of the fat body culture system. The effect of different concentrations of the labelled amino acid, leucine, is shown in Fig. 3. A plateau, presumably representing saturation at a limiting step, is approached at 30 μm leucine, and this concentration was adopted for routine use. Time-course studies (Fig. 4) showed that labelled vitellogenin first appeared in the medium about 1 hour after addition of isotope, the intervening time presumably being required for intracellular synthesis, processing and transport of the protein. Three hours was selected as a standard incubation time in the linear phase of incorporation.

Using this system as the assay, we examined the dose-response relationship of JH applied to allatectomized female locusts (Fig. 5). Topical application in oil or acetone to the abdominal cuticle proved more effective than injection, and, after weak effects were obtained from single applications of JH (cf. Engelmann *et al.*, 1971) we turned to a regime of multiple applications at 8 hour intervals. Presumably, this counteracts rapid inactivation of the hormone. Under these conditions, a remarkably sharp end-point was obtained (Fig. 5), with half-maximal effect at a total dose of about 75 μg, and maximal effect above 120 μg of dl-JH-I. With the saturating dose, 40-50% of the labelled protein output from the fat body was precipitated by the antiserum, which is close to the proportion of about 60% obtained with fat body from intact egg-maturing locusts. These quantities of JH are surprisingly large in relation to the doses of the order of 1 μg found effective for vitellogenesis in cockroaches (Engelmann, 1974; Kunkel, 1973) or Monarch butterflies (Pan and

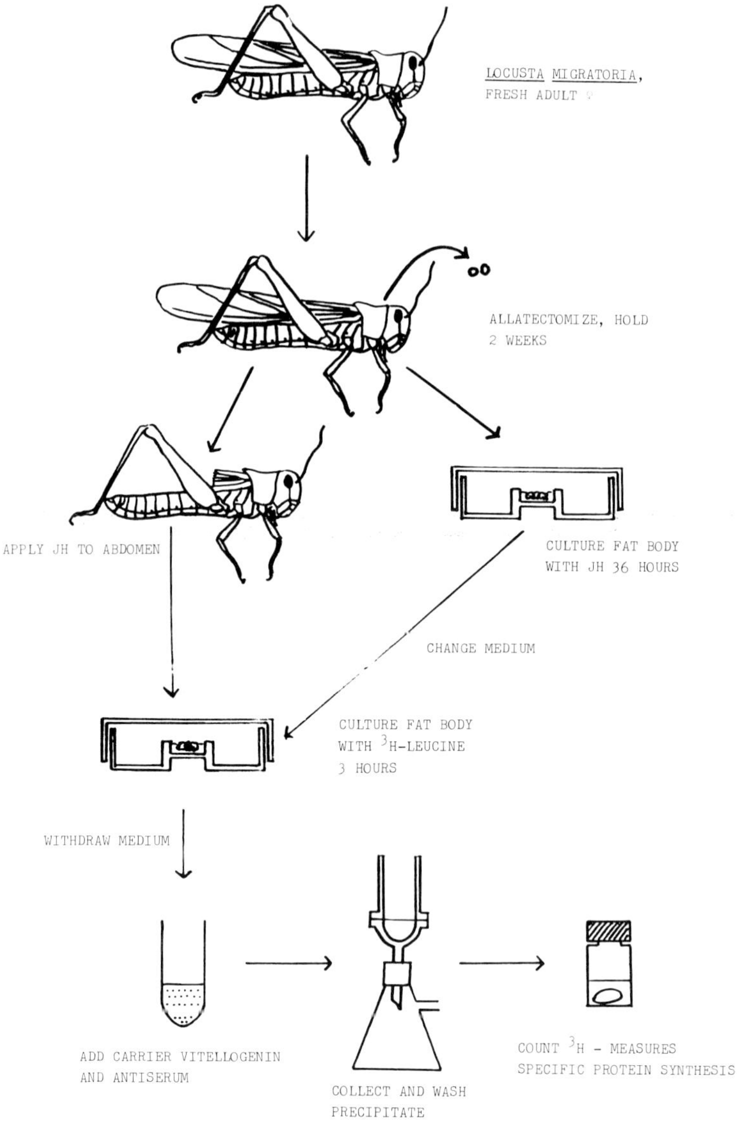

Fig. 2. Procedure for assay of JH-induced vitellogenin synthesis in locust fat body *in vitro.* Freshly emerged adult female locusts are allatectomized and kept for 2 weeks; JH in 10 µl acetone is applied topically to the abdomen, or, alternatively, fat body is placed in culture in M-14 medium (Marks and Reinicke, 1965) with JH; after appropriate time, fat body is placed in synthetic medium (Table 1) with ^{3}H-leucine for 3 hours; medium is collected, to 0.2 ml samples purified vitellogenin (20 ug) is added, one sample is precipitated with anti-vitellogenin serum and another sample with 10% trichloroacetic acid (TCA); both are collected on Millipore filters, washed with 5% TCA, ethanol and ether, and counted by liquid scintillation.

TABLE 1.

Medium for culture of *Locusta migratoria* fat body

Component	mg/100 ml	mM	Component	mg/100 ml	mM
Amino Acids			Amino Acids (cont'd)		
Alanine	44.6	5	Tyrosine	18.1	1
Arginine-HCl	105.4	5	Valine	23.4	2
Asparagine.H_2O	75	5	Lysine-HCl	91.5	5
Aspartic Acid	66.5	5	Leucine	0.39	0.03
Cysteine-HCl.H_2O	35.1	2	Inorganic Salts		
Glutamic Acid	73.5	5	NaCl	585	100
Glutamine	73	5	KCl	112	15
Glycine	75	10	$MgCl_2.6H_2O$	160	8
Histidine-HCl.H_2O	314	15	$CaCl_2$	44	4
Isoleucine	26.2	2	Na_2HPO_4	71	5
Methionine	29.8	2	Carbohydrates		
Phenylalanine	33	2	D-Glucose	90	5
Proline	57.5	5	Trehalose	756	20
Serine	105	10	Antibiotics		
Threonine	23.8	2	Streptomycin ($SO_4^{=}$)	2.5	
Tryptophan	40.8	2	Penicillin G (Na^+)	1.5	

The medium was adjusted to pH 7.2 with 1 N NaOH and sterilized by filtration. L-Amino acids were used. The leucine was labelled with ^3H.

TABLE 2.

Induction of vitellogenin synthesis in locust fat body by juvenile hormone or corpora allata *in vitro.*

Fat body from allatectomized female locusts was cultured in the medium described in Table 1 for 36 hours. JH was added as 30 µg dl-JH-I dissolved in 1 ml medium as described in the text, and the tissue was transferred to fresh JH-containing medium after 12 and 24 hours. Corpora allata came from reproductively mature males, 5 pairs were present in 1 ml medium, and the tissue was transferred to fresh medium with corpora allata after 18 hours.

	Cpm/0.2 ml medium/3 h		Antibody ppt/TCA ppt, %	
Treatment	Antibody ppt.	TCA ppt.	Raw data	Corr. for control
Control	355	2266	16	–
+ JH-I	862	1731	50	29
+ Corpora allata	978	2258	43	28

Each value is the average from 2 incubations.

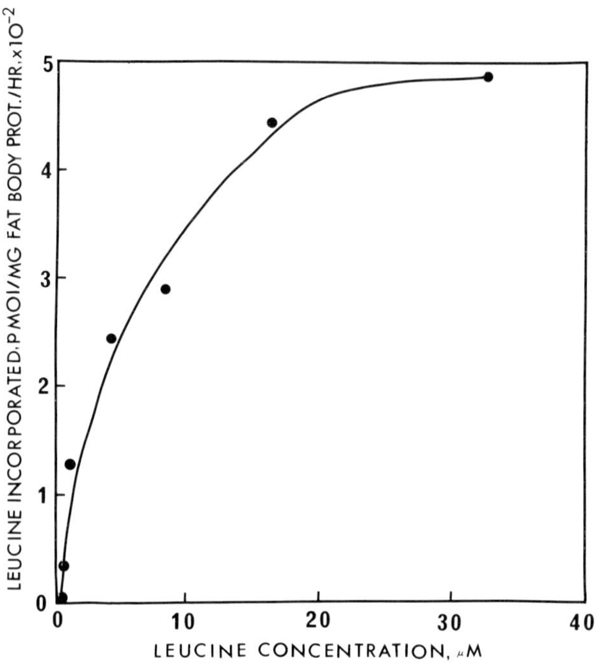

Fig. 3. Effect of leucine concentration on incorporation into total protein by fat body *in vitro.* Fat body was incubated for 3 hours in synthetic medium containing ^{3}H-leucine plus unlabelled leucine to give the concentrations shown, medium was treated with TCA and the protein precipitate washed and counted.

Wyatt, 1976). The effects of JH analogs and mimics are being studied, as well as further analysis of time and dose relationships.

Next, we wished to see whether JH would stimulate when applied directly to tissue *in vitro*. Despite its lipid character, JH is appreciably soluble in water (Kramer *et al.*, 1974). We therefore evaporated a solution of the hormone in cyclohexane in the culture vessel, added the culture medium and subjected this to sonic oscillation in order to bring the hormone into solution, before adding fat body from allatectomized female locusts. As an alternative source of JH, corpora allata dissected from mature males were cultured along with fat body. The results from an experiment of this type are shown in Table 2. The cultures containing JH (either synthetic or released from corpora allata) showed about 2.5 times as much incorporation in the specific precipitate as did controls, with no increase in incorporation into total protein. The radioactivity in the immuno-precipitate in control cultures is believed to include a blank due to inadequate washing, for a modified procedure has recently lowered this value to 60-80 cpm. When corrected for the control value, the results from the JH-treated cultures show almost 30% of protein output in the specific product. Although the results varied quantitatively, several experiments of this type each gave greater relative incorporation in the antibody precipitate from JH-treated cultures than from controls. Therefore, although these results are preliminary and in need of repetition and extension, we believe that they are valid, representing the induction of specific protein synthesis in an insect tissue by treatment with JH

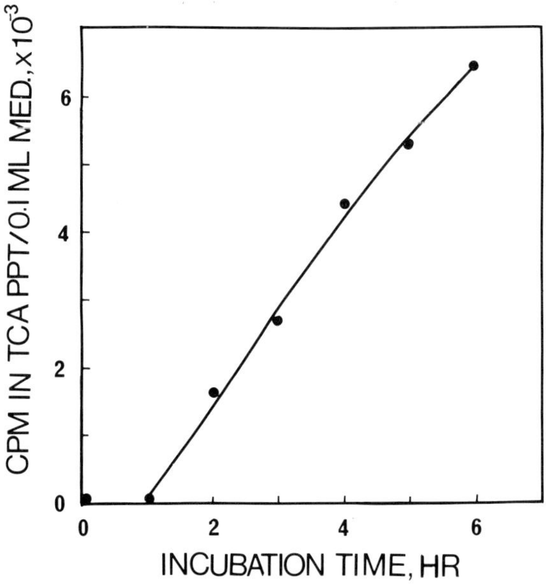

Fig. 4. Release of newly-synthesized protein by fat body *in vitro* with time. Fat body was incubated in synthetic medium with ^{3}H-leucine (30 μM) for different times, then the medium was collected, treated with TCA, and the protein precipitate was washed and counted.

in vitro. Recently, several interesting morphological effects of JH applied to insect tissue *in vitro* have been reported (Oberlander and Tomblin, 1972; Riddiford, 1976). The response of locust fat body in culture of JH may perhaps be improved by adding the hormone along with a hemolymph carrier protein which may facilitate delivery to the cells and protect against inactivation by esterase (Sanburg *et al.*, 1975).

Other experiments indicate that the action of juvenile hormone on adult locust fat body is accompanied by substantial changes in RNA, including ribosomal RNA and vitellogenin messenger RNA (cf. Engelmann, 1974, for *Leucophaea*). We are optimistic that continued studies with this system may contribute to understanding of the action of juvenile hormone at the molecular level .

IV. References

Benz, J., Girardie, A. and Cazal, M. (1970). *J. Insect Physiol. 16,* 2257.

Chen, T.T., and Wyatt, G.R. (1976). In preparation.

Dejmal, R.I., and Brookes, V.J. (1972). *J. Biol. Chem. 247,* 869.

Doane, W.W. (1973). In Insects - Developmental Systems (S.J. Counce and C.H. Waddington, ed.) Academic Press, London, Vol. 2, pp. 291-497.

Engelmann, F. (1974). *Amer. Zool. 14,* 1195.

Engelmann, F., and Friedel, T. (1974). *Life Sci. 14,* 587.

Engelmann, F., Hill, L., and J.L. Wilkens (1971). *J. Insect Physiol. 17,* 2179.

Hagedorn, H.H. (1974). *Amer. Zool. 14,* 1207.

Hagedorn, H.H., O'Connor, J.D., Fuchs, M.S., Sage, B., Schlaeger, D.A., and Bohm, M.K. (1975). *Proc. Nat. Acad. Sci. U.S.* (in press).

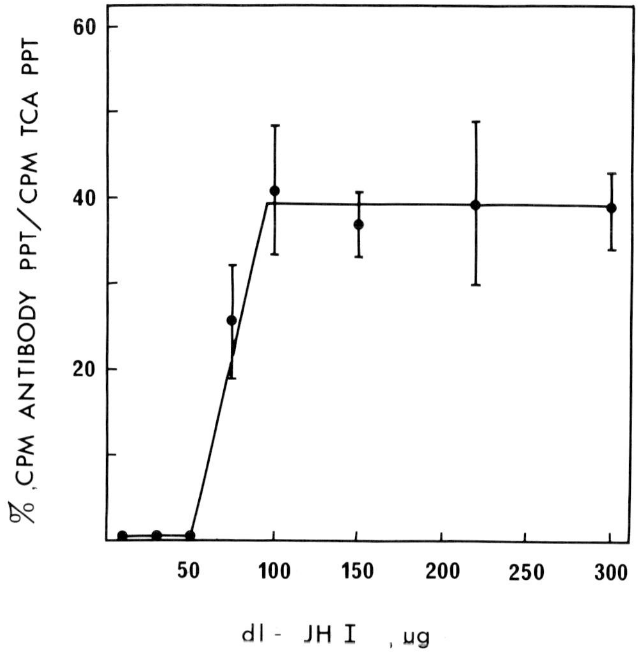

Fig. 5. Dose-response curve for induction of vitellogenin synthesis by JH *in vivo.* The procedure was as in Fig. 1, using dl-JH-I (cecropia C-18 JH; Eco-Control, Inc., Cambridge, Mass.). Each point shows the mean and S.E.M. from 6 test animals.

Hill, L. (1965). *J. Insect Physiol. 11,* 1605.

Ilan, J., Ilan, J., and Patel, N. (1970). *J. Biol. Chem. 245,* 2175.

Kramer, K.J., Sanburg, L.L., Kezdy, F.L., and Law, J.H. (1974). *Proc. Nat. Acad. Sci. U.S. 71,* 493.

Kunkel, J.G. (1973). *J. Insect Physiol. 19,* 1285.

Kunkel, J.G., and Pan, M.L. (1976). *Comp. Biochem. Physiol.* (in press).

Minks, A.K. (1967). *Arch. Neerl. Zool. 17,* 175.

Oberlander, H., and Tomblin, C. (1972). *Science 177,* 441.

Pan, M.L., and Wallace, R.W. (1974). *Amer. Zool. 14,* 1239.

Pan, M.L., and Wyatt, G.R. (1971). *Science 174,* 503.

Pan, M.L. and Wyatt, G.R. (1976). In preparation.

Riddiford, L. (1976). In: Invertebrate Tissue Culture: Application in medicine, biology and agriculture (E. Kurstak and K. Maramorosch, eds). Academic press, New York.

Sanburg, L.L., Kramer, K.J., Kezdy, F.J., Law, J.H., and Oberlander, H. (1975). *Nature 253,* 266.

Wallace, R.A., and Bergink, E.W. (1974). *Amer. Zool. 14,* 1159.

Williams, C.M. and Kafatos, F.C. (1971). *Mitt. Schweiz. Ent. Ges. 44,* 151.

Willis, J.H. (1974). *Ann. Rev. Entomol. 19,* 97.

Wyatt, G.R. (1972). In: The Hormones (G. Litwack, ed.) Academic Press, New York. Vol. 2, pp. 385-490.

Chapter 16

IN VITRO ANALYSIS OF FACTORS REGULATING THE JUVENILE HORMONE TITER OF INSECTS

J. Nowock and L. I. Gilbert

I. Introduction

Insect development can be viewed as a progressive sequence of cyclic events that is controlled in part by the titer of juvenile hormone (JH). A general decline in endogenous JH, with modulations within each cycle, is in large measure responsible for the transformation of the larva to the pupa and then to the adult. In the female adults of many insect species the corpora allata (source of JH) are reactivated and the JH secreted during adult life then controls aspects of the reproductive cycle. Past studies in which exogenous JH was applied to insects at inauspicious times revealed that normal development could be disrupted and suggested that in addition to the importance of synthesis and release of JH by the corpora allata in determining the temporal pattern of JH concentration, metabolic degradation and excretion were also crucial parameters to be considered (e.g. Gilbert and Schneiderman, 1960).

Recent studies on hemolymph proteins have demonstrated that specific high molecular weight lipoproteins can bind JH and juvenoids (*Tenebrio molitor*, Trautmann, 1972; *Hyalophora gloveri*, Whitmore and Gilbert, 1972; *Locusta migratoria*, Emmerich and Hartmann, 1973). In the tobacco hornworm *Manduca sexta*, a smaller protein (MW ∽34,000) exhibits high specificity for JH (Kramer *et al.*, 1974; Goodman and Gilbert, 1974) although a lipoprotein with lower binding affinity is also present.

In vitro studies demonstrated that one role of the low molecular weight binding protein (BP) is maintenance of the circulating JH titer by preventing the non-specific uptake of the hormone into lipophilic compartments (Hammock *et al.*, 1975). Furthermore, the BP provides protection for JH from the general esterases in the hemolymph, except for a specific esterase which can hydrolyze JH while it is a constituent of the JH-BP complex (Sanburg *et al.*, 1975a,b; Hammock *et al.*, 1975). These findings suggest that BP and esterases are important components of the control system that regulates the JH titer and thus regulates insect development. Although the BP is present in the hemolymph during all developmental stages studied, there is some evidence for temporal fluctuations during the life of *M. sexta (*W. Goodman, personal communication). Developmental studies on hemolymph esterases revealed alterations in general activity and the appearance of a JH-esterase at a specific stage (Weirich *et al.*, 1973; Sanburg *et al.*, 1975a, b). To obtain a comprehensive picture of JH regulation it is necessary to understand the control of synthesis, release and turnover of the BP and esterases. Data pertaining to the origin of the hemolymph BP and esterases are requisite for this understanding. The present paper summarizes our *in vitro* experiments on that subject.

II. Site of synthesis of JH binding proteins

For a preliminary screening of JH binding activity in *Manduca sexta* larval salivary glands, fat body, carcass and gut-Malpighian tubules complex were cultured separately for 6-12 hr in a medium based on the amino acid and cation content of mature last instar larvae (Table 1). The resulting culture media and 105,000 g supernatants of tissue homogenates (free of microsomal epoxide hydratase activity) were treated for 12 hr with diisopropyl fluorophosphate (DFP) to inhibit esterases, subsequently charged with [^{3}H]-JH and submitted to Sephadex G-25 gel filtration. Protein-bound JH should be eluted in the exclusion volume whereas unbound hormone which forms monomeric solutions at the concentrations applied, should be recovered in the inclusion volume. Of the various samples tested, only media and tissue extracts from fat body and carcass contained label associated with one or more macromolecules. Culture medium in which isolated corpora allata had been secreting JH did not yield a JH binding fraction.

These data were corroborated by experiments in which media and tissue extracts were analyzed by a micro-double diffusion test against serum containing BP-antibodies. Only media and homogenate supernatants from fat body and carcass formed precipitin arcs (Fig. 1).

Although the arc patterns suggest that the reactants shared immunochemically similar antigens, they did not prove conclusively that BP was present since the antiserum was not totally specific. That is, the antigen preparation used for immunization contained two minor contaminants in addition to the BP. These two contaminants however, were devoid of binding activity.

When [^{3}H]-JH was introduced as a marker, the antiserum could be used to detect BP. To distinguish between tissue sequestering of BP and *de novo* synthesis, fat body and carcass were cultured *in vitro* in the presence of 2 μCi/ml (5.7 x 10^{-6}M) L-[U-^{14}C] leucine in medium devoid of cold leucine. After incubation, [^{3}H]-JH was added to both the DFP-treated media and 105,000 g supernatants. Antiserum was then added and the precipitates were radioassayed. The data revealed that the [^{3}H]-JH was bound most efficiently to the immunoprecipitates derived from the fat body medium and tissue extract (Table 2).

A small amount of binding was noted in the carcass 105,000 g supernatant whereas essentially no [^{3}H] was associated with the precipitate from the carcass medium. In addition, carcass incorporated [^{14}C]-leucine to a much lower extent than fat body. Although rates of uptake and precursor pools were not determined, the data demonstrate that the fat body had the greater capacity for protein synthesis, thus making it the more probable candidate for the site of synthesis of BP.

To further analyze both the presence of BP in the immunoprecipitate and the distribution of [^{14}C]-leucine between the antigens, fat body and carcass media and

TABLE 1

Composition of Medium for the Culture of MANDUCA SEXTA Tissues

Substance	Concentration	
	mg/liter	mole/liter
KCl	3,000	40.24
KH_2PO_4	1,360	9.99
$NaHCO_3$	420	5.00
$MgCl_2.6\ H_2O$	2,050	10.08
$MgSO_4$	1,200	9.97
$CaCl_2.2\ H_2O$	800	5.44
$Glucose.H_2O$	4,000	20.18
$Trehalose.2\ H_2O$	2,000	5.29
Sucrose	9,120	26.64
L-Alanine	420	4.71
L-Arginine HCl	350	1.66
L-Aspartic acid	400	3.00
L-Cysteine HCl	260	1.48
L-Glutamic acid	600	4.08
L-Glutamine	1,200	8.21
L-Glycine	200	2.66
L-Histidine HCl	710	3.12
L-Isoleucine	230	1.75
L-Leucine	250	1.91
L-Lysine HCl	400	2.19
L-Methionine	500	3.35
L-Phenylalanine	200	1.21
L-Proline	500	4.34
L-Serine	200	1.90
L-Threonine	260	2.18
L-Tryphophan	100	0.49
L-Tyrosine	180	0.99
L-Valine	800	6.83
B_{12}	0.02	0.00008
Biotin	0.01	0.00004
D-Ca pantothenate	0.02	0.00042
Choline HCl	0.20	0.00127
Folic acid	0.02	0.00004
Inositol	0.05	0.00030
Niacin	0.02	0.00002
Pyridoxine HCl	0.02	0.00017
Riboflavin	0.05	0.00013
Thiamine HCl	0.02	0.00006
Penicillin	100 I.U.	
Streptomycin	100	

pH 6.6; 340 mosmol/kg

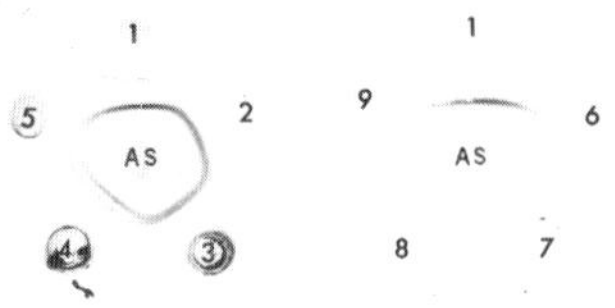

Fig. 1. Immunodiffusion analysis of culture media and homogenate supernatants against anti-binding protein serum. (1) hemolymph; (2) fat body medium; (3) fat body extract; (4) carcass extract; (5) carcass medium; (6) gut-Malpighian tubule medium; (7) gut-Malpighian tubule extract; (8) salivary glands extract; (9) salivary glands medium; anti-serum (AS) in center well.

TABLE 2

Immunoprecipitation of Double Labeled Culture Media and Homogenate Supernatants

Tissue	Specific activity [^{14}C] (dpm/mg protein)	Immunoprecipitable radiolabel [^{14}C] dpm	[^{3}H] dpm
Fat body			
Medium	312,186	2,864	3,199
Homogenate supernatant	20,271	1,021	3,114
Carcass			
Medium	4,428	96	88
Homogenate supernatant	3,792	260	618

Tissues were incubated *in vitro* in the presence of [^{14}C]-leucine (2 μCi/ml). Media and 105,000 g supernatants of tissue homogenates were charged with [7-ethyl-1,2- H(N)] JH (C_{18}). Free JH was removed by charcoal adsorption. Aliquots (100 μl) were combined with 300 μl antiserum against bindinggprotein. The immunoprecipitates were washed and radio-assayed. The total [^{14}C]-label added per 100 μl aliquot was 106,143 dpm for fat body medium, 67,010 dpm for fat body homogenate, 1,488 dpm for carcass medium and 14,521 dpm for carcass homogenate. All samples contained 214,400 dpm [^{3}H]-JH before the charcoal treatment.

extracts were subjected to immunoelectrophoresis on agarose films. Only fat body medium and extract produced a precipitin arc that corresponded to BP. Autoradiography of these pherograms resulted in significant exposure at the sites of the precipitin arcs, indicating that the BP was indeed synthesized *de novo*. Carcass medium and 105,000 g supernatant formed one and two arcs respectively. However, these arcs can be attributed to non-BP antigens. Their autoradiographic reaction was very weak as would be expected from the immunoprecipitation data.

To confirm our findings that the fat body has the ability to synthesize and release BP, [^{14}C]-leucine containing fat body preparations were charged with [^{3}H]-JH and subjected to slab gel electrophoresis. The proteins thus separated were analyzed for electrophoretic mobility and binding capacity. Using hemolymph as a standard (Fig. 2), [^{3}H]-label was found to be associated with 2 proteins. The faster migrating (R_f = 0.62) band represents the high affinity, low molecular weight BP while the second (R_f = 0.14) was tentatively characterized as a lipoprotein on the basis of its staining behavior with lipid crimson. Carrierfree JH remained at the origin while polar metabolites migrated with the front. These two proteins, both labeled with [^{3}H] from JH and [^{14}C] from leucine, were present in the fat body preparations. In contrast, electrophoretic analysis of carcass homogenate supernatant failed to demonstrate the presence of any protein with binding activity.

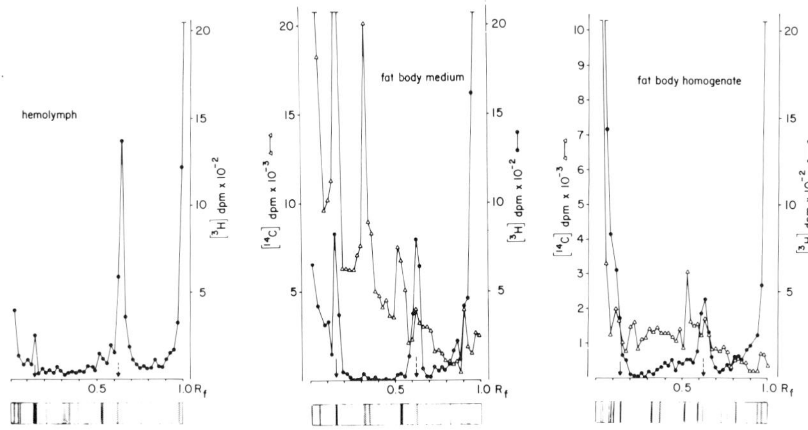

Fig. 2. Radiolabel distribution pattern of samples separated by gel electrophoresis. Fat body was cultured in medium containing 5 μCi/ml L-[^{14}C]-leucine. Homogenate supernatant and medium were charged with [^{3}H]-JH and separated in duplicate on a slab gel. [^{3}H]-JH treated hemolymph served as a standard. One track was sliced and the sections were combusted in a Packard sample oxidizer. The parallel track was stained for proteins. Cathode on the left. Solid arrow depicts position of the lipoprotein while dashed arrow depicts position of the binding protein.

Thus, the immunological and electrophoretic analyses demonstrated the *de novo* synthesis of BP by the fat body and its rapid release into the culture medium. Since the binding capacity of the tissue extracts never exceeded that of the corresponding medium, there was no indication of storage of BP. Although some [^{3}H]-JH was found associated with the immunoprecipitate from the carcass extract, no binding activity could be detected using slab gel or immunoelectrophoresis. The former result could be due to minor contamination by the hemolymph and the sensitivity of the immunoprecipitation technique. The extremely low incorporation of [^{14}C]-leucine into carcass protein and the essential absence of JH binding activity in the medium indicate that the carcass is not the site of synthesis of BP.

From the gel electrophoresis analyses it is apparent that the high molecular weight protein with relatively low JH affinity is also synthesized and released by the fat body. The fact that this protein is the only lipoprotein of the 5 present in the hemolymph (and the culture medium) which binds JH is suggestive of some specificity and perhaps a physiological role. The lipoprotein may serve as a sink during times of intense JH secretion by the corpora allata.

III. Site of origin of JH esterases.

Fluctuations in esterase activity during the development of *Manduca sexta* were first reported by Weirich *et al.* (1973) who determined the hydrolytic activity of hemolymph using both α-naphthyl acetate (α-NA) and JH as substrates. The differences in the activity patterns as a function of developmental stage suggested that two esterase populations were present in the hemolymph. In extending these studies, Sanburg *et al.* (1975a,b) found that these two populations can be distinguished by their sensitivity to DFP. The general esterases are rapidly inactivated by low concentrations of DFP whereas the JH-specific esterases are fairly resistant to the inhibitor. Furthermore, the JH-BP complex protects the hormone from enzymatic attack by the general esterases but does not afford protection from the JH-specific esterases (Sanburg *et al.*, 1975a,b; Hammock *et al.*, 1975). Preliminary information regarding the origin of the hemolymph esterases arose from our *in vitro* studies on the metabolic degradation of JH by the fat body (Hammock *et al.*, 1975). These data indicated that fat body from several larval stages released esterases into the medium which hydrolyzed free JH but not JH bound to BP. Only 3-day-old last larval instar fat body released an enzyme(s) that attacked the bound JH.

If the hemolymph esterases actually originate in the fat body, one should expect some correlation between the activity patterns of these two compartments. Therefore, the hydrolytic activities of hemolymph and fat body 105,000 g supernatants were compared during the last larval instar of *M. sexta* using α-NA and JH as substrates. The developmental pattern of the esterase activity in the hemolymph exhibited a maximum at days 3 and 4 for both substrates (Fig. 3).

This is in general accord with the findings of Weirich *et al.* (1973) and Sanburg *et al.* (1975a,b) and any slight differences probably reflect variations in rearing conditions. Fat body homogenate supernatants showed a similar pattern with a sharper peak at day 4. However, the complexity of the *in vivo* situation due to the simultaneous presence of esterases produced for export and those which are constituents of the fat body, as well as possible differences in rates of synthesis and release etc., led us to *in vitro* analyses.

Fat body was incubated for 24 hr since preliminary experiments revealed that protein release is linear during this period of time. The media were then analyzed for α-NA, total, and specific JH hydrolytic activity. Fig. 4 shows the secretory capacity of fat body from different stages and demonstrates that JH hydrolytic activity peaks at day 3. Total and specific JH hydrolytic activity exhibited essentially the same pattern indicating that JH-ester cleavage is determined primarily by the activity of the JH-specific esterases. The increase in JH-esterase activity initiated on the second day suggests the possibility of an induction phenomenon. α-NA hydrolytic activity exhibited a completely different pattern, indicating a separate control for this population of esterases.

Supporting evidence for the premise that hemolymph esterases originate in the fat body was obtained from studies on the effect of inhibitors of RNA and protein

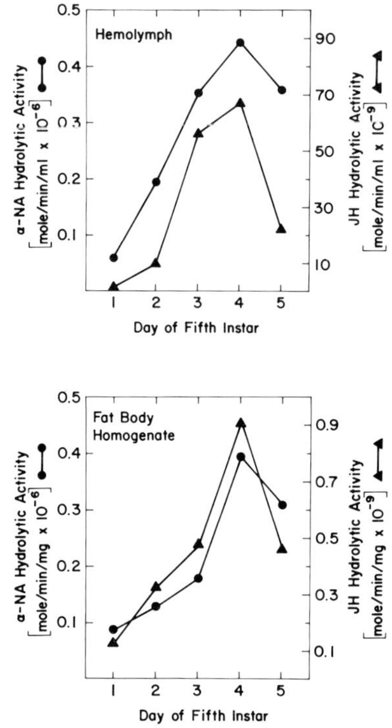

Fig. 3. α-NA and JH hydrolytic activity of hemolymph and fat body 105,000 g supernatant during the last larval instar. "General esterase" was assayed spectrophotometrically with α-NA as substrate. JH hydrolytic activity was determined with [7-ethyl-1,2-^{3}H(N)]JH (C_{18}) as substrate and quantitated by extraction of the reaction mixture with ethyl acetate after incubation. This was followed by TLC separation and liquid scintillation counting of the appropriate zones of the silica gel plates (Weirich *et al.*, 1973). Three different enzyme concentrations were assayed for each point to insure that measurement was in the linear range.

synthesis on the release of these enzymes. Aliquots of fat body from 3-day-old fifth instar larvae were cultured in the presence or absence of puromycin, cycloheximide, actinomycin D or α-amanitin. After a 3 hr incubation period during which pre-existing proteins were released into the medium, the tissues were transferred to fresh medium and cultured for another 6 hr under the same conditions. The media were then analyzed for released protein and esterase activity. In conjunction with this experiment, the inhibition of macromolecular synthesis was also studied: After 3 hr of preincubation, fat bodies were transferred to media containing [^{3}H]-uridine triphosphate and/or [^{14}C]-leucine, and incubation was allowed to continue for another 3 hr. Previous studies demonstrated that the incorporation of [^{14}C]-leucine into protein was linear over this period of time.

The data from these studies (Table 3) revealed that when puromycin or cycloheximide were utilized as inhibitors there appeared to be a correlation between protein

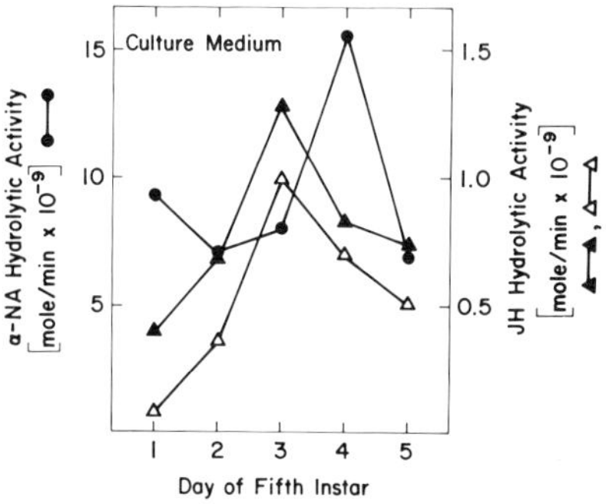

Fig. 4. Release of esterases from fat body derived from last instar larvae of differing ages. Approximately equal amounts of tissue (~450 mg fresh weight) were cultured *in vitro* for 24 hr and the media were assayed for esterase activity. In contrast to the procedure in Fig. 3, JH hydrolytic activity was determined with [^{3}H]-methoxy-JH(C_{18}) as substrate. Quantitation was achieved by partitioning the generated labeled methanol and the unreacted JH in a two-phase system. Methanol:chloroform (20:50) was mixed with the reaction medium (22) to yield a 20:50:22 mixture. 80 ± 2% of the methanol appears in the aqueous phase. The parition of unreacted JH into this phase is negligible. Closed triangles indicate total JH hydrolysis. Short treatment with DFP (10^{-4}M, 15 min) revealed DFP-resistant (i.e. JH-specific) activity (open triangles). For comparison, all values are expressed on the basis of 1 mg fat body protein.

TABLE 3

Effect of Inhibitors on Macromolecular Synthesis and Protein Release by M. SEXTA Fat Body

	Percent Inhibition			
	Puromycin (50 µg/ml)	Cycloheximide (1 µg/ml)	Actinomycin D (1 µg/ml)	α-Amanitin (0.5 µg/ml)
RNA synthesis			100	32
Protein synthesis	95	58	28	22
Protein release	41	22	<1	2
α-NA hydrolytic activity	21	24	21	24
JH hydrolytic activity:				
Total	30	15	19	9
DFP resistant	34	12	12	6

Aliquots of fat body from 3-day-old last instar larvae were incubated with and without inhibitors and the media were analyzed for released proteins and α-NA, total and specific JH hydrolytic activity. Quantitation as in legend to Fig. 4. For determination of RNA and/or protein synthesis, uridine-5, 6-[^{3}H] 5'-triphosphate (10 µCi/ml) and/or L-[U-^{14}C] leucine (0.5 µCi/ml) were added. The concentration of cold leucine in the medium was reduced to 0.1mM. At the end of the incubation period tissues were homogenized in the medium and TCA was added to 10%. The washed TCA precipitable material was collected on glass fiber filters and radioassayed. Percent inhibition was calculated on the basis of specific activities ([^{3}H] or [^{14}C] dpm/mg protein).

synthesis and release, although the latter is inhibited to a lesser degree. A substantial amount of protein was detected in the medium even when the incorporation of [^{14}C]-leucine into protein was almost totally inhibited. This suggests that the period required for the transit of synthesized proteins into the surrounding medium is greater than 3 hr. Protein synthesis, and consequently protein release, were only slightly affected by treatment with actinomycin D or α-amanitin. This indicates that at least some of the mRNA has a half life greater than 3 hr and is in accord with previous studies of stable mRNA in eukaryotes (Kafatos and Gelinas, 1974). That α-amanitin was less effective than actinomycin D in inhibiting the incorporation of [^{3}H]-UTP (both affecting protein synthesis to a similar degree) was not surprising since α-amanitin is specific for RNA polymerase II. The release of esterases into the medium with the capacity to hydrolyze JH, both total and DFP resistant, was inhibited to a lesser degree than total protein release, although the relative extent of inhibition appeared to be similar when protein synthesis inhibitors were utilized. Actinomycin D and x-amanitin differed in their effectiveness in inhibiting the release of enzymes capable of hydrolyzing JH. α-NA hydrolytic activity was generally reduced by all inhibitors tested.

Since studies utilizing inhibitors of RNA and protein synthesis cannot distinguish between effects on the synthesis of specific proteins (i.e. esterases) or of proteins that activate existing proenzymes, conclusions from such experiments must be drawn with caution. Therefore we cannot exclude the possibilities that the function of the fat body in hemolymph esterase production is to activate pre-existing esterases previously synthesized by the fat body, or synthesized at other sites and sequestered by the fat body. Notwithstanding these possibilities, the inhibitor experiments and the developmental study on the secretory capacity of fat body do demonstrate that this organ is engaged in the release of active esterases. The inhibitor studies further indicate that this release is not due to simple leakage from damaged cells but is a controlled process. This is supported by the fact that fat body homogenates and incubation media exhibit different esterase patterns when analyzed by gel electrophoresis and stained with α-NA and fast blue RR.

IV. Conclusion

The titer of JH within an insect is critical for normal development and is determined by the rate of JH biosynthesis and release by the corpora allata, the presence of "protective" hemolymph proteins, and the activity of enzymes that degrade the hormone. The data presented here reveal that the hemolymph binding protein which protects JH from nonspecific esterases is synthesized in the fat body and released into the hemolymph. The fat body also releases two populations of esterases, one of which can degrade JH that is complexed to binding protein. The appearance of these JH specific esterases is stage specific and suggests the presence of a precise control mechanism (e.g. enzyme induction). It should be emphasized that we did not survey all the organs and tissues of *M. sexta* for hemolymph esterase biosynthesis. Therefore, although we now consider the fat body to be the primary site of origin, other sites may also exist. The fat body however, does play a major albeit indirect role in regulating the JH titer, although the means by which it in turn is controlled is presently a matter of conjecture.

Acknowledgments

These studies were supported by Grants AM-02818 from the National Institutes of Health and GB-27574 from the National Science Foundation. J. Nowock is grateful to the Deutsche Forschungsgemeinschaft for post-doctoral fellowship support. We thank R. Bell (U.S.D.A.) for supplying the *M. sexta* eggs; K. Judy (Zoecon Corp.l for providing the corpora allata culture medium; and W. Goodman (Northwestern University) for the binding protein antiserum. The [^{3}H]-methoxy-JH was a generous gift from K.H. Trautmann, (Dr. R. Maag, A.G.)

V. References

Emmerich, H., Hartmann, R. (1973). *J. Insect Physiol. 19,* 1663.

Gilbert, L. I., and Schneiderman, H.A. (1960). *Trans. Amer. Microsc. Soc. 79,* 38.

Goodman, W., and Gilbert, L.I. (1974). *American Zool. 14,* 1289.

Hammock, B., Nowock, J., Goodman, W., Stamoudis, V., and Gilbert, L.I. (1975). *Molec. Cell. Endocrinol.* (in press).

Kafatos, F.C., and Gelinas, R. (1974). In: Biochemistry of Cell Differentiation (Kornberg, H.L. and Phillips, D.C., eds.) *9,* 223-264.

Kramer, K.J., Sanburg, L.L., Kezdy, F.J., and Law, J.H. (1974). *Proc. Nat. Acad. Sci. USA 71,* 493.

Sanburg, L.L., Kramer, K.J., Kezdy, F.J., Law, J.H., and Oberlander, H. (1975a). *Nature 253,* 266.

Sanburg, L.L., Kramer, K.J., Kezdy, F.J., and Law, J.H. (1975b). *J. Insect Physiol. 21,* 873.

Trautmann, K.H. (1972). *Z. Naturforsch. 27b,* 263.

Weirich, G., Wren, J., and Siddall, J.B. (1973). *Insect Biochem. 3,* 397.

Whitmore, E., and Gilbert, L.I. (1972). *J. Insect Physiol. 18,* 1153.

Chapter 17

IN VITRO ACTION OF ECDYSONE AND JUVENILE HORMONE ON EPIDERMAL COMMITMENT IN THE TOBACCO HORNWORM

L. M. Riddiford

I. Introduction

One of the most important questions in insect endocrinology today is what is the cellular mode of action of the two insect hormones concerned with growth and development---i.e. ecdysone and juvenile hormone? To study the mode of action of hormone, ideally one needs a single target tissue which can be isolated in large amounts and on which the hormone can act *in vitro*. For ecdysone, there are several such systems, most notably the salivary gland chromosomes of higher Diptera (Ashburner, 1970; Ashburner *et al.* 1973) and imaginal discs of various insect species (Fristrom, 1973; Oberlander, 1972, 1975; Yund and Fristrom, 1975). But for juvenile hormone (JH) there have been only a few reports of its action *in vitro* (Oberlander and Tomblin, 1972; Chihara and Fristrom, 1973; Davey and Huebner, 1974; Ittycheriah *et al.*, 1974; Sanburg *et al.*, 1975). In the morphogenetic system JH had its usual inhibitory effect, blocking either leg evagination and subsequent differentiation (Chihara and Fristrom, 1973) or pupal cuticle deposition (Oberlander and Tomblin, 1972; Sanburg, *et al*, 1975), and effects were only seen at very high doses of Cecropia C18JH (10^{-4} M or higher). Davey and Huebner (1974) showed a gonadotropic effect of C18JH *in vitro* at the more physiological concentration of 3×10^{-7} M; this concentration was found to be optimal for the induction of patency (spaces between the follicular cells) in the *Rhodnius* ovary which in turn allows increased uptake of the vitellogenin.

Although the *Rhodnius* ovary may be a good system in which to look at the cellular mode of action, this gonadotropic action of JH may be quite different from its morphogenetic action. Therefore, our aim in the past few years has been to find a system in which one could look at the morphogenetic action of JH *in vitro*---a system in which JH would have its *status quo* effect (Williams, 1961) but in which the tissue would show a response, both in the presence and in the absence of JH.

In the tobacco hornworm *Manduca sexta* the physiology of both the larval-larval molt and the larval-pupal molt has been worked out in some detail (Truman, 1972; Truman and Riddiford, 1974; Nijhout and Williams, 1974a,b). In the fourth larval instar the JH titer in the hemolymph is high until after the beginning of the molt to the final larval instar (Fain and Riddiford, 1975). Prothoracicotropic hormone (PTTH) release which initiates ecdysone secretion and thus the molt occurs at a specific time of day as defined by the photoperiod (Truman, 1972). Then after ecdysone has acted on the epidermal cells, the JH titer declines until the time of ecdysis when it increases again (Fain and Riddiford, 1975). In the 5th (final) instar the JH titer remains elevated until the animal attains a weight of about 5 grams (on the 2nd day after ecdysis) at which time the corpora allata are inactivated and the JH titer gradually declines (Nijhout and Williams, 1974b; Nijhout, 1975). During the succeeding day at a time determined by photoperiod, PTTH and ecdysone are released to initiate metamorphosis--a change in appearance and behavior from the feeding stage to the wandering stage (Truman and Riddiford, 1974). This small amount of ecdysone (Bollenbacher *et al.*, 1975) also irrevocably commits the epidermal cells to a program of pupal differentiation (Truman, *et al.*, 1974) but it does not elicit pupal cuticle synthesis. This synthesis is initiated two days later by a second, larger release of ecdysone (Truman and Riddiford, 1974; Bollenbacher *et al.*, 1975).

Since JH prevents pupal cuticle synthesis by the epidermis only when applied before the first release of ecdysone (Truman, *et al.*, 1974), this switchover in commitment seemed to be an ideal system in which to look at JH action. Here one is unencumbered by the molecular events leading to cuticle synthesis and rather concerned only with those events triggered by ecdysone which lead to a permanent switchover in cellular commitment to pupal differentiation. This paper is concerned with the establishment of an *in vitro* system in which ecdysone triggers the switchover in cellular commitment and in which JH will prevent this switchover but will allow the cells to continue to produce larval cuticle.

II. *In vitro* culture procedures

Manduca sexta were reared on an artificial diet as previously described (Truman, 1972) under a 12L:12D photoperiod at 25°C. Donor larvae were selected to weigh between 6.3 and 7 gms at mid-day on the third day of the fifth larval instar.

After surface sterilization in 0.05% $HgCl_2$ in 50% ethanol and thorough rinsing, the dorsal integument of abdominal segments 4 to 6 was removed. The fat body and much of the muscle was carefully dissected away from the epidermis in a specially prepared *Manduca* saline (Cherbas, 1973) containing several drops of an alcoholic solution of phenylthiourea (recrystallized) (PTU). The cleaned epidermis was cut into small pieces and rinsed twice in chemically defined Grace's medium (GibCo), the first rinse containing 200 µg/ml streptomycin sulfate and PTU, the second only PTU. The tissue was then incubated in 0.7 ml defined Grace's medium in plastic culture wells (Linbro Disposotrays).

β-ecdysone (Rohto Pharmaceutical Co.) was dissolved in 10% isopropanol and sterilized through a Millipore filter. C18JH (JHI) (Ro6-9550 from R Maag Co.) and epoxygeranylsesamole (Eco-control) (EGS) were dissolved in sterile Grace's medium by sonication using siliconized (Siliclad) glassware.

The wells were placed in sterile petri dishes and surrounded by sterile *Manduca* saline to maintain high humidity. The dish was continuously perfused with 95% O_2–5% CO_2 mixture and was gently shaken (Thomas Rotating Apparatus) at 24-25°C.

III. *In vivo* assay for epidermal commitment

To assay the commitment of the epidermal cells, the epidermis and attached cuticle was removed after the incubation in Grace's medium with or without hormones, cut into smaller pieces, and implanted into fourth instar *Manduca* larvae before PTTH release. The host larva was anesthetized, submerged in *Manduca* saline, and the tip of the proleg was cut off. Two to three pieces of epidermis and cuticle were implanted into the hemocoele. A few crystals of 1: 1 mixture of PTU: streptomycin were added, and the proleg tied off with a loop of thread.

After implantation into the host larva, the epidermal cells of the implant grew around their cuticle and formed an epidermal cyst just as Piepho (1938) originally described for implants of *Galleria* integument. During the subsequent molt of the host, the epidermal cyst also molted and formed a new cuticle. Figure 1 shows the types of cuticle that were produced. Larval cuticle was distinguished by its thick translucent appearance, its rubbery consistency, and the many tiny papillae seen in Fig. 1a. Also, it was covered with densely pigmented epidermal cells. Pupal cuticle was hard and brittle, usually tan or sometimes white, opaque and shiny, and covered by colorless epidermis.

When epidermis was taken from a larva on the fourth day of the fifth instar during the time of the ecdysone-induced switchover of commitment or after culture with insufficient ecdysone or with ecdysone and low amounts of JH, cysts were found to produce both larval and pupal cuticle (Fig. 1b). Often pupal patches covered large areas, but sometimes small fingers or tiny sports of distinct pupal cuticle were scattered amidst otherwise normal larval cuticle. Thus, each epidermal cell changes its commitment individually and independently.

Occasionally, a thin untanned layer of smooth transparent cuticle, often with some sculpturing, was formed. This cuticle was classified as pupal since it had none of the characteristics of larval cuticle and usually contained one or two tan patches. It was most prevalent after incubation with high doses of β-ecdysone (5μg/ml).

Thus, in this *in vivo* assay, the type of cuticle produced by the implanted epidermis during the larval-larval molt is not determined by the hormonal milieu of the host, but rather reflects the commitment of the epidermis at the time of implantation. Once the *Manduca* epidermis was committed to pupal differentiation by an exposure to ecdysone in the absence of JH, then the JH in the hemolymph of the fourth instar larval host had no detectable effect on the type of cuticle produced. In these experiments the duration of exposure to the JH of the host ranged from one to three days before initiation of the molt.

IV. Ecdysone-induced switchover in epidermal commitment.

Since PTTH and ecdysone release which initiate metamorphosis occur on the fourth day of the last larval instar in *Manduca* (Truman and Riddiford, 1974), epidermis from larvae which weigh >6 gms on the third day of the instar is still committed to larval differentiation. Yet by this time the JH titer in the hemolymph has begun to decline (Nijhout and Williams, 1974b). Therefore, this epidermis was incubated in Grace's medium with different amounts of β-ecdysone for 20-23 hours to determine the concentration necessary to effect a complete switchover in commitment to pupal development. The circles (●) in figure 2 show the % which formed completely pupal cuticle in the *in vivo* assay as a function of the logarithm of the dose of β-ecdysone. With 1 μg/ml β-ecdysone 93% of the epidermis completed the change before implantation into the larval host. Recovery of implanted cysts ranged between 50 and 80%, and the cuticle formed was hard and brittle and usually tan or beginning to tan. When the concentration of β-ecdysone was increased to 5 μg/ml, 100% of the cysts formed pupal cuticle. But the percentage of epidermal cysts recovered from the molted 5th instar host was low (<20%) due to encapsulation of the implanted piece by the blood cells of the host. Moreover, many of the cysts formed thin untanned cuticle described above. Apparently, this dose of ecdysone was excessive for effecting the switchover in commitment although it is necessary for the production of pupal cuticle by these same cells two days later (Mitsui and Riddiford, manuscript in preparation).

Fifty percent complete transformation to pupal commitment was effected by 0.1 μg/ml β-ecdysone (2 x 10^{-7} M) (Fig. 2). But a still lower dose of β-ecdysone was all that was necessary for a partial switchover. Figure 3 shows that although 0.01 μg/ml β-ecdysone only effected complete pupal commitment in 11% of the 18 cysts, it caused at least some cells to change their commitment in 50% of the cysts. As the dose of β-ecdysone was increased, the % partial and complete switchover increased. The dose-response curve for cysts showing at least 50% pupal cuticle is given by the triangles (▲) in Figure 2, and extrapolation gives a 50% response at 6 x 10^{-8} M β-ecdysone.

To confirm the larval commitment of the donor epidermal cells, Fig. 3 indicates that none of the 62 cysts which had been incubated in Grace's medium without ecdysone formed pupal cuticle.

These *in vitro* experiments have confirmed our previous findings with the intact animal (Truman *et al.*, 1974)--namely, that the first secretion of ecdysone which causes the change in behavior from feeding to wandering also effects a permanent change in cellular commitment from larval to pupal. Moreover, the dose of β-ecdysone to effect this switchover in 50% of the epidermal cells is in the same range as that found by Ashburner (1973) for induction of chromosome puffs in *Drosophila melanogaster,* Fristrom *et al* (1973) for evagination of *Drosophila* imaginal discs, and Fain and Riddiford (1973; Fain, 1975) for the formation of larval crochets. If one considers the cysts which showed at least 50% pupal cuticle, the dose necessary is four-fold less and more in line with the doses of estrogen and progesterone necessary to effect a physiological response in chickens and mammals (Jensen and DeSombre, 1973; O'Malley and Means, 1974). Furthermore, the dose of 1 μg/ml to effect the complete switchover agrees amazingly well with the amount of ecdysone found in the whole animal at the peak of the first pulse of ecdysone (Bollenbacher *et al.,* 1975).

V. Prevention of the ecdysone-induced switchover by juvenile hormone

Since 1 μg/ml β-ecdysone effected nearly complete switchover to pupal commitment in at most 20 hours and apparently caused little damage to the epidermal cells, this dose was used for the JH experiments. In the intact animal JH had to be applied 6-10 hours before the time of ecdysone release, as determined by ligation experiments (Truman *et al,* 1974). Therefore, since the epidermis was explanted from larvae with a very low hemolymph titer of JH (Nijhout and Williams, 1974b), the epidermis was pretreated with the JH for 3-4 hours before the ecdysone was added to the culture. The epidermis then was further incubated with both hormones for 20 hours before implanation into the assay larva.

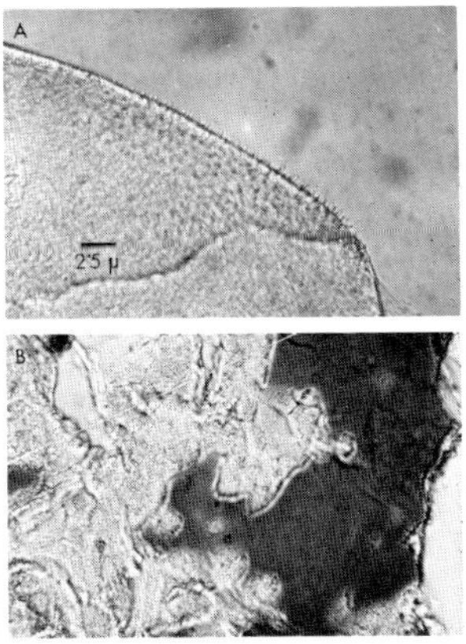

Fig. 1. Types of cuticle formed by implanted epidermal cysts after *in vivo* molt.
a) Larval cuticle. Note papillae on edge.
b) Cuticle from one cyst showing both translucent larval (left) and tan pupal (right) cuticle.

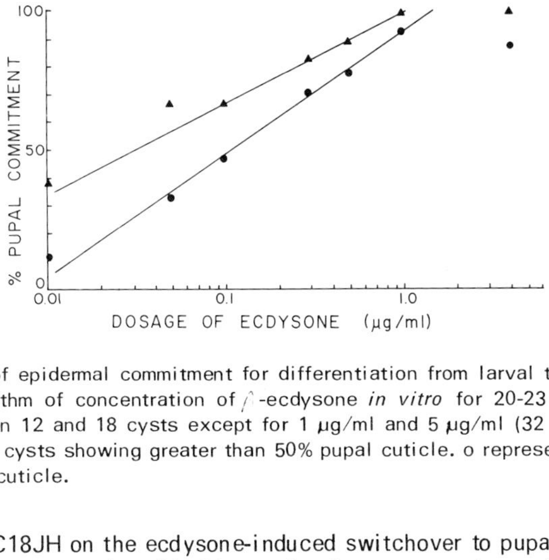

Fig. 2. Change of epidermal commitment for differentiation from larval to pupal as a function of the logarithm of concentration of β-ecdysone *in vitro* for 20-23 hours. Each point represents between 12 and 18 cysts except for 1 μg/ml and 5 μg/ml (32 and 23 respectively). ▲ represent % cysts showing greater than 50% pupal cuticle. o represent % cysts forming completely pupal cuticle.

The effects of C18JH on the ecdysone-induced switchover to pupal commitment are seen in Table 1. Five μg/ml C18JH completely prevents the switchover to pupal commitment and does not interfere with the subsequent expression of larval commitment during the *in vivo* assay. As the concentration of C18JH is reduced, the percentage of epidermal pieces that become commited to pupal differentiation in response to the ecdysone increases. But even at 0.005 μg/ml only 50% show complete pupal commitment. The dose-response curve is linear down to 0.05 μg/ml and shows a 50% complete inhibition of the switchover (i.e., 50% completely larval) at 0.28 μg/ml (9.5×10^{-7} M). If, however, those that remain at least 50% larval are included, then the 50% inhibition dose is approximately 7×10^{-8} M.

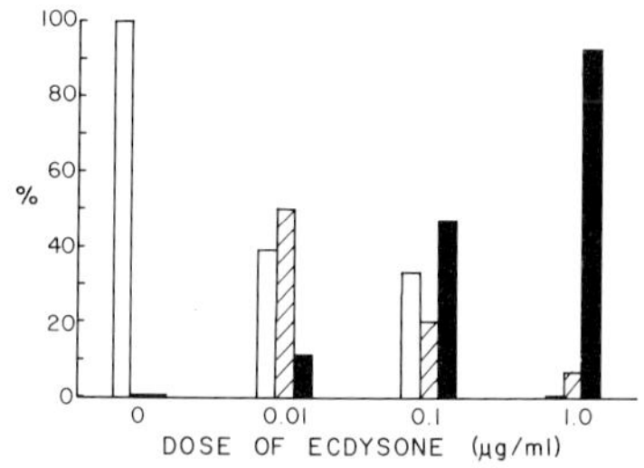

Fig. 3. Types of cuticle formed by cysts after exposure to selected concentrations of β-ecdysone *in vitro.* Number of cysts per concentration: 0 μg/ml, 58; 0.01 μg/ml, 18; 0.1 μg/ml, 15; 1.0 μg/ml, 32. Open bars - larval cuticle; crosshatched bars - both larval and pupal cuticle; solid bars - pupal cuticle.

When the JH mimic epoxygeranylsesamole (EGS) was used in similar experiments, a similar response was seen. Three μg/ml completely inhibited the switchover, and with 0.03 μg/ml only 33% showed complete pupal commitment.

Thus, JH was able to prevent the ecdysone-induced switchover in commitment of the epidermal cells. Moreover, it was able to effect this at near physiological levels which did not interfere with subsequent expression of the larval commitment of these cells. In the 5th instar of *Manduca,* the hemolymph titer of JH in C18JH equivalents is about 0.8 ng/ml (3 x 10^{-9} M) (Fain and Riddiford, 1975; Nijhout, 1975); but the titer at the time of PTTH and ecdysone release which initiates the 4th to 5th larval molt is about seven-fold higher (2 x 10^{-8} M). Therefore, the concentration of ca. 10^{-6} M necessary to retain complete larval commitment is still quite high. But the C18JH (Ro6-9550) used in these experiments is a mixture of stereoisomers including about 14% of the natural isomer (Pawson *et al.,* 1972); in various bioassays it ranges from 10%-50% of the activity of the natural isomer. Studies are presently in progress with a C18JH preparation which contains >95% the natural *trans, trans, cis* isomer.

Although the morphogenetic action of JH *in vitro* in the prevention of imaginal disc development has previously been reported (Chihara and Fristrom, 1973; Oberlander and Tomblin, 1972; Sanburg *et al,* 1975), the effective doses have always been about 10^{-4} M or higher. Also, in these systems the responses have been a complete inhibition of differentiation and cuticle fomation, a fact not surprising since imaginal discs only grow and do not make cuticle in the presence of JH in the larva. Yet in such a system one cannot be sure that the inhibition seen at these high doses is not just phamacological, although Chihara and Fristrom (1973) report immediate reversibility of the inhibition upon removal of the JH. In the *Manduca* epidemal system described in this paper the effect of JH is also inhibition as it prevents the ecdysone-induced onset of pupal commitment. But the fact that the epidermis is not adversely affected by the JH is shown by its ability to make larval cuticle when exposed to an *in vivo* molt. Furthermore, preliminary data indicate that when the epidermis is exposed to ecdysone in the presence of JH for 20 hours, then removed to hormone-free media, it will form a definitive larval cuticle in about 2 days (Riddiford, unpublished). But when exposed only to ecdysone for 20 hours, then to hormone free media, no cuticle is fomed. Rather a second exposure to ecdysone is necessary and results in pupal cuticle formation (Mitsui and Riddiford, manuscript in preparation).

Sanburg *et al.,* (1975) have suggested that the JH-binding protein isolated from the hemolymph of fifth instar *Manduca* larvae plays an important role not only in protecting JH from degradation by the general esterases byt also in the action of JH on its target tissue. Their experiments which show the enhancement of the JH inhibition of cuticle deposition in *Plodia* wing discs upon addition of the binding protein (Sanburg *et al.,* 1975) tend to support this conclusion. However, even with the binding protein the dose of C18JH necessary for 50% inhibition was $>10^{-4}$ M. In the experiments reported here there was no protein in the medium and no binding protein added. Although the complete absence of blood cells is unlikely because of their adherence to insect tissue, their number was minimal due to the extensive washing. The absence of fat body in the cultures insures the virtual absence of binding protein since Nowock and Gilbert (this conference) have now reported the production of the binding protein by the fat body *in vitro.* Therefore, one must conclude that JH can also act on the epidermis without the intermediary of the binding protein. It remains to be seen if the addition of the JH-binding protein would substantially lower the effective dose of JH required for its *status quo* effect.

VI. Timing of DNA synthesis in the epidermis during the larval-pupal transformation.

In the classical view of the endocrine control of metamorphosis, ecdysone is thought to cause apolysis from the overlying cuticle, DNA synthesis, and then secretion of a new cuticle (for reviews see Wyatt, 1972; Gilbert and King, 1973; Willis, 1974) Since in many developing systems DNA synthesis is thought to be necessary to commit the cells to a particular path of differentiation (Holtzer *et al,* 1972), the prevalent idea in the insect literature has been that ecdysone induces this critical round of DNA synthesis and the level of JH determines whether the larval, pupal, or adult gene-set will be read to synthesize the cuticle (Schneidermen, 1972; Williams and Kafatos, 1972). Consequently, the morphogenetic action of JH has been assumed to be somehow involved with this DNA synthesis, but Willis (1974) argues persuasively against this narrow view.

Since the ecdysone-induced change in commitment from larval to pupal differentiation occurs at a very specific time in *Manduca,* it seemed to be an ideal system in which to look at this problem. Concurrently with the development of the *in vitro* system for the study of the hormonal control of epidemal commitment, DNA synthesis in the intact animal was monitored by autoradiography. Fifth instar larvae of known age were injected with 5 μc/gm ^{3}H-thymidine (specific activity 20 Ci/mmole, New England Nuclear Corp.) and then sacrificed 2 hours later. The integument from various regions was fixed, sectioned, coated with NTB-3 Nuclear track emulsion (Eastman Kodak), and exposed for 4 weeks at -20°C.

Table 2 compares the timing of the switchover of commitment of the epidermal cells as assayed by implantation into 4th instar larvae to that for DNA synthesis. As is readily apparent, no DNA synthesis occurs during the day that the switchover to pupal commitment occurs. Even the following day there is only DNA synthesis in the proleg region but none in the dorsal epidermis. But on day 1 of the wandering phase when the second ecdysone secretion is beginning, all of the epidermis is heavily labeled. By the following day this burst of DNA synthesis is completeand cuticle secretion begins.

These data thus indicate that the ecdysone-induced change in commitment of the epidermal cells occurs nearly 2 days before DNA synthesis is seen. Thus, it is

TABLE 1

Commitment of epidermis from 6.5 ± 0.3 gm fifth instar MANDUCA larvae after IN VITRO exposure to ecdysone and juvenile hormone.

Hormone (μg/ml)			*Type of cuticle produced (%)*		
C18JH	*Ecdysone*	*Number*	*Larval*	*Larval & Pupal*	*Pupal*
0	0	59	100	0	0
0	1	39	0	7	93
0.005	1	8	12	38	50
0.05	1	21	10	52	38
0.5	1	15	60	33	7
5	1	6	100	0	0

TABLE 2

DNA synthesis in MANDUCA epidermis during the larval-pupal transformation.

	*Epidermal Commitment (%)**			*Epidermal DNA Synthesis***		
Stage	*Larval*	*Larval & Pupal*	*Pupal*	*Dorsal*	*Ventral*	*Proleg*
Feeding 5th instar						
3rd day						
p.m.	100	0	0	0	0	0
4th day						
a.m.	30	40	30	0	0	0
p.m.	9	16	78	0	0	0
Wandering stage						
Day 0						
a.m.	0	0	100	0	+/ –	++
p.m.	––	––	––	0	0	+
Day 1						
a.m.	––	––	––	++++	++	+++
p.m.	––	––	––	++++	++	++
Day 2						
a.m.	––	––	––	0	0	0

* Epidermis was removed at the indicated time and its commitment was directly assayed by the *in vivo* larval assay described in section III.

** Autoradiographs of epidermis from two larvae at each time point were scanned, and the amount of labeling was scored as follows:

0: no detectable label
+/–: label in a few cells in only one of the two animals
+: < 10 cells labeled per section
++: 10-30 cells labeled per section
+++: 30-50 cells labeled per section
++++: > 50 cells labeled per section

not DNA synthesis itself that effects the irreversible switchover. Rather it is some cellular event, perhaps the synthesis of a regulatory protein, that occurs well before this synthesis. Later the final round of DNA synthesis occurs just prior to the expression of the differentiated state, i.e., cuticle formation.

VII. Conclusion.

Manduca larval epidermis thus provides an excellent system in which to study the mode of action of both ecdysone and juvenile hormone *in vitro*. It can be easily cultured in Grace's medium and upon exposure to β-ecdysone *in vitro* will change its commitment for differentiation from larval to pupal. If, however, JH is also present during this *in vitro* exposure to ecdysone, its larval commitment will be maintained. The beauty of this system is that the commitment to differentiation and the expression of this commitment, i.e. cuticle synthesis, occur at two distinctly separate times, thus requiring two pulses of ecdysone. The cellular event involved in the switchover of commitment in this system is not DNA synthesis but some prior happening. Investigations of the nature of this event are continuing with this *in vitro* system.

Acknowledgements

I thank Ms. Janice Moore for technical assistance, Drs. C. Gunthart and W. Vogel of R. Maag for the gift of Ro6-9550, Dr. Margery Fain for advice on culture techniques, and Dr. James W. Truman for a critical reading of the manuscript. Supported by grants BMS74-02781-A02 from NSF and RF73019 from the Rockefeller Foundation.

VIII. References

Ashburner, M. (1970). *Adv. Insect Physiol. 7,* 2.

Ashburner, M. (1973). *Develop. Biol. 35,* 47.

Ashburner, M., Chihara, C., Metlzer, P., and Richards, B. (1973). *Cold Spring Harbor Symposium 38,* 655.

Bollenbacher, W.E., Vedeckis, W.V., Gilbert, L.I., and O'Connor, J.D. (1975). *Develop. Biol. 44,* 46.

Cherbas, P.T. (1973). Biochemical studies of insecticyanin. Ph.D. thesis, Harvard University.

Chihara, C.J. and Fristrom, J.W. (1973). *Develop. Biol. 35,* 36.

Davey, K.G. and Huebner, E. (1974). *Can. J. Zool. 52,* 1407.

Fain, M.J. (1975). Endocrine physiology of larval molting in the tobacco hornworm. Ph.D. thesis, Harvard University.

Fain, M.J. and Riddiford, L.M. (1973). *Amer. Zool. 13,* 1272.

Fain, M.J. and Riddiford, L.M. (1975). *Biol. Bull.,* in press.

Fristrom, J.W. (1972). *Results and Problems in Cell Differentiation 5,* 109.

Fristrom, J.W., Logan, W.R., and Murphy, C. (1973). *Develop. Biol. 33,* 441.

Gilbert, L.I. and King, D.S. (1973). In: The Physiology of Insecta (M. Rockstein, ed.), Vol. I, pp. 249-370, Academic Press, New York.

Holtzer, H., Weintraub, K., Mayne, R., and Mochan, B. (1972). *Curr. Topics Devel. Biol. 7,* 229.

Ittycheriah, P.I., Marks, E.P., and Quraishi, M.S. (1974). *Ann. Ent. Soc. Am. 67,* 595.

Jensen, E.V. and De Sombre, E.R. (1973). *Science 182,* 126.

Nijhout, H.F. (1975). Biol. Bull. (in press).

Nijhout, H.F. and Williams, C.M. (1974a). *J. Exp. Biol. 61,* 481.

Nijhout, H.F. and Williams, C.M. (1974b). *J. Exp. Biol. 61,* 493.

Oberlander, H. (1972). *Results and Problems in Cell Differentiation 5,* 155.

Oberlander, H. (1975). *In Vitro* (in press).

Oberlander, H. and Tomblin, C. (1972). *Science 177,* 441.

O'Malley, B.W. and Means, A.R. (1974). *Science 183,* 610.

Pawson, B.A., Scheidl, F., and Vane, F. (1972). In: Insect Juvenile Hormones, Chemistry and Action (J.J. Menn and M. Beroza, eds.), pp. 191-216, Academic Press, New York.

Piepho, H. (1938). *Biol. Zentralbl. 58,* 356-366.

Sanburg, L.L., Kramer, K.J., Kezdy, F.I., Law, J.H., and Oberlander, H. (1975). *Nature 253,* 266.

Schneiderman, H.A. (1972). In: Insect Juvenile Hormones, Chemistry and Action (J.J. Menn and M. Beroza, eds.), pp. 3-27, Academic Press, New York.

Truman, J.W. (1972). *J. Exp. Biol. 57,* 805.

Truman, J.W. and Riddiford, L.M. (1974). *J. Exp. Biol. 60,* 371.

Truman, J.W., Riddiford, L.M., and Safranek, L. (1974). *Develop. Biol. 39,* 247.

Williams, C.M. (1961). *Biol. Bull. 121,* 572.

Williams, C.M. and Kafatos, F.C. (1972). In: Insect Juvenile Hormones, Chemistry and Action (J.J. Menn and M. Beroza, eds.), pp. 29-41, Academic Press, New York.

Willis, J.H. (1974). *Ann. Rev. Entomol. 19,* 97.

Wyatt, G.R. (1972). In: Biochemical Actions of Hormones (G. Litwack, ed.), Vol. 2, pp. 385-490, Academic Press, New York.

Yund, M.A. and Fristrom, J.W. (1975). *Devel. Biol. 43,* 287.

Chapter 18

METABOLISM OF MOLTING HORMONE ANALOGS BY CULTURED COCKROACH TISSUES

E.P. Marks

I. Introduction

Treatment of leg regenerates of cultured cockroaches, *Leucophaea maderae* (F.), with β-ecdysone will induce the deposition of a chitin-containing cuticle (Marks and Leopold, 1971). Treatment with α-ecdysone will also induce cuticle deposition, but only when the ecdysone is incubated with the leg regenerate for several days (Marks, 1973a). Recently, King and Marks (1974) demonstrated that cultured leg regenerates slowly convert α-ecdysone to β-ecdysone. In addition, Kaplanis *et al.* (1969) reported that 22, 25-dideoxyecdysone was converted to α- and β-ecdysone in *Manduca* pupae and that it terminated pupal diapause though it produced only erratic and inclusive molting activity on cockroach leg regenerates (unpublished data). Since then King (1972) has reported that fat body from *Manduca* is able to convert 22,25-dideoxyecdysone to β-ecdysone *in vitro*. Because this compound may be a precursor in the biosynthesis of β-ecdysone, I decided to determine whether incubation with *Leucophaea* fat body would enhance the ability of 22,25-dideoxyecdysone to induce cuticle formation by cultured leg regenerates. The compound 22-isoecdysone, which had given negative results in earlier experiments (Marks and Leopold, 1971), was included in my experiments.

II. Methods

Late-instar nymphs of *L. maderae* were removed from the colony immediately after molting. After 24 hours, the middle legs were removed at the coxotrochanteral joint, and the insects were held for an additional 25 days. Then the leg regenerates were removed from the coxal stumps, and about 10 cu mm of fat body were removed from the abdomen. These tissues were placed under dialysis strips in Rose multipurpose tissue chambers (Rose *et al.*, 1958). Plastic coverslips were used, and the chambers were filled with M20 culture medium with 7% fetal bovine serum (Marks, 1973b).

The test compounds were weighed and dissolved in dimethyl sulfoxide (DMSO) to give a concentration of 10 $\mu g/\mu l$. One μl of the solution was injected through the gasket and dialysis membrane directly into the fat tissue or, in the absence of fat body, was placed under the dialysis strip alongside the leg regenerate. The chambers were incubated for 14 days at 26°C and then examined by phase contrast microscopy for evidence of cuticle deposition (see Marks and Leopold, 1971).

III. Results

In the first series of experiments, four compounds -- α-ecdysone, β-ecdysone, 22-isoecdysone, and 22,25-dideoxyecdysone -- were tested on leg regenerates in the absence of fat body. β-ecdysone gave the best response: 93 percent of the regenerates produced cuticle. The 82 percent response to α-ecdysone was only slightly lower (Table 1). The response to 22,25-dideoxyecdysone, as earlier, was erratic: 27 percent of the leg regenerates produced cuticle. The response to 22-isoecdysone was effectively at the control level.

TABLE I

Induction of cuticle deposition by cockroach leg regenerates caused by ecdysone and ecdysone analogs in the presence or absence of fat body

	Tissue			
	Regenerate		Regenerate + fat body	
Compound	N	%	N	%
Untreated control	10	10	10	0
β-ecdysone	14	93		
α-ecdysone	17	82		
22,25-dideoxyecdysone	22	27	22	91
22-isoecdysone	12	8	10	0
22-isoecdysone + 22,25-dideoxyecdysone			10	100

In the second series of experiments, either 22,25-dideoxyecdysone or 22-isoecdysone was injected directly into the fat body of chambers containing both tissues. After 14 days, 91 percent of the leg regenerates in chambers treated with 22,25-dideoxyecdysone showed evidence of cuticle deposition, but none of the regenerates treated with 22-isoecdysone responded.

In the third series of experiments, 22,25-dideoxyecdysone and 22-isoecdysone were injected together into the fat body. In this series, all the leg regenerates responded by producing cuticle.

IV. Discussion

The conversion of 22,25-dideoxyecdysone to β-ecdysone *in vivo* (see Robbins *et al.*, 1971) and *in vitro* (see King, 1972) involves at least three steps (Fig. 1).

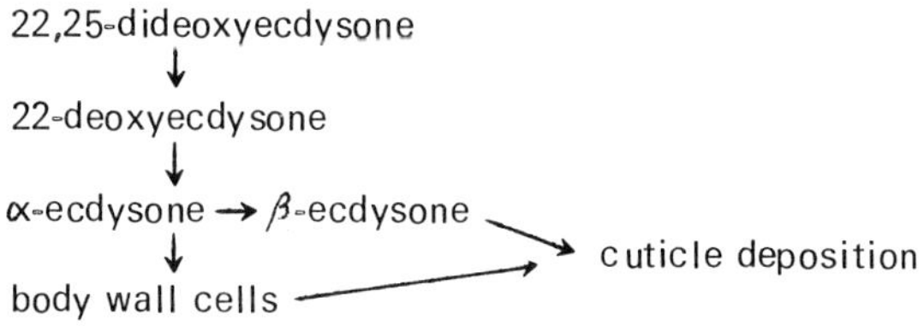

Fig. 1. Tentative schema for the conversion of 22,25-dideoxyecdysone to β-ecdysone and the deposition of cuticle (after King, 1972).

The results of the experiment in which the compounds were tested against leg regenerates alone were substantially in agreement with the findings of King (1972), Marks (1973a), and King and Marks (1974): body wall tissue could carry out the final conversion of α- to β-ecdysone and thus could respond to either compound. Furthermore, King (1972) found that *Manduca* body wall tissue could not carry out the first two steps -- conversion of 22,25-dideoxyecdysone. This, in turn, explains why in earlier experiments, cockroach leg regenerates responded so poorly. Apparently, the tissues of leg regenerates, which are primarily epidermal, convert 22-25-dideoxyecdysone only poorly, if at all. However, some explants contain large numbers of adipohemocytes, which may account for the occasional positive responses that do occur. Oberlander (1974) used imaginal discs from *Plodia* and found that none of his test compounds induced cuticle deposition.

The results of the experiment in which the fat body was co-cultured with leg regenerates indicated that cockroach fat body was able to metabolize 22,25-dideoxyecdysone, at least to α-ecdysone, but that it could not convert 22-isoecdysone to any active form. These results, again, were in agreement with the finding by King (1972) that fat body from *Manduca* could metabolize 22-25-dideoxyecdysone to β-ecdysone *in vitro*. On the other hand, Oberlander (1974) obtained no cuticle-forming response with this compound when the fat body was present. In the latter case, the lack of response must have resulted from failure of the tissues to carry out the conversion to β-ecdysone.

The last experiment was conducted to determine whether any interference by 22-isoecdysone with the processes involved in the conversion could be detected. The negative results indicated that the structure of 22-isoecdysone is such that it is not converted to α- or β-ecdysone nor does it inhibit the conversion of 22,25-dideoxyecdysone by the fat body.

A comparison of my results with those of King (1972) and Oberlander (1974) led to two inferences. First, the results obtained by King with *Manduca* and by Marks with *Leucophaea* differed from those obtained by Oberlander when he used *Plodia* more because of the differences in the tissues involved than because of the species of insect. Apparently, *Manduca* and *Leucophaea* epidermal tissues have hormonal responses that are more like each other than they are like the imaginal disc tissue of *Plodia*. Thus Oberlander *et al.* (1973) showed that *Plodia* wing discs do not metabolize α- to β-ecdysone *in vitro* so if the fat body converted 22,25-dideoxyecdysone only as far as α-ecdysone, the discs would not form cuticle. This situation differs from that with *Manduca* and *Leucophaea* in which the conversion of α- to β-ecdysone can be carried out by either the fat body or the body wall.

The second inference is that it may be possible to exploit the differences in biochemical capacity of fat body and leg regenerate tissue. They could be used to investigate the metabolism of additional ecdysone precursors and could be used to investigate substances that might interfere with the metabolism of such precursors. These possibilities are currently being investigated.

V. References

Kaplanis, J., Robbins, W., Thompson, M., and Baumhover, A. (1969). *Science 166,* 1540.

King, D.S. (1972). *Gen. Comp. Endocrinol. Suppl. 3,* 221.

King, D., and Marks, E. (1974). *Life Sci., 15,* 147.

Marks, E. P. (1973a). *Gen. Comp. Endocrinol. 21,* 472.

Marks, E. P. (1973b). In: "Methods and Applications in Tissue Culture" (P. F. Kruse and M. K. Patterson, eds), pp. 153-156, Academic Press, New York.

Marks, E. P., and Leopold, R. A. (1971). *Biol. Bull. 140,* 73.

Oberlander, H. (1974). *Experientia 30,* 1409.

Oberlander, H., Leach, C., and Tomblin, C. (1973). *J. Insect Physiol. 19,* 993.

Robbins, W., Kaplanis, J., Svoboda, J., and Thompson, M. (1971). *Ann. Rev. Entomol. 6,* 53.

Rose, G., Pomerat, C., Shindler, T., and Trunnel, J. (1958). *J. Biophys. Biochem. Cytol. 4,* 761.

Chapter 19

STAGE AND TISSUE-SPECIFIC HEMOGLOBIN SYNTHESIS IN AN INVERTEBRATE

H. Laufer, G. Bergtrom and R. Rogers

I. Introduction

In any study of comparative biochemistry one is immediately impressed with the variety and ubiquity of the hemoglobins (Hbs). These unique molecules vary considerably in structure and physiology throughout the animal kingdom and in the plant species in which they are found. However, Hbs have a number of features in common. These are an iron protoporphyrin IX prosthetic group, varying degrees of alpha-helical structure in the globin moiety, the typical hydrophobic infolding containing the heme group and, of course, the ability to reversibly bind O_2. These properties may be taken as a definition of Hb.

Hemoglobins are present in quantity in the vertebrates and because they are easily assayed, they are among the most thoroughly investigated proteins. The physiological properties are well characterized and the sites of synthesis are known. Studies of the mechanisms and regulation of Hb synthesis in the vertebrates are contributing much to our understanding of transcriptional and translational processes in general. Despite the intensity of the research in these areas, surprisingly little is known about the function and synthesis of Hb in invertebrates.

This report reflects our concern with the site of Hb synthesis in an invertebrate, the insect *Chironomus*, and developmental factors affecting changing patterns of Hb synthesis. The data which we present are the first which show that Hbs are synthesized from amino acids and that the globins are conjugated with heme derived from porphyrin precursors. They also demonstrate that Hb synthesis in *Chironomus* is a powerful tool with which to study development at the molecular level.

II. Non-vertebrate hemoglobins - general considerations

1. *Distribution and function*

Hbs are widely distributed among non-vertebrate phyla. Perhaps most curious is the occurrence of Hbs in some yeasts and molds as well as in the root nodules of N_2-fixing legumes (Kubo, 1939; Oshino *et al.*, 1973; Keilin, 1953). The function of these Hbs is not clear, though Wittenberg (1974) suggests that Hbs in legumes facilitate O_2 transport into the symbiotic rhizobia (bacteroids), thereby enhancing nitrogenase (N_2-fixing) activity. Yeasts, however, show no changes in respiratory function or cell multiplication rate when stripped of their Hb by ethyl hydrogen peroxide treatment.

Among the invertebrates, Hbs are found in certain Protozoa (Keilin *et al.*, 1953; Smith *et al.*, 1962), Nemathelminthes (Fox, 1955; Fox and Taylor, 1955; Crompton *et al.*, 1963; Young *et al.*, 1973), Annelids (Lankester, 1872; Bloch-Raphael, 1939; Vinogradov *et al.*, 1970; Hoffman and Svedberg, 1933; Mangum, 1970; Seamonds *et al.*, 1970; Padlan *et al.*, 1974; Patel *et al.*, 1963), Molluscs (Fox and Vevers, 1960; Sminia *et al.*, 1972), Arthropods (Fox, 1948; 1955; 1949; Manwell, 1966; Braunitzer *et al.*, 1968), and Echinoderms (Lemberg *et al.*, 1949). For a comprehensive review of the occurrence of Hbs and other porphyrin compounds, see Rimington and Kennedy (1962).

Hbs in the higher invertebrates are found either free in the coelomic fluid (hemolymph) , in blood corpuscles, or as in some insects, in cells associated with tracheae (Fox, 1955; Sminia *et al.*, 1972; Dinulescu, 1932; Lankester, 1872). The function of these Hbs is more difficult to define than in vertebrates which have a closed vascular system. In the latter, the necessity for O_2 transport by Hbs to the various body organs from a specific port of O_2 entry is obvious. Hbs evolved as a mechanism to sequester O_2 at low O_2 tensions. In the absence of a closed vascular system it is necessary to measure O_2 tensions at the level of cells ultimately receiving this O_2 in order to assess the importance of the O_2 transport function of Hb. The obstacles involved in such measurements limit the ability to show experimentally a clear function for invertebrate Hbs. Hb function in invertebrates may in fact be one of O_2 transport. Other possibilities include O_2 storage, or as a metabolic source of porphyrins. In some aquatic diving insects, O_2 may be bound by Hbs in order to maintain a sufficient internal O_2 tension without increasing buoyancy (Miller, 1964).

2. *Biochemical properties*

Numerous techniques have focussed on the characterization of the Hb molecule in animals. The vertebrate hemoglobins (and myoglobin) have been so well studied that a summary of their biochemical properties can be found in many textbooks of biochemistry. In recent years, however, numerous invertebrate Hbs have been isolated and their molecular weight, structure, and other physical properties determined.

The molecular weights of invertebrate Hbs range from 15-16000 in *Chironomus* to 18000-3.00 x 10^6 in some annelids (Svedberg, 1933; Scheler *et al.*, 1959; Shlom *et al.*, 1973; Braun *et al.*, 1968; Thompson *et al.*, 1968). Legume Hb has a molecular weight of 15-∽17000 (Broughton *et al.*, 1971, Wittenberg *et al.*, 1972 and others). The molecular weight of crustacean Hb was estimated at ∽200,000-600,000 (Horne *et al.*, 1974; Hoshi *et al.*, 1966).

The molecular weight of *Chironomus* Hb was originally shown to be 33,000 (Svedberg *et al.*, 1934). Braun *et al.* (1968) demonstrated the presence of Hbs in *C. thummi* with molecular weights of 16,000 and 32,000. Heme (protoporphyrin IX) is bound stoichiometrically as one heme per 15-16,000 molecular weight unit in *Chironomus* (Amiconi *et al.*, 1972). Thus, the basic Hb unit in *Chironomus* is a 15000-16000 molecular weight monomer.

The high molecular weight Annelid Hbs can be separated into components ranging from 12000-52000 molecular weight (Shlom *et al.*, 1973) using mercaptoethanol or SDS. However, not all of the low molecular weight components appear to bind heme. High molecular weight Hbs (>30000) in general probably result from the aggregation of smaller units.

From this brief discussion it is apparent that the quaternary structure of Hbs varies considerably between species. In addition, several types of Hb can co-exist in the same organism, as is the case for *Chironomus* and some Annelids (Braunitzer *et al.*, 1968; Manwell, 1966; Thompson, 1968; Vinogradov *et al.*, 1970; Seamonds *et al.*, 1971). In the latter case, there are at least two Hbs in the coelomic corpuscles in addition to those in the coelomic cavity of *Glycera*. As many as 10-14 different Hbs are known for *Chironomus*, including monomers and dimers. Hbs in *Chironomus* represent up to 90% of the hemolymph proteins (English, 1969).

The tertiary structure of monomeric Hbs from Annelids, *Chironomus* and vertebrates have been compared by x-ray crystallography (Padlan *et al.*, 1974). The results demonstrate a striking similarity in the carbon backbones between all species examined. All of the molecules have the myoglobin- like fold in which the heme group is inserted. In all cases, hydrophobic amino acid residues surround the heme moiety. Such residues are required for O_2 exchange. Without them, O_2 would be rapidly discharged to the surrounding medium.

In spite of the overall similarity of some monomeric Hbs from widely divergent species, the secondary and particularly the primary structures can vary considerably. Thus the alpha-helical content of Vertebrate Hb, sperm whale myoglobin, *Chironomus* Hb and *Glycera* Hb are not identical (Huber *et al.*, 1968; Padlan *et al.*, 1974). Amino acid analyses and sequencing (where available) show that there is up to 93% non-homology of amino acid residues and only two residues are in homologous positions in all Hbs sequenced (Padlan *et al.*, 1974). Even among the numerous Hbs isolated from *Chironomus thummi* hemolymph, there are many dissimilarities in amino acid content and N- and C-termini (Braun *et al.*, 1968).

Apparently certain Key positions in the different Hb molecules are occupied by similar amino acid residues which serve to break or alter the helical structure of the polypeptide chains in just the right places, binding the molecule to assume the overall similarity of tertiary structure. Differences in the hydrophobic groups which seclude the heme from an otherwise hydrophillic environment probably result in the differences in O_2 (and other ligand) binding affinities observed for these Hbs. Other subtle differences in primary, secondary and tertiary structure may determine the ability of Hbs to dimerize or aggregate and the related cooperativity or non-cooperativity of O_2 binding (i.e., the binding of the first O_2 molecule generally facilitates binding of the succeeding molecules in polymeric Hbs (Haurowitz, 1963).

3. Location and source of hemoglobins in non-vertebrates

The synthesis and origin of Hbs has been demonstrated in the root nodules of legumes. Studies of isotopic precursor incorporation into Hbs show that protoporphyrin IX is synthesized by the bacteroid symbiont, while the globin is determined by the plant genome (Dilworth, 1969; Cutting *et al.*, 1971; 1972; Broughton *et al.*, 1971). In yeasts and protozoa, Hb is a component of the cytoplasm. Generally, Hb is found free in the body fluids of invertebrates (Lankester, 1872; Shlom *et al.*, 1973). The coelomic cells of some polychaete annelids also contain Hb (Vinogradov *et al.*, 1970). Cells in the body wall of flatworms appear to contain Hb (Fox, 1955; Fox *et al.*, 1955). Hb-containing cells (e.g., pore cells) are found in Molluscs (Sminia *et al.*, 1972) using Hb-specific staining procedures. Among crustaceans (e.g., *Daphnia*) Hb is found in cells of the nervous system, in eggs and in fat cells (Fox, 1955b). Insect Hbs are sometimes found in groups of cells richly penetrated by tracheae (e.g., *Buenoa*, *Gastrophilus*) as well as in cells in the gut and nervous tissue (e.g., *Chironomus*, personal observation). These "tracheal cells" seem ontogenically to be derived from cells of the fat body (Dinolescu, 1932; Bare, 1928). In addition to these Hbs, Myoglobin has been found associated primarily with Mollusc and Annelid muscles (Fox 1955a; Fanelli *et al.*, 1958; Tentori *et al.*, 1972).

The cytological or histological localization of Hbs in cells and organs of invertebrates has led to considerable speculation regarding the source of these pigments in the organism. The main lines of evidence supporting such speculations stem from observed differences in the physiochemical properties of various invertebrate and vertebrate Hbs. Differences in absorption spectra, molecular weights, ligand binding affinities, etc. suggested that different organisms elaborated their own Hbs, presumably in the tissues where they were localized. However, Hbs and porphyrins are certainly available to all invertebrates from dietary sources. Clearly, in *Rhodnius*, a blood sucking insect, the localization of Hb in the eggs and midgut is not indicative of Hb synthesis by these cells (Wigglesworth, 1943). The presence of Hbs within cells in an invertebrate may suggest a site of synthesis, storage, breakdown, or utilization. It is the purpose of this study to expand a preliminary report (Bergtrom *et al.*, 1975b) on the site of hemoglobin synthesis, assembly and secretion, and the synthesis of stage-specific Hbs in *Chironomus* by the use of short-term tissue and organ culture.

III. Synthesis of insect hemoglobins - site of synthesis

An organ may manufacture a macromolecule from smaller precursor molecules, or it can sequester it in whole or in part from dietary sources. Organ culture experiments were designed to distinguish between these two alternatives in the case of *Chironomus* Hb.

1. Organs and culture procedures

Tissues from larvae in the fourth instar of *C. thummi* were dissected and cultured in Cannon's modified medium (Ringborg & Rydlander, 1971), supplemented with either ^{3}H-δ-amino levulinic acid (δ-ALA, a porphyrin precursor) or a mixture of tritiated L-amino acids (5.050 Ci/mM and 1.3 Ci/mM, respectively). In double label experiments, ^{3}H-δ-ALA and ^{14}C-L-amino acids (5.050 Ci/mM and 49 Ci/mM, respectively) were used. Salivary glands from 35 larvae, or the gut complex (including the malphigian tubules) or fat body from 12 larvae were cultured in 50-75 μl of medium containing the isotopes.

The morphology of the gut complex and salivary glands has been described (Miall and Hammond, 1900; Kloetzel and Laufer, 1969). Fat body is an irregular sheet of cells several layers thick suspended in the body cavity of *Chironomus*. Another component of the larval fat body, lying between the epidermis and the musculature of the body wall, could not be isolated without difficulty and was therefore, not included in the cultures. Fat body cells adhere loosely to each other, and in phase contrast microscopy, are highly vacuolated. Some of these vacuoles are pink while other are yellow, or contain irregular greenish inclusions. This was the major cell type found in the cursory examination of the fat body used in organ cultures. Fat body-like cells were also the major cell component found floating in the hemolymph of *C. thummi*, based on phase contrast microscopy.

The various tissues were each cultured with isotopes for 24 hours. The ability of the fat body, gut complex and salivary gland to synthesize, assemble and secrete Hbs was determined by examining the protein released into the culture medium during the incubation period.

2. Detection of hemoglobin synthesis and secretion

The medium, from which the tissues were removed by low speed (121 x g) centrifugation, was dialyzed against several changes of 0.009M PO_4 (Na) buffer (0.04M NaCl, 0.04% KCN, pH 8.1) overnight to remove unincorporated counts. In some cases, the same result was achieved by Sephadex G-25 gel filtration. The radioactivity of an aliquot of each sample was determined and a known amount (20,000-50,000 cpm) was prepared for gel electrophoresis. Preparation involved the addition of 10 μl fresh unlabelled fourth instar hemolymph as marker, sucrose to increase sample density, and bromophenol blue. Discontinuous acrylamide gels were prepared and run at 125-200V for 2-4 hours according to procedures described elsewhere (Bergtrom *et al.*, 1975a).

Nine hemoglobins were visible as red bands after electrophoresis (Bergtrom *et al.*, 1975a, b). The Hb pattern is comparable to that obtained by Braunitzer *et al.* (1968), except that these workers detect an additional Hb band. The band positions were recorded and the gels were sliced, digested in H_2O_2 at 90°C and counted in Aquasol (or a fluor made from Toluene, PPO & POPOP, and Triton X-100) (Bergtrom *et al.*, 1975a). The positions of the Hb bands and the details of the experiment are presented in the figures.

3. Isotope controls

Hbs were labelled either with ^{3}H-δ-ALA or ^{3}H-amino acids. After acid-acetone treatment of the sample (Bergtrom *et al.*, 1975a) approximately 86% of the label from the ^{3}H-amino acids is found in the protein fraction, while 87% of the label from the ^{3}H-δ-ALA is removed by the acid-acetone. This indicates that there is little or no significant interconversion of porphyrin and polypeptide precursors.

When labelled Hb bands are individually eluted and re-electrophoresed along with cold hemolymph markers, radioactivity is found migrating only with the band originally eluted, and, in smaller amounts, its adjacent bands (Bergtrom *et al.*, 1975a).

This suggests that there is no heme exchange detectable between labelled and unlabelled Hbs.

4. Results demonstrating the site of hemoglobin synthesis in the fourth larval instar

Fat body was cultured with ^3H-δ-ALA and ^{14}C-amino acids for 24 hours. An electropherogram of the medium is shown in Figure 1. The ^{14}C-amino acid incorporation profile demonstrates that the nine Hb bands present in fourth instar hemolymph comigrate with the label. In addition, there are several amino acid peaks between bands 1 and 2. These are presumably non-heme proteins or globins, since the ^3H-δ-ALA profile shows no incorporation in this region. Tritiated δ-ALA is incorporated into the nine Hb bands, however, indicating that the amino acid incorporation into the bands represents globin synthesis. Thus, all nine visible Hbs are synthesized and secreted by fat body in organ culture.

Although these results can be explained by artifacts such as the loss of terminal amino acids during Hb purification or incomplete purification, the labelling technique used here does reveal that there may be at least 2 and 3 components in Hb bands 6 and 4, respectively. In some experiments, as many as 14 δ-ALA-labelled heme proteins migrating with visible red regions of the gel can be detected. Braunitzer *et al.* (1971) suggested that there may be more Hbs than those resolved by electrophoresis or ion exchange chromatography.

Although salivary glands do incorporate amino acids in 24 hours into proteins on gel electrophoresis, none of these proteins incorporate ^3H-δ-ALA (Figures 2a, 2b, on next page). Thus the salivary gland does not synthesize and secrete Hbs. Based on a 24-hour incorporation of both ^3H-δ-ALA and ^3H-amino acids into gut tissue (including the Malphigian tubules) there is a low level of Hb synthesis and secretion. This synthesis amounted to about 5% of the total Hb synthesis by fat body and gut complex from the same larvae. If the fat body between the body wall musculature and the epidermis also synthesized Hb, the Hb synthesized by the gut may be still less than this figure. We have found that the body wall in the absence of the gut complex, salivary glands, and dissectable fat body, is still capable of Hb synthesis.

These experiments demonstrate that a) isolated larval tissues in organ culture synthesize heme and globin from small precursor molecules, b) the heme and globin are assembled and secreted by these tissues, c) the fat body is the major site of synthesis of Hbs in the fourth larval instar and d) there may be more than 9 (or 10) Hbs in *C. thummi* hemolymph separable by acrylamide gel electrophoresis.

IV. Hemoglobins in development

Changing patterns of hemoglobin synthesis during vertebrate development have been recognized. Mouse embryonic hemoglobins, distinct from adult hemoglobins, have been characterized by electrophoresis and chromatography (Fantoni *et al.*, 1967). Fantoni *et al.*, (1968) demonstrated that the synthetic rates of the three distinct embryonic mouse hemoglobins change during fetal development leading to alterations in hemoglobin patterns. The adult alpha-globin chain is also present during fetal life, and contributes to fetal tetrameric hemoglobins (designated E_I, E_{II} and E_{III}). The changeover from embryonic to adult hemoglobins reflects an alteration in the site of hemoglobin synthesis, liver-derived erythroid cells producing adult hemoglobins which replace the yolk sac derived erythroid cells which were responsible for the elaboration of fetal hemoglobins. This changeover occurs about

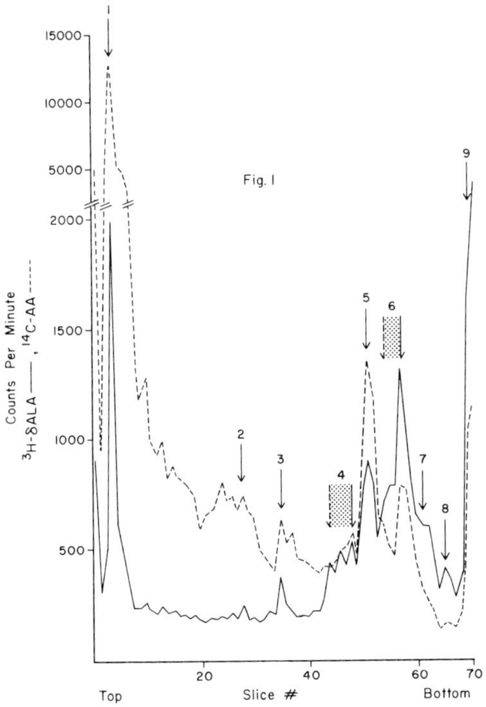

Fig. 1. Labelled precursor incorporation into products of fat body synthesis and secretion in organ culture for 24 hours. Incorporation of ^{3}H-δ-ALA (1 mCi/ml) and ^{14}C-amino acids (0.5 mCi/ml). Fifteen per cent of the dialyzed sample were co-electrophoresed with cold fourth instar hemolymph markers. Positions of the Hb markers are indicated by arrows. (Adapted from Bergtrom *et al.*, 1975).

midway through embryonic development (Rifkind *et al.*, 1969; Moore *et al.*, 1970; Marks *et al.*, 1972; Ingram, 1972). In frogs, the shift from larval to adult hemoglobins also occurs during development (Weber, 1967), a process mediated by thyroid gland hormones.

Chironomids, whose life cycle include a series of aquatic larval stages, contain several electrophoretically distinguishable hemoglobins in the hemolymph (Braunitzer *et al.*, 1968); Laufer and Poluhowich, 1971; Tichy, 1970; Plagens *et al.*, 1970). The electrophoretic pattern and number of hemoglobins present in any given instar is specific and changes with development (Wulker *et al.*, 1969; Manwell, 1966; Schin *et al.*, 1974; Bergtrom *et al.*, 1975b).

1. Procedures for detecting in vivo hemoglobin synthesis

Hemolymph was collected from larvae of *C. thummi* in the mid-third instar and early and mid-fourth instar and diluted 1:1 with KCN-PO_4 buffer (above). Twenty μl of this solution were electrophoresed on acrylamide gels prepared in a slab gel apparatus with ingredients identical to those used in the tube gels in the previous experiments, in order to confirm the stage specificity of Hbs observed by other workers and determine the earliest stage at which changes are detectable.

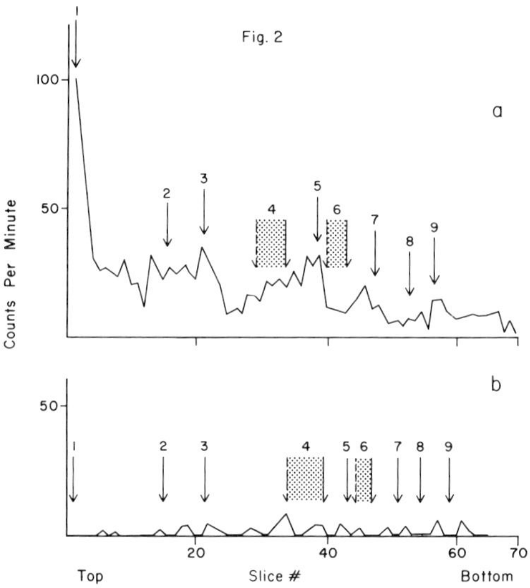

Fig. 2. Labelled precursor incorporation into products of salivary gland synthesis and synthesis and secretion in organ culture for 24 hours.
a) incorporation of ^{3}H-amino acids (1.5 mCi/ml)
b) incorporation of ^{3}H-δ-ALA (1.0 mCi/ml)
In each experiment the total sample, after dialysis, was co-electrophoresed with cold fourth instar hemolymph markers. (Adapted from Bergtrom *et al.*, 1975).

The synthesis of Hbs by third and fourth instar larvae was examined *in vivo*. Thirty and 20 larvae (respectively) were placed in a glass vial (1" in diameter) containing 0.75 µl of tap water supplemented with ^{3}H-amino acids or ^{3}H-δ-ALA and ∽1 mg of food (Laufer and Wilson, 1960) and aerated. The larvae were rinsed thoroughly in isotope-free running water, blotted and bled. The hemolymph collected was centrifuged (12,000 g, 10') to remove cell particulates and debris, and the supernatant dialyzed in the same manner as the organ culture medium samples. An aliquot of sample with 20,000-50,000 cpm (sufficient for electrophoretic resolution) was made dense with a sucrose crystal and electrophoresed. The gels were fixed and sliced for scintillation counting.

To determine if Hbs labelled only in the fourth and not in the third instar were synthesized *de novo*, the larvae still in the third instar after a 24-hour exposure to ^{3}H-amino acids were collected and washable radioactivity was removed by thorough rinsing. The incubation was then continued for 10 days with aeration in approximately 100 ml of isotope-free water containing food. At the end of this "chase" period, larvae had advanced 4-7 days into the fourth instar. These were bled. The hemolymph collected was centrifuged, dialyzed and electrophoresed as described. Fourth instar larvae were subjected to similar "pulse-chase" treatments. In this case the chase period was shortened to five days in order to avoid the onset of metamorphosis.

2. Results demonstrating stage-specific hemoglobin synthesis

The electrophoretic patterns of Hbs collected from larvae in different developmental stages are compared in Figure 3 (schematic electropherogram). Hb in bands 2 and 3 appear in the mid-fourth instar and are not present in sufficient quantity to be detectable in either third instar or "red-head" fourth instar larvae. Hbs in bands 1 and 4-9 are common to all three stages. In gels stained with Coomassie brilliant blue, a faint sharp band from third and early fourth instar hemolymph migrates with an R_f near band 2 Hb from mid-fourth instar hemolymph.

To determine if different stage larvae could synthesize the Hbs found in their hemolymph *in vivo*, larvae were allowed to incorporate ^{3}H-amino acids for 24 hours *in vivo* during the mid-third and fourth instars. Larvae in the fourth instar incorporated label into material migrating with Hb bands 1-9 on electrophoretic gels (Figure 4). Results of similar experiments using ^{3}H-δ-ALA instead of amino acids suggest that fourth instar larvae synthesize all of the visible Hbs found in their hemolymph. In addition, amino acid labelled proteins migrating between Hb bands 1 and 2 are probably not heme proteins.

Fig. 3

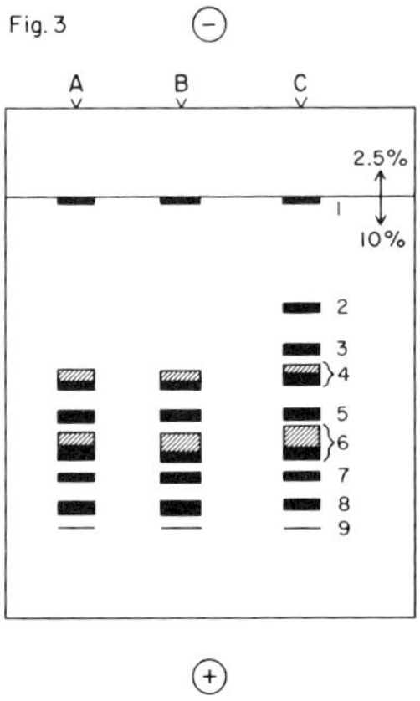

Fig. 3. Slab gel electrophoresis of whole hemolymph from third (A), early fourth (B, "redhead"), and mid-fourth (C) instar larvae. Unstained gels showing resolution of red hemoglobin bands. Note boundary between 2.5% stacking and 10% running acrylamide gels. Hb bands are numbered consecutively from 1 to 9.

When mid-third instar larvae incorporate ^{3}H-amino acids *in vivo*, bands 1 and 4-9 are most heavily labelled (Figure 5). Label is also incorporated into material migrating between bands 1 and 4, which on the basis of experiments with ^{3}H-δ-ALA, are non-heme proteins.

To determine if Hb bands 2 and 3 are synthesized *de novo* in the fourth instar, third instar larvae were fed ^{3}H-amino acids for 24 hours, and then "chased", in the absence of isotope, into the fourth instar. The results (Figure 6) show that only bands 1 and 4-8 (and possibly 9) remain heavily labelled after the chase. Since the larvae were bled in the fourth instar all nine Hbs are present on the gel. In addition, the label in non-heme proteins between bands 1 and 4 is near background levels, suggesting that non-heme proteins are degraded during the chase

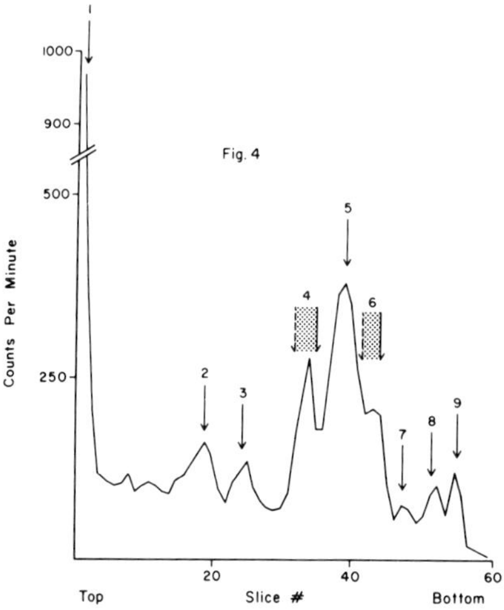

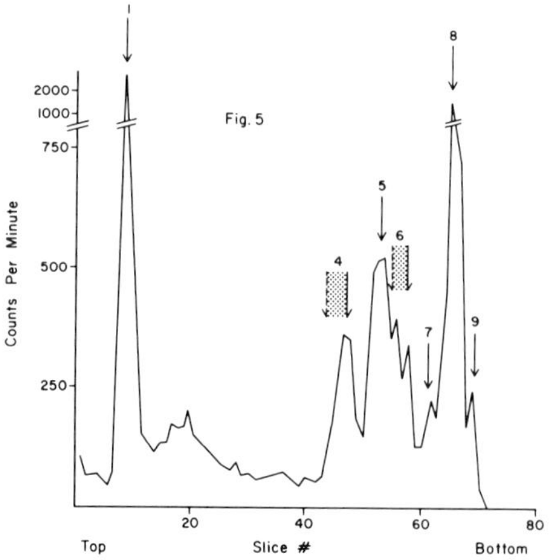

Fig. 4. Disc gels of labelled hemolymph proteins from fourth instar larvae incubated *in vivo* with ^{3}H-amino acids (0.1 mCi/ml) for 24 hours. Incorporation profile of stacking gel is not shown. Approximately 30% of the sample was electrophoresed. Arrows indicate Hb band positions.

Fig. 5. Disc gels of labelled hemolymph proteins from third instar larvae incubated *in vivo* with ^{3}H-amino acids (0.1 mCi/ml) for 24 hours. Approximately 30% of the sample was electrophoresed.

period. Two minor peaks in this region (2X background) may reflect the reutilization of the products of degradation of non-heme proteins in the synthesis of bands 2 and 3. The possibility exists, however, that Hbs 2 and 3 are synthesized in small quantities during the third instar.

Thus far, the results indicate that particular Hbs are synthesized at specific developmental stages of *Chironomus*, and suggest that Hbs 2 and 3, present in sufficient quantity to be detected without staining only in fourth instar larvae, are only synthesized in significant amounts in the fourth instar and are probably "new" proteins, not aggregates or alterations of Hbs present already in the third instar. Further, "pulse-chase" experiments suggest that Hbs synthesized in the third instar are not turned over as rapidly as non-heme proteins synthesized during the same 24-hour pulse. The latter observation was confirmed in pulse-chase experiments conducted on larvae entirely within the fourth instar (Figure 7).

After the 24 hour labelling period in either third or fourth instar larvae, the "chase" periods included at least 5-7 days in the fourth instar. Thus Hbs synthesized during the label incubation in both instars are exposed to fourth instar hemolymph for similar lengths of time. This supports the conclusion that Hbs 2 and 3 are synthesized *de novo* in the fourth instar. Further, amino acid analyses of Hbs 1-9 suggest that each is a unique protein (Braun *et al.*, 1968). However, variations in the hormonal milieu in different periods of the fourth instar affecting the selective or specific aggregation of Hbs 4-9 to result in the appearance of Hbs 2 and 3 cannot be ruled out.

The observation that fat body synthesizes Hb is not unreasonable in light of the fact that insect fat body secretes many blood proteins (Shigmatsu, 1958; Laufer, 1960). Evidence, though inconclusive, suggest that fat cells of crustaceae (e.g., *Daphnia*) may synthesize the heme of Hb (Smaridge, 1956). At least two Hbs (electrophoretic bands 2 and 3) are not synthesized in the mid-third instar, or are synthesized in insignificant amounts. The synthesis of these two Hbs, present in the fourth instar, is presumably induced or stimulated by a factor(s) acting on the site of synthesis. Since their appearance is correlated with the last larval molt, it is possible that ecdysones play a role, direct or indirect, in this process. It is of interest that Hbs are degraded, and Hb synthesis is presumably turned off at metamorphosis (Laufer and Poluhowich, 1971; Schin *et al.*, 1974). These processes may also be mediated by ecdysones.

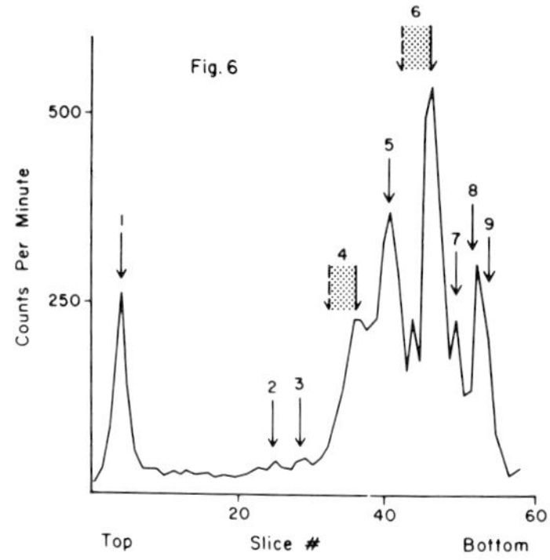

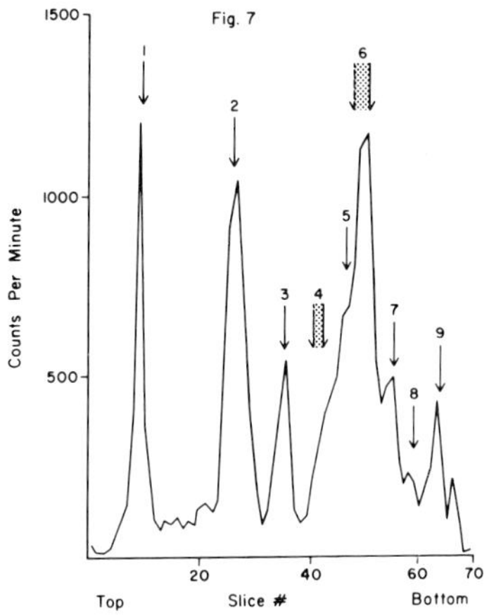

Fig. 6. Disc gels of labelled hemolymph proteins from fourth instar larvae after a 24-hour pulse exposure to ^{3}H-amino acids (0.1 mCi/ml) during the third instar and a 240-hour chase in the absence of isotope.

Fig. 7. Labelled hemolymph proteins from fourth instar larvae after a 24-hour pulse exposure to ^{3}H-amino acids (0.1 mCi/ml) during the fourth instar and a 120-hour chase in the absence of isotope.

V. Discussion and conclusions

The first case of hemoglobin synthesis by an invertebrate is demonstrated both by the *in vivo* incorporation of labelled amino acids and δ-ALA by *Chironomus* and their incorporation into Hbs in organ cultures. Fat body is the major site of Hb synthesis in fourth instar larvae and is capable of synthesizing and secreting all of the electrophoretically distinguishable Hbs normally synthesized by fourth instar larvae *in vivo*. The labelling technique indicates that several of the visible Hbs on acrylamide gels may be made up of 2-3 components. It is alleged (Braunitzer *et al.*, 1971) on the basis of terminal amino acid analyses, that there may be as many as 40 different Hbs in *Chironomus* (see above). Also, after 24 hours of isotope incorporation, the level of labelling differs among the hemoglobins. These differences also vary between experiments, perhaps due to differences in the physiological or developmental state of the larvae. Differences in peak height, may be the result of differential rates of synthesis of the Hbs or may be due to differences in the specific activities of the amino acids in the isotope mixture, or in the globin polypeptides.

The change in the electrophoretic pattern of Hbs in *Chironomus* during development originally demonstrated by Wulker *et al.*, (1969) and Manwell (1966) has been confirmed. Hemoglobin bands 2 and 3 are present only in mid-fourth instar larvae, and are apparently not synthesized in the third instar. "Pulse-chase" experiments suggest that Hb bands 2 and 3 are synthesized *de novo* in the fourth instar. The temporal correlation between the synthesis and appearance of the two additional Hbs and molting from the third to the fourth instar strongly suggests that certain aspects of Hb synthesis may be under hormonal (e.g., ecdysone) regulation.

Low levels of Hb synthesis and secretion by gut tissues can be explained in two ways. The gut may have been active in Hb synthesis at an earlier larval stage, exhibiting only residual Hb synthesis in the fourth instar. Alternatively, there may be a population of cells common to both gut and fat body, but present in greater quantity in the latter. These cells might synthesize Hb. The presence of free floating cells in the hemolymph resembling fat body cells have been noted in *Chironomus*. These or other cell types (e.g., oenocytes) may adhere to various organs in the body cavity. Fat body cells have in fact been found associated with the gut (K. Judy and E. Marks, personal communication).

In view of the presence of Hbs in various invertebrate tissues and cells, it would be useful to apply radioisotope techniques to determine if such cells are sites of Hb synthesis. Of particular interest are the specialized cells derived from fat body in *Castrophilus* (tracheal cells) which contain Hb (Dinolescu, 1932). Similar Hb-containing cells are found in the non-parasitic notonectid *Buenoa* (Bare, 1928). Our findings in *Chironomus* suggest that Hb in these other species may be synthesized by fat body-derived cells.

VI. Acknowledgment

The authors gratefully acknowledge the hospitality of Professors J.E. Edstrom and H.O. Halvorson of the Karolinska Institute and Brandeis University, respectively, where the early phases of this work were initiated. We also acknowledge the technical assistance of Thomas Goralski and Henry Danziger.

This research was supported in part by grants from the NSF, the University of Connecticut Research Foundation, an NIH special fellowship and a NATO senior fellowship.

VI. References

Amiconi, G., Antonini, E., Brunori, M., Formaneck, H. and Huber, R. (1972). *Eur. J. Biochem. 31*, 52.

Bare, C.O. (1928). *The University of Kansas Science Bulletin 18*, 265.

Bergtrom, G., Rogers, R. and Laufer, H. (1975a). *Manuscript submitted for publication to J. Cell Biol.*

Bergtrom, G., Laufer, H. and Rogers, R. (1975b). *Am. Zool. Abstract in press.*

Bloch-Raphaël, C. (1939). *Ann. Inst. Oceanogr. Monaco 19,* 1.

Braun, V., Crichton, R.R. and Braunitzer, G. (1968). *Hoppe-Seyler's Z. Physiol. Chem. 349,* 197.

Braunitzer, G., Glossman, H. and Horst, J. (1968). *Hoope-Seyler's Z. Physiol. Chem. 349,* 1789.

Braunitzer, G., Craig, S., Buse, G. and Plagens, U. (1971). *Limnologica (Berlin) 8,* 119.

Broughton, W.J., Dilworth, M.J. (1971). *Bioch. J. 125,* 1075.

Crompton, D.W.T., and Smith, M.H. (1963). *Nature (Lond.) 197,* 1118.

Cutting, J.A. and Schulman, H.M. (1971). *Biochim. Biophys. Acta 229,* 58.

Cutting, J.A. and Schulman, H.M. (1972). *Biochim. Biophys. Acta 261,* 321.

Dilworth, M.J. (1969). *Biochim. Biophys. Acta 184,* 432.

Dinulescu, G. (1932). *Ann. Sci. Nat. Zool. 15,* 1.

English, D.S. (1969). *J. Embryol. Exp. Morph. 22(3),* 465.

Fanelli, A. Rossi, Antonini, E. and Povoledo, D. (1958). I.U.P.A.C. Symposium on Protein Structure (Nevberger, A., ed.) p. 144, Methuen, London.

Fantoni, A., Bank, A. and Marks, P.A. (1967), *Science 157,* 1327.

Fantoni, A., De la Chapelle, A. and Marks, P.A. (1968). *J. Biol. Chem. 244,* 675.

Fox, H.M. (1948). *Proc. Roy, Soc. B 135,* 196.

Fox, H.M. (1949), *Nature 164,* 59.

Fox, H.M. (1955a). *Proc. Roy. Soc. B 143,* 203.

Fox, H.M. (1955b). *Bull soc. zool. France 80,* 288.

Fox, H.M. and Taylor, A.E.R. (1955). *Proc. Roy. Soc. B 143,* 214.

Fox, H.M. and Vevers, H.G. (1960). In: The Nature of Animal Colours, Sidgwick and Jackson, London.

Havrowitz, F. (1963). In: The Chemistry and Function of Proteins, pp. 256-279, Academic Press, New York.

Hoffman, R.J. and Mangum, C.P. (1970). *Comp. Biochem. Physiol. 36,* 211.

Horne, F.R. and Beyenbach, K.W. (1974). *Arch. Biochem. Biophys. 161,* 369.

Hoshi, T., Sugano, H. and Masuguchi, A. (1966). *Sci. Rep. Niigata Univ. Ser. D 3,* 1.

Huber, R., Formanek, H. and Epp, O. (1968). *Naturwiss, 55,* 75.

Ingram, V.M. (1972). *Nature 235,* 338.

Keilin, D. (1953). *Nature (Lond.) 172,* 390.

Keilin, D. and Ryley, J.F. (1953). *Nature 172,* 451.

Kloetzel, J. and H. Laufer. 1969, *J. Ultrastructure Res. 29,* 15-36.

Kubo, H. (1939). *Acta Phytochim. (Tokyo) 11,* 195.

Lankester, E.R. (1972). *Proc. Roy. Soc. (Lond.) 21,* 70.

Laufer, H. (1960). *Ann. N.Y. Acad. Sci. 89,* 490.

Laufer, H. and Wilson, M. (1970). In: General and Comparative Endocrinology, (R.E. Peter and A. Gorbman, eds.) pp. 135-200, Prentice-Hall Inc., Englewood Cliffs, New Jersey.

Laufer, H. and Poluhowich, J. (1971). *Limnologica (Berlin) 8,* 125.

Lemberg, R. and Legge, J.W. (1949). In: Haematin Compounds and Bile Pigments, Interscience, New York.

Manwell, C. (1966). *J. Embryol. Exp. Morph. 16,* 259.

Marks, P.A. and Rifkind, R.A. (1972). *Science 175,* 955.

Miall, L.C. and Hammond, A.R. (1900). The Structure and Life History of the Harlequin Fly *(Chironomus).* Oxford, At the Clarendon Press.

Miller, P.L. (1964). *Proc. R. Ent. Soc. Lond. (A) 39,* 166.

Moore, M.A.S. and Metcalf, D. (1970). *Brit. J. Haematol. 18,* 279.

Oshino, R., Oshino, N., Chance, B. and Hagihara, B. (1973). *Eur. J. Biochem. 35,* 23.

Padlan, E.A. and Love, W.E. (1974). *J. Biol. Chem. 249,* 4067.

Patel, S. and Spencer, C.P. (1963). *J. Mar. Biol. Ass. U.K. 43,* 167.

Plagens, U., Fittkass, E.J., Jonasson, P.M. and Braunitzer, G. (1972). Deutschen Akadem. Wiss. Berlin, 184.

Rifkind, R.A., Chui, D. and Epler, H. (1969). *J. Cell Biol. 40,* 343.

Rimington, C. and Kennedy, G.Y. (1962). In: Comparative Biochemistry - A Comprehensive Treatise, Vol. IV, (M. Florkin and H.S. Mason, eds.) pp. 558-614, Academic Press, New York.

Ringborg, U. and Rydlander, L. (1971). *J. Cell Biol. 51,* 355.

Scheler, W. and Schneiderat, L. (1959). *Acta Biol. Med. Ger. 3,* 588.

Schin, K.S., Poluhovich, J., Gamo, T. and Laufer, H. (1974). *J. Ins. Physiol. 20,* 561.

Seamonds, B., Forster, R.E. and Gottlieb, A.J. (1971). *J. Biol. Chem. 246,* 1700.

Shigematsu, H. (1958). *Nature 182,* 880.

Shlom, J.M. and Vinogradov, S.N. (1973). *J. Biol. Chem. 248,* 7904.

Smaridge, M.W. (1956). *Quart. J. Microscop. Sci. 97,* 205.

Sminia, T., Boer, H.H. and Niemantsverdriet, A. (1972). *Z. Zellforsch. 135,* 563.

Smith, M.H., George, P. and Preer, J.R. (1962). *Arch. Bioch. Biophys. 99,* 313.

Svedberg, T. (1933). *J. Biol. Chem. 103,* 311.

Svedberg, T. and Eriksson-Quensel, I.-B. (1934). *J. Amer. Chem. Soc. 56,* 1700.

Tentori, L., Vivaldi, G., Carta, S., Marinucci, M., Massa, A., Antonini, E. and Brunori, M. Int. J. Peptide Protein Res. *5,* 187.

Thompson, P., Bleecker, W. and English, D.S. (1968). *J. Biol. Chem. 243,* 463.

Tichy, H. (1970). *Chromosoma 29,* 131.

Vinogradov, S.N., Machlik, C.A. and Chao, L.L. (1970). *J. Biol. Chem. 245,* 6533.

Wever, R. (1967). In: The Biochemistry of Animal Development (R. Weber, ed.) Vol. 2, pp. 227-301, Academic Press, New York.

Wigglesworth, V.B. (1943). *Proc. Roy. Soc. B 131,* 313.

Wittenberg, B.A., Wittenberg, J.B. and Noble, R.W. (1972). *J. Biol. Chem. 247,* 4008.

Wittenberg, J.B., Bergerson, F.J., Appleby, C.A. and Turner, G.L. (1974). *J. Biol. Chem. 249,* 4057.

Wülker, W., Maier, W. and Bertau, P. (1969). *Z. Naturforschg. 246,* 110.

Young, J.O. and Harris, J.H. (1973). *Freshwater Biol. 3,* 85.

Chapter 20

DISSOCIATION AND REAGGREGATION OF FAT BODY CELLS DURING INSECT METAMORPHOSIS

H. OBERLANDER

I. Introduction

Insect fat body undergoes a dramatic transformation during metamorphosis. The connective tissue sheath supporting the fat body is destroyed, and the compact strands of larval fat body become loosely cohesive. Many of the dissociated cells then reassemble to form new compact adult tissue(Walker, 1966; Walters, 1969, 1972).

Reaggregation of dissociated cells has been the subject of scientific inquiry since Wilson published his observations of sponges in 1907. Specific cell aggregation in sponges continues to be studied with *in vitro* technology (Humphreys, 1970). Similarly, the *in vitro* reaggregation of dissociated vertebrate embryonic cells has been investigated intensively (See reviews by Zwilling, 1968; and Roth, 1973). Hence, it seemed likely that dissociation and reaggregation of insect fat body cells should also be amenable to study *in vitro.* Some progress has been made in this direction. Walters and Williams (1966) demonstrated that dissociated pupal fat body cells of saturniid moths reaggregate when cultured in pupal blood. Furthermore, Judy and Marks (1971) observed that fat body adhering to hindgut of *Manduca sexta* dispersed *in vitro.*

In the course of investigating the action of beta-ecdysone on imaginal disks of *Plodia interpunctella* (Hübner) *in vitro*, we noted that "the fat body cells dissociated after several days *in vitro* and surrounded the disks..." (Dutkowski and Oberlander, 1973). Thus, there was an opportunity to study for the first time the effects of hormones on fat body dissociation *in vitro.* This paper reports on the effects of beta-ecdysone and juvenile hormone on the dissociation of *P. interpunctella* larval fat body. In addition, the specificity of the reaggregation of fat body cells on the cultured imaginal disks was examined.

II. Influence of beta-ecdysone

The sensitivity of the fat body to hormone was studied by examining the effects of different concentrations of beta-ecdysone. Fat body was dissected from final instar-larvae (weighed 15 to 18 mg) and cultured (5 sheaths/dish) in 1 ml of modified Grace's medium as previously described (Oberlander *et al.,* 1973). In all experiments the fat body was cultured 24 hours before the addition of hormone. Also, each expe-ment included matched controls to which solvent (10% ethanol) was added. In no case did the control fat body dissociate. The degree of dissociation is based on estimates from examination with a dissecting microscope.

As reported in Table 1, concentrations of hormone as low as 0.005 µg/ml were effective in stimulating fat body dissociation, although concentrations of 0.05 µg/ml or greater had a more pronounced effect (See Fig. 1). Extensive dissociation was not observed for fat body incubated with 0.05 µg/ml (or greater) beta-ecdysone until the second week of culture; a full response to 0.005 µg/ml took three weeks.

TABLE 1

Concentration of Beta-ecdysone and Fat Body (15-18 mg larvae) dissociation.

Concentration beta-ecdysone (µg/ml)	Degree of dissociation* 1 week	2 weeks	3 weeks
0	0	0	0
0.005	+	+	+++
0.05	+	+++	++
0.2	+	+++	+++
0.5	+	+++	+++
5.0	+	++	+++

* + denotes 10-25%,
++ denotes 25-50%,
+++ denotes 50-75%.
Each treatment was replicated 3 times.

The preceding experiments demonstrated that the fat body is sensitive to low concentrations of beta-ecdysone (0.005 µg/ml), although the full response took a considerable time to develop. In view of this latent period, the effects of hormone treatments of limited duration were examined. The minimum effective duration of hormone treatment (0.5 µg/ml) was 15 hours. A 48-hour exposure was required to obtain a pronounced effect, although even 48 hours was less effective than continuous exposure to hormone (See Table 2). When the fat body was cultured with beta-ecdysone longer than 48 hours, it became fragile and broke up during handling.

III. Effect of larval age

A comparison of the degree of dissociation of fat body from final-instar larvae of different ages is shown in Table 3. The fat body was fully competent to dissociate in response to beta-ecdysone at any time during the final larval instar. Furthermore,

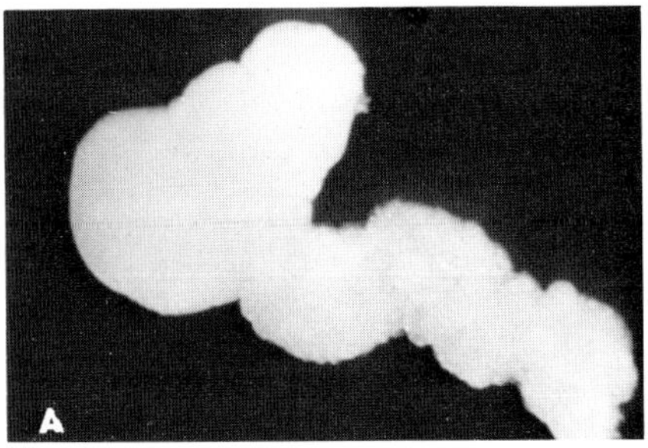

Fig. 1A. A sheath of fat body cultured for two weeks in modified Grace's medium.

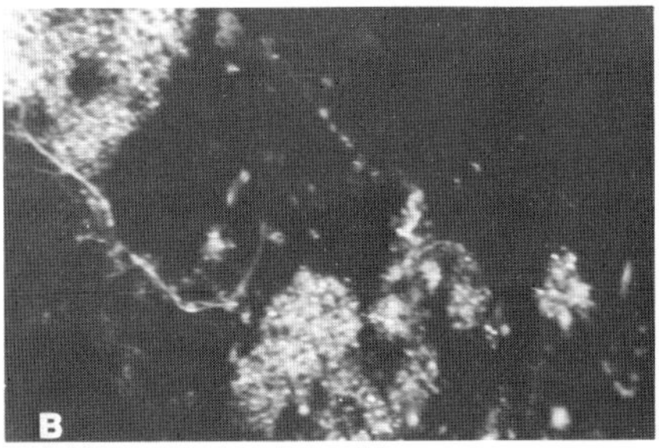

Fig. 1B. Fat body cultured two weeks in medium containing 0.5 μg/ml beta-ecdysone.

TABLE 2

Duration of Exposure to Beta-ecdysone (0.5 μg/ml) and Fat Body (15-18 mg larvae) Dissociation.

Exposure to beta-ecdysone (hours)	Degree of dissociation* 1 week	2 weeks	3 weeks
0	0	0	0
2	0	0	0
6	0	0	0
15	+	+	+
24	+	+	+
48	+	+	++
Continuous	+	+++	+++

* + denotes 10-25%
++ denotes 25-50%
+++ denotes 50-75%
Each treatment was replicated 3 times.

TABLE 3

Effect of Fat Body Age on Beta-ecdysone (0.5 µg/ml)-Induced Dissociation.

Larval Age: Days after egg laying	Degree of dissociation* 1 week	2 weeks	3 weeks
11 (8 - 11 mg)	+	++	+++
12 (12 - 15 mg)	+	++	+++
13 (15 - 18 mg)	+	+++	+++
14 (wandering)	+	++	+++

* + denotes 10-25%

++ denotes 25-50%

+++ denotes 50-75%

Each treatment was replicated 3 times. Matched control cultures of fat body at each age did not dissociate in the absence of beta-ecdysone.

fat body from wandering-stage larvae had not received a sufficient hormonal stimulus *in vivo* because it did not dissociate *in vitro* unless additional beta-ecdysone was present. This is consistent with the timing of fat body dissociation during metamorphosis.

IV. Influence of juvenile hormone

Larval fat body does not dissociate during the larval molt cycle, though fat body from newly molted last-instar larvae is responsive to beta-ecdysone *in vitro*. Experiments were therefore conducted with *Cecropia* juvenile hormone (mixed isomers of methyl-10,11-epoxy-7-ethyl-3,11-dimethyl-2,6-tridecadienoate) to determine whether juvenile hormone is directly responsible for preventing histolysis of larval fat body.

Larvae that weighed 12-15 mg were used as a source of fat body because at this stage the imaginal disks are sensitive to juvenile hormone (Oberlander and Tomblin, 1972). In a control experiment fat body was incubated in culture medium with either juvenile hormone (200 µg/ml) or an equivalent amount of solvent (DMSO). Neither treatment was toxic, nor was dissociation induced. Next, fat body and imaginal wing disks were cultured in a glass microdish containing 200 µl of modified Grace's medium, 12 µg juvenile hormone and 160 µg of binding protein (isolated from *Manduca sexta;* courtesy of Prof. John Law, University of Chicago). [The binding protein enhances the effect of juvenile hormone, probably by retarding its degradation (Sanburg *et al.*, 1975)]. After 24 hours, 0.1 µg beta-ecdysone was added to the cultures. By three weeks the fat body cultured with beta-ecdysone alone had dissociated; there was little or no dissociation in the cultures treated with juvenile hormone. The imaginal disks served as an internal control for the effectiveness of juvenile hormone; 80% produced cuticle in response to beta-ecdysone alone, but none made cuticle in the presence of beta-ecdysone and juvenile hormone.

V. Specificity of fat body adhesion to imaginal disks

Larval fat body that dissociates *in vitro* reaggregates spontaneously at the surface of the liquid along the edge of the culture dish. If imaginal disks are present, they become a focus for fat body reaggregation. The specificity of adhesion of fat body (15-18 mg larvae) to the imaginal disks was therefore assessed by incubating with beta-ecdysone (0.5 µg/ml) and (a) wing disks; (b) wing disks that were fixed for 5 minutes in 70% ethanol; (c) pieces of midgut; or (d) chips of agar. In these experiments the fat body dissociated by the second week and by the third week were adhering to the midgut or normal wing disks. There was no adhesion to the agar or the fixed wing tissue. Hence, dissociated fat body cells will eventually adhere to each other and to other live tissue that may be present as a focus for reaggregation.

VI. Conclusions

The experiments reported in this paper demonstrate that fat body of *P. interpunctella* larvae dissociates in response to beta-ecdysone, but is inhibited from doing so in the presence of juvenile hormone. The developmental behavior of *P. interpunctella* fat body *in vitro* is distinguished from that of wing imaginal disks because full competence to respond to beta-ecdysone is present throughout the last larval instar; in the case of wing disks it appears only at mid-instar(Oberlander and Tomblin, 1972). Therefore, we may conclude that larval fat body does not dissociate until metamorphosis begins because juvenile hormone prevents the fat body from responding to beta-ecdysone, even though it is competent to do so.

The low concentration (0.005 µg/ml) of beta-ecdysone to which fat body responds makes this one of the most sensitive ecdysone responses examined *in vitro*. However, the long exposure required to achieve a full effect indicates that the action of beta-ecdysone on this tissue is consistent with the theory of Ohtaki *et al.* (1968): "covert" effects of the hormone accumulate within the target tissue until threshold is reached for an "overt" response. Thus, the fact that during metamorphosis wing disk development proceeds in advance of fat body histolysis can be explained by the different sensitivities to beta-ecdysone demonstrated in the *in vitro* experiments. Imaginal disks require a shorter exposure to beta-ecdysone *in vitro* than does fat body. Also, disks from wandering larvae begin to metamorphose *in vitro* without additional hormone, but fat body from wandering larvae does not dissociate without ecdysone.

The ultrastructural changes that accompany fat body histolysis *in vivo* have been described (Walker, 1966), but biochemical changes have not been studied. Walters (1974) showed that fat body could be dissociated by treatment with trypsin, pronase or collagenase. However, trypsin digestion was incomplete, and pronase damaged the cells. Collagenase dissociated the fat body completely without damaging the cells. Walters suggested that a tissue collagenase may be responsible for fat body histolysis during metamorphosis. Manifestly, the action of beta-ecdysone on the activity of such enzymes should be investigated. Further, it is possible that the inhibitory action of juvenile hormone on fat body dissociation can be explained in terms of suppressing such enzyme activities.

Walters (1969) also found that reaggregation of fat body cells occured either in the presence of hemocytes or with gentle agitation. In the *P. interpunctella* cultures the fat body was washed several times and cultured without blood. Presumably, few hemocytes accompanied the fat body into culture. The cultures were not agitated, except as necessary for periodic examination. This probably accounts for the slow pace of dissociation and reaggregation observed in the *P. interpunctella* cultures.

An insect plasma factor was required for reaggregation of pupal fat body cells in Walter's (1969) experiments. The factor had some properties of proteins, but various sources of vertebrate proteins did not serve as substitutes. In the *P. interpunctella* cultures insect plasma protein was not required. Since it is well known that the fat body itself adds proteins to tissue culture medium (e.g. Pan *et al.*, 1969), this may account for the difference in observations, since the *P. interpunctella* cultures were initiated with intact larval fat body rather than dissociated pupal cells.

The adhesion of *P. interpunctella* fat body cells to each other or to imaginal disks and midgut *in vitro* in the absence of hemocytes or insect plasma protein makes it possible to study the reconstitution of adult fat body under controlled conditions. Clearly, fat body dissociation and reaggregation can be duplicated *in vitro* and provides a favorable system for studying both hormone action and cell adhesion in insects.

VII. References

Dutkowski, A., and Oberlander, H. (1973). *J. Insect Physiol. 19,* 2155.

Humphreys, T.D. (1970). *Transpl. Proc. 2,* 194.

Judy, K. J. and Marks, E. P. (1971). *Gen. Comp. Endocrinol. 17,* 351.

Oberlander, H., Leach, C. E., and Tomblin, C. (1973). *J. Insect Physiol. 19,* 993.

Oberlander, H., and Tomblin, C. (1972). *Science 177,* 441.

Ohtaki, T., Milkman, R., and Williams, C. M. (1968) *Biol. Bull.* Woods Hole *135,* 322.

Pan, M. L., Bell, W. J., and Telfer, W. H. (1969). *Science 165,* 393.

Roth, S. (1973). *Quart. Rev. Biol. 48,* 541.

Sanburg, L. L., Kramer, K. J., Kezdy, F. J., Law, J. H., and Oberlander, H. (1975). *Nature, 253,* 266.

Walker, P. A. (1966). *J. Insect Physiol. 12,* 1009.

Walters, D. R. (1969). *Biol. Bull.,* Woods Hole *137,* 217.

Walters, D. R. (1972). *Am. Zool. 12,* 102.

Walters, D. R. (1974). *J. Insect Physiol. 20,* 49.

Walters, D. R., and Williams, C. M. (1966). *Science 154,* 516.

Wilson, H. V. (1907). *J. Exp. Zool. 5,* 245.

Zwilling, E. (1968). *Develop. Biol. Suppl. 2,* 184.

III

NUTRITIONAL REQUIREMENTS AND ESTABLISHMENT OF CELL LINES

Chapter 21

THE DEVELOPMENT OF AN INSECT TISSUE CULTURE MEDIUM

G.R. Wyatt and S.S. Wyatt

I. Introduction

From the remarkably diverse and successful applications of insect cell and tissue culture reported at this meeting, it is quite clear that this field of endeavour can now be properly described as having come of age. So it may perhaps be excusable to take this opportunity to describe some experiments done just 21 years ago which, as events have turned out, played a significant role in the embryonic development of insect tissue culture.

II. The establishment of an culture medium.

In the early 1950's, we were both employed at the recently established Laboratory of Insect Pathology of the Canadian Department of Agriculture at Sault Ste-Marie, Ontario, where one of the major projects was research on insect-pathogenic viruses, with a view to possible applications in selective insect control. S.S.W., as a graduate student, undertook the project of attempting to produce insect tissue cultures for use in virus studies. The state of insect tissue culture at the time was quite primitive. Since the pioneering work of Goldschmidt (1916) described by Williams at this meeting, there had been only a few other contributions, mostly not very successful. A remarkable paper for its time, however, was that of Trager (1935). Using fragments of larval silkworm *(Bombyx mori)* ovaries explanted into a simple synthetic medium, Trager had obtained considerable cell outgrowth, including some mitoses. After adding to these cultures some hemolymph from virus-diseased silkworms he observed the infective process, leading to production of large numbers of the polyhedral inclusion bodies typical of this class of virus. By 1951 this highly promising report had never been followed up. S.S.W. succeeded in repeating Trager's culture method, but the outgrowing cells never appeared really healthy. Although their numbers increased for several days, already by 24 hours the cells were granular (by phase optics) and their activity was declining. Many attempts were made to improve these cultures by adjustments in the medium and additions of tissue extracts and protein digests, but initially with quite limited success.

G.R.W. became interested in the project, and we collaborated on the problem of an insect tissue culture medium. We assumed that an appropriate medium should, in its main features, resemble the blood or body fluids of the species being used, so we began to collect information on silkworm hemolymph. To supplement what was in the literature, we performed, with the aid of Dr. Crossley Loughheed, some new analyses (Wyatt, Loughheed and Wyatt, 1956). Then using the available data together with a generous addition of guesswork and trial and error, we formulated a new physiological solution for silkworm tissue culture (Table 1; S.S. Wyatt, 1956).

TABLE I

Physiological solution for culture of tissue from the silkworm (Bombyx mori) (S.S. Wyatt, 1956).

Component	mg/100 ml	mM	Component	mg/100 ml	mM
Inorganic salts			Amino acids		
$NaH_2PO_4 \cdot H_2O$	110	8	L-Arginine HCl	70	3.3
$MgCl_2 \cdot 6H_2O$	304	15	DL-Lysine HCl	125	6.9
$MgSO_4 \cdot 7H_2O$	370	15	L-Histine	250	15.7
KCL	298	40	L-Aspartic acid	35	2.63
$CaCl_2$	81	7.2	L-Asparagine	35	2.65
			L-Glutamic acid	60	4.08
Sugars			L-Glutamine	60	4.11
Glucose	70	3.9	Glycine	65	8.66
Fructose	40	2.2	DL-Serine	110	10.5
Sucrose	40	1.1	DL-Alanine	45	5.05
			β-Alanine	20	2.25
Organic acids			L-Proline	35	3.0
Malic	67	5	L-Tyrosine	5	0.27
α-Ketogluaric	37	2.5	DL-Threonine	35	2.94
Succinic	6	0.5	DL-Methionine	10	0.67
Fumaric	5.5	0.5	L-Phenylalanine	15	0.9
			DL-Valine	20	1.7
			DL-Isoleucine	10	0.77
			DL-Leucine	15	1.14
			L-Tryptophan	10	0.49
			L-Cystine	2.5	0.1
			Cysteine HCl	8	0.5

L-amino acids were used if available; otherwise, DL-forms were used. The solution was adjusted to pH 6.35 and used with addition of 10% of heat-treated silkworm hemolymph. (Some arithmetical errors in the original table have been corrected.).

The inorganic components were based on the excellent systematic analyses of Bialascewicz and Landau (1938). We used their data on hemolymph of fully grown feeding larvae, the stage used for tissue culturing, but lowered the calcium content somewhat to allow for the probability that a proportion of this ion in hemolymph was protein-bound. The balancing anions posed a problem, since the published analyses showed chloride equivalent to less than 20% of the total cation titer and most of the remainder was unidentified. We included a low level of inorganic phosphate, based on analyses, and some sulfate, on the supposition that variety would be a good thing, and made up most of the balance with chloride, for lack of better ideas. Since then it has become known that major anionic components of the hemolymph of higher insects include a number of organic phosphates, bicarbonate and sometimes citrate and other organic acids (Wyatt, 1961). We had an indication from the early work of Tsuji (1909), as cited in the useful tabulations of Yamafuji (1937), that silkworm hemolymph contained substantial levels of several di- and tri-carboxylic acids related to the citric acid cycle. From empirical testing in silkworm ovarian cultures, it appeared that a combination of four of these gave some stimulation of growth, and accordingly this mixture was included in the culture medium.

The sugars presented a problem since there were no published data on sugars in silkworm hemolymph that appeared to be reliable, and, indeed, very few on the hemolymph of any other insect. While some authors had reported glucose from tests that measured reducing power, more critical work by others had shown that the compounds responsible were in fact largely not sugars. We carried out analyses of silkworm blood sugars by paper chromatography and found evidence for glucose, fructose and sucrose, but only at trace levels (Wyatt *et al.*, 1956). [However, determination of total carbohydrate in hemolymph extract by means of the anthrone reaction showed far higher levels, of a glucose derivative which was later identified as trehalose (Wyatt and Kalf, 1958)]. For the formulation of a culture medium, it seemed that some carbohydrate was desirable as a sustained energy source, so we provided what seemed a reasonable level by simply elevating the levels of the three sugars detected on chromatograms to a total of 150 mg/100 ml.

For the amino acids, we obtained fairly satisfactory analytical data by quantitative paper chromatography (Wyatt *et al.*, 1956), and also during the course of the work received the paper of Sarlet, Duchateau and Florkin (1952), reporting determinations of 14 amino acids in silkworm hemolymph by microbiological assay. We formulated a mixture based on our analyses from the late larval stage, with quantities rounded off for convenient weighing. Knowing little about the pathways of protein synthesis and wishing to omit nothing that might be essential, we included both cystine and cysteine, as well as tryptophan, which the chromatograms from hemolymph did not show. The chromatograms did, however, regularly reveal β-alanine, so we included this in the medium, having no idea of its function.

The mixture was adjusted to pH 6.35, corresponding to our best estimate for silkworm hemolymph, and had an osmotic pressure (measured as freezing point depression of 0.53°C) close to reported values for silkworm blood.

To provide for unknown needs of the cells, it seemed appropriate to add a proportion of silkworm hemolymph itself, as Trager (1935) had done. This confronted us with the problem of the phenol oxidase activity, well known in insect blood, which, upon exposure to air, leads to the production of melanin and various toxic intermediates. Of several compounds tested to inhibit phenoloxidase, only phenylthiourea gave permanent inhibition, but there was an indication that this substance might itself be harmful to the cells. Eventually, taking a cue from the experiments of Levenbook (1950) on *Gastrophilus* hemolymph, we found that heating of the hemolymph to 60°C for 5 minutes caused coagulation of some protein including the phenoloxidase, which could then be removed by centrifuging. Heat-treated silkworm hemolymph was added to the medium, usually at a concentration of 10%.

Of course a new culture medium was not formulated in a single step. But it is our recollection that the first experiments with a medium designed on the basis of the composition of hemolymph showed marked and obvious improvement over those done previously. From then on, modifications were readily reflected in the appearance and number of the cells, so that progressive improvement became possible. Under the

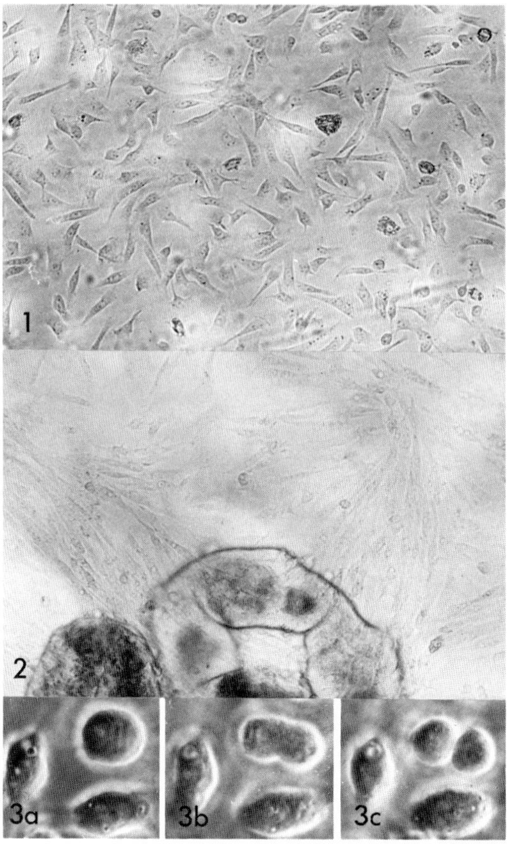

Fig. 1-3. Cell growth in hanging-drop cultures of larval silkworm ovarian tissue, printed from the negatives of S.S. Wyatt (1956). Phase contrast.
Fig. 1. Typical growth of dispersed cells, 4 days after explanation of ovarian fragments. x168.
Fig. 2. Less usual type of out-growth of cells in contact; a portion of an ovariole can be seen. 5 days. x175.
Fig. 3. Mitosis in a 4-day old culture. x735. *a*, 0 min; *b*, 9 min; *c*, 20 min.

conditions finally achieved in this project, outgrowth continued in hanging drop cultures for as long as a week and in small roller tube cultures, in which the medium could be changed, for more than 2 weeks. The cells appeared healthy, with clear cytoplasm and frequent mitosis. Figure 1 shows a hanging drop culture at 4 days of development, with the migrating, separated, fibroblast-like cells that were regularly obtained. Figure 2 shows a slower-growing continuous outgrowth from the explant which was observed in occasional cultures. Figure 3 shows three stages of mitosis in a 4 day culture.

In 1954, quite soon after these promising results were obtained, we both left the laboratory where this work was done. S.S.W. turned to raising a family, and G. R.W. to studies on insect biochemistry, beginning with some problems of hemolymph chemistry that arose from the work just described. There was no opportunity for further refinement of the tissue culture medium, or serious attempts at long-term culturing or establishment of a cell line.

As is well known, these latter problems were soon taken up by others. Tom Grace, in Australia, improved the medium, by the addition of group of water-soluble vitamins and some adjustment of ionic ratios, pH and osmotic pressure (Grace, 1962;1967). In 1962 he reported the landmark step of the establishment of the first continuously growing insect cell line, from the Australian wild silkmoth *Antheraea eucalypti*, soon to be followed by others. Imogene Schneider (1964, 1966) modified the medium for use with *Drosophila* by alteration of the sodium: potassium ratio to make sodium the dominant cation, as in the hemolymph of flies. Marks and Reinicke (1965; Marks, 1973) from the same basis developed media suitable for culturing tissues of cockroaches and other Orthoptera. Many others have since contributed to the development and modification of culture media suited to a wide variety of insect cells (Vago, 1971).

Reviewing the various insect tissue culture media that have been described, it is interesting to see how the inheritance from our early experiments has been passed down. Among the amino acids, for example, we included β-alanine in the medium simply because we found it in silkworm hemolymph. β-Alanine is, of course, not one of the amino acids of proteins and is unlikely to assist the growth of insect cells; more recently, it has been shown to have a role in the tanning of insect cuticle (Bodnaryk, 1971). Yet it was included in the media of Grace (1962), Schneider (1966), and Shields and Sang (1970), among others Landureau (1966), included β-alanine in his first culture medium for cockroach tissues, but, after failing to find evidence for utilization of it, omitted it from subsequent modifications (Landureau and Jollès, 1969; Landureau and Grellet, 1972). Now, at this meeting, Mitsuhashi (1976) has shown that β-alanine may even have some detrimental effect on the growth of cell cultures from butterfly pupal ovaries. It would seem that β-alanine should henceforth be omitted from insect tissue culture media.

The proportions of the protein amino acids that we used, which are continued in Grace's (1962) medium, were based on the analysis of silkworm hemolymph, but they may not be necessary or optimal for insect cell growth. These proportions have been varied by recent workers; Marks and Reinecke (1965) and Landureau and Grellet (1972) developed formulations for cockroach tissue culture with more nearly equimolar quantities of the different amino acids. That the proportions of the amino acids are not at all critical is also suggested by the success with which lactalbumin hydrolysate has been used in many culture media. One must, however, be on guard for the effects of interactions between medium components. Histidine, for example, the most abundant amino acid in our original formulation, is an effective chelator for divalent cations, and its presence in large amounts may help to prevent the precipitation of the rather insoluble phosphates of calcium and magnesium. Lowering of histidine levels should, therefore, perhaps be accompanied by reduction of the levels of these ions; alternatively, the inclusion of citrate, which is often abundant in insect hemolymph (Levenbook and Hollis, 1961) and also forms complexes with them, might be valuable. The mixture of Krebs cycle acids which we introduced, or some variant of them, has been included in many insect tissue culture media, and as there is some evidence that they improve cell growth (Wyatt, 1956) continuation of this practice may be justified.

The three sugars, glucose, fructose and sucrose, added to our original medium on the slender grounds that have been described, are still included in the medium of Grace (1972) and some others. After trehalose was discovered as the major blood sugar of most insects, it was included in some insect tissue culture media, for example, that of Schneider (1966) and Landureau (1966). While this seemed reasonable, and trehalose can support the growth of insect cell cultures (Stockdale and Gardiner, 1976), it has never been shown to have any superiority over glucose in this role. Sucrose is probably generally not utilized by insect cells (Stockdale and Gardiner, 1976); its consumption by *Antheraea eucalypti* cells reported by Clements and Grace (1967) may perhaps have been due to the presence of some sucrase in the hemolymph included in their culture medium. It seems reasonable, therefore, to include merely glucose as an energy source, as Landureau and Grellet (1972) have done, while sucrose may be added if a relatively inert ingredient is required for adjustment of osmotic pressure.

With regard to the proportions of inorganic cations, our assumption that these should resemble those found in the hemolymph of the species being cultured seems in general to be born out. Whereas lepidopteran cells flourish in high-potassium media, cockroach tissues prefer media with a sodium:potassium ratio greater than 1 (Ting and Brooks, 1965; Marks and Reinecke, 1965). Very little effort, however, has been put into systematic testing of the inorganic components, and we are still ignorant of the optima or limits for growth of various types of insect cells.

Now that insect tissue culture has become established, through developments during the past 20 years, as a successful and valuable technique, it is encouraging to see some workers undertaking the critical re-examination of the composition of culture media (e.g. Landureau and Grellet, 1972; Mitsuhashi, 1976; Hinks, 1976). Such efforts may be expected to yield media which are more rationally composed, simpler, less expensive and at least as successful as those currently in use.

III. References

Bialascewicz, K., and Landau, C. (1938). *Acta Biol. Exp. 12*, 307.

Bodnaryk, R.P. (1971). *J. Insect Physiol. 17*, 1201.

Clements, A. and Grace, T.D.C. (1967). *J. Insect Physiol.*, *13*, 1327.

Goldschmidt, R. (1916). *Biol. Zentr. 36*, 161.

Grace, T.D.C. (1962). *Nature 195*, 788.

Grace, T.D.C. (1967). *In Vitro 3*, 100.

Hink, W.F., and Strauss, E. (1976). In: Invertebrate Tissue Culture: Application in Medicine, Biology and Agriculture (E. Kurstak and K. Maramorosch, eds), Academic Press, New York.

Landureau, J.C. (1966). *Exp. Cell Research 41*, 545.

Landureau, J.C., and Grellet, P. (1972). *C.R. Acad. Sci., Ser. D. 274*, 1372.

Landureau, J.C., and Jolles, P. (1969). *Exp. Cell Research 54*, 391.

Lenvenbook, L. (1950). *Biochem. J. 47*, 336.

Levenbook, L. and Hollis, W.W. (1961). *J. Insect Physiol.*, *6*, 52.

Marks, E.P. (1973). In: Tissue Culture Methods and Applications (P.J. Kruse and M.D. Patterson, eds), Academic Press, New York, 153-156.

Marks, E.P., and Reinicke, J.P. (1965). *J. Kansas Entomol. Soc. 38*, 179.

Mitsuhashi, J. (1976). In: Invertebrate Tissue Culture: Application in Medicine, Biology and Agriculture. (E. Kurstak and K. Maramorosch, eds), Academic Press, New York.

Sarlet, H., Duchateau, G., and Florkin, M. (1952). *Arch. Internat. Physiol. 60*, 126.

Schneider, J. (1964). *J. Exp. Zool. 156*, 91.

Schneider, J. (1966). *J. Embryol. Exp. Morphol. 15*, 271.

Shields, G., and Sang, J.H. (1970). *J. Embryol. Exp. Morphol. 23*, 53.

Trager, W. (1935). *J. Exp. Med. 61*, 501.

Ting, K.Y., and Brooks, M.A. (1965). *Ann. Entomol. Soc. Amer. 58*, 197.

Tsuji, C. (1909). *Sanji Hokoku 35*, 1.

Stockdale, H. and Gardiner, G.R. (1976). In: Invertebrate Tissue Culture: Application in Medicine, Biology and Agriculture (E. Kurstak and K. Maramorosch, eds). Academic Press, New York.

Vago, C. (ed.) (1971). Invertebrate Tissue Culture. Vol. 1. Academic Press, New York.

Wyatt, G.R. (1961). *Ann. Rev. Entomol. 6*, 75.

Wyatt, G.R., and Kalf, G.F. (1958). *J. Gen. Physiol. 40*, 833.

Wyatt, G.R., Loughheed, T.C., and Wyatt, S.S. (1956). *J. Gen. Physiol. 39*, 853.

Wyatt, S.S. (1956). *J. Gen. Physiol. 39*, 841.

Yamafuji, K. (1937). Tabulae Biologicae (W. Junk, C., Oppenheimer and Weisbach, eds). W. Junk, The Hague, Vol. 14, pp. 36-50.

Chapter 22

INSECT CELL LINE: AMINO ACID UTILIZATION AND REQUIREMENTS

J. Mitsuhashi

I. Introduction

Recently synthetic media have been introduced for culturing insect cells. The nutritional requirements of insect cells *in vitro*, however, have been studied only to a limited extent. The utilization of amino acids by insect cell lines has been examined in cultures of the *Antheraea eucalypti* (Ae) (Grace and Brzostowski, 1966), *Periplaneta americana* (EPa) (Landureau and Jollès, 1969), and *Carpocapsa pomonella* cell lines (CP 1268 and CP 169) (Hink *et al*, 1973). The sugar utilization has been studied in Ae (Grace and Brzostowski, 1966; Clements and Grace, 1967), and CP (Hink *et al*, 1973). The vitamin requirements have been studied only in EPa (Landureau, 1969). To know the nutritional requirements of a certain cell line is not only profitable for the improvement of the culture medium but also important for the characterization of the cell line. In the present study, the utilization and requirements of amino acids by the *Papilio xuthus* cell line (Px 58) were investigated.

II. Materials and methods

A mixed cell population of Px 58 cell line, which has been established from pupal ovaries of the swallow tail, *Papilio xuthus* (Mitsuhashi, 1973), was used. The stock colony has been maintained in MGM-431 medium (Mitsuhashi, 1972) at 25°C by subculturing twice a week.

For the quantitative determination of amino acid changes in the medium, replicate tissue cultures were set up in roller tubes. Five ml of the cell suspension, containing about 1.0×10^5 cells per ml, was distributed to each tube. The cultures were maintained at 25°C. Sampling was made on the 0, 4th and 8th days. At one time, 3 tubes were harvested, and the amino acid analysis was made for each tube. In order to prepare the samples for the analysis, Hink *et al's* procedure (Hink *et al*, 1973) was adopted. First the cells were spun down at 2,000 r.p.m. for 10 minutes. Four ml of the supernatant were mixed with an equal volume of 70% ethyl alcohol. This mixture was kept at 5°C for 2 hours. Then the precipitate was removed by centrifugation at 3,500 r.p.m. for 15 minutes. The supernatant was slightly turbid, and was passed successively through membrane filters of 1.2 µm, 0.6 µm, and 0.45 µm. The resulting filtrate was diluted with lithium citrate buffer, pH 2.73. The final dilution rate of the sample was 50 times. For the amino acids which were in small amounts in the medium, only 5 times diluted samples were prepared. The analysis was made by Na-Li two column method in a Japan Electron Optics Laboratory 6AH automatic amino acid analyser.

For the determination of the essential amino acids for the growth of Px 58 cells, experiments were made in which amino acids were deleted from MGM-431 medium one at a time. By the replicate tissue culture method the cell number was recorded at 0 day and the 8th day. The replicate tissue cultures were prepared by distributing 2 ml of cell suspension (1.5×10^5 cells per ml) into each roller tube. The culture was kept at 25°C for 8 days without changing the culture medium. At the beginning and at the end of cultivation, 3 tubes were sampled, and the cell enumeration was done by the use of Thoma's hemocytometer. The growth rate of the cells in each test medium was expressed as the number of cells after 8 days of cultivation.

III. Results

1. Utilization of amino acids by the cell line.

During 8 days of cultivation, the cell number increased as shown in Fig. 1, and most of the amino acids in the medium decreased (Fig. 2). The rate of the decrease, however, varied in different amino acids. Lysine, histidine, tryptophan, arginine, threonine, serine, glutamic acid, and β-alanine, did not decrease markedly. The amino acid which showed the greatest decrease was glutamine. Asparagine, cystine, aspartic acid, methionine, tyrosine, proline, leucine, isoleucine, and valine decreased 10 - 40 per cent during 8 days cultivation. α-alanine was the only amino acid which increased during cultivation. In addition to the amino acids, ammonia was noticed to increase during cultivation.

2. Requirements of amino acids by the cell line.

MGM-431 media lacking one amino acid were prepared. Since MGM-431 medium contained 10 per cent fetal bovine serum, and the serum contained free amino acids, it was impossible to delete any one of the amino acids completely. An analysis, however, showed that the free amino acids derived from fetal bovine serum were almost negligible in MGM-431 medium (Table 1).

The growth rate of the cells in the media lacking asparagine, α-alanine, glutamic acid, glycine, and phenylalanine were not significantly different from that in the complete MGM-431 medium (Fig. 3). In the medium lacking β-alanine, the cells grew much faster compared with the control. Deletion of the other amino acids resulted in the reduction of the growth rate of the cells compared with the control. Among them, the deletion of arginine, aspartic acid, cystine, and tryptophan still permitted limited growth of the cells, while the deletion of glutamine, histidine, isoleucine, leucine, lysine, methionine, proline, serine, threonine, tyrosine, and valine stopped the cell growth completely or even killed the cells.

The long-term cultivation of Px 58 cells was performed with media lacking one of amino acids. In the media lacking asparagine, α-alanine, β-alanine, glutamic acid, glycine or phenylalanine, the cells could be subcultured for an infinite period. These amino acids could be removed altogether without deleterious effect on the

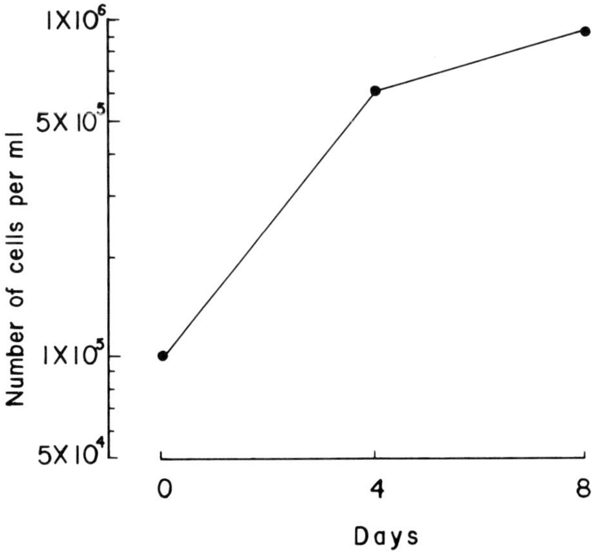

Fig. 1. The growth curve of the Px 58 cells during 8 days of experimental culture. Medium, MGM-431. Temperature, 25°C.

TABLE 1

Composition of amino acids, added to MGM-431 medium, and free amino acids derived from fetal bovine serum added to MGM-431 medium (mg per liter MGM-431 medium).

Amino acids	A.M.*	F.B.S.**	Amino acids	A.M.*	F.B.S.**
α-alanine	263	13.7	Leucine	63	5.7
β-alanine	167	trace	Lysine	521	3.3
Arginine	583	nil	Methionine	42	0.2
Aspartic acid	293	0.9	Phenylalanine	125	3.1
Asparagine	293	nil	Proline	292	2.9
Cystine	21	nil	Serine	917	4.7
Glutamic acid	500	14.7	Threonine	146	3.7
Glutamine	500	5.5	Tryptophan	83	nil
Glycine	542	8.4	Tyrosine	42	2.0
Histidine	2083	0.8	Valine	83	5.5
Isoleucine	42	1.6			

* amino acids in MGM-431 medium.

** free amino acids derived from fetal bovine serum (Microbiological Associates Inc., Bethesda, U.S.A.). (After Mitsuhashi, 1976).

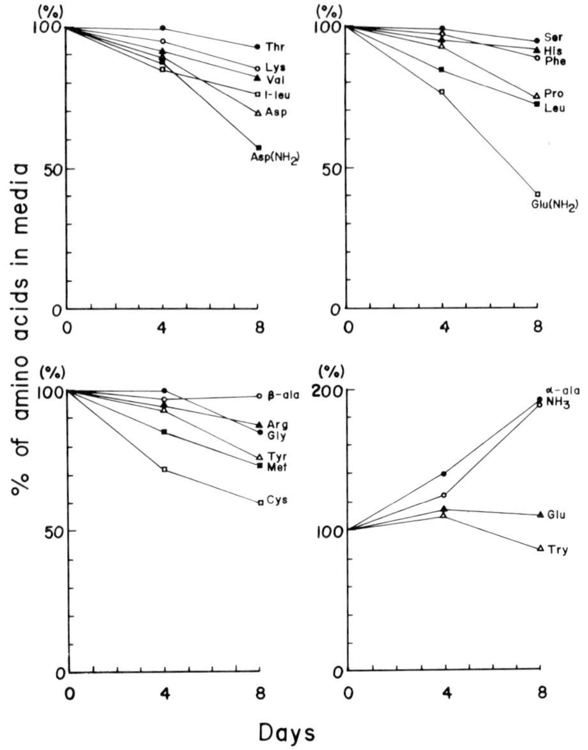

Fig. 2. Quantitative changes in amino acids in MGM-431 medium during 8 days cultivation of Px 58 cells.

cell growth. In media lacking other amino acids, the cells deteriorated sooner or later. The cells, however, could adapt to the medium lacking serine. In this medium the cells stayed alive for a long time, gradually multiplied, and finally the growth rate of the cells became comparable to the control. In the medium lacking glutamine, the cells attached themselves to the glass, and stayed alive for several months, but apparently did not multiply.

IV. Discussion

The utilization of amino acids by insect cell lines has been reported in Ae cells (Grace and Brzostowski, 1966), EPa cells (Landureau and Jollès, 1969), and CP cells (Hink *et al*, 1973). In the present study, it became evident that the use of amino acids by Px 58 cells followed different patterns from those reported before. In the culture of Px 58 cells, most of the amino acids decreased as the cultures increased in age. The decreasing pattern of each amino acid, however, varried.

The three basic amino acids, lysine, histidine, and arginine in MGM-431 medium did not decrease markedly. This does not mean that these amino acids are not important. If any one of these amino acids was deleted, the growth of the cells was severely impaired. These three amino acids, therefore, are essential for the

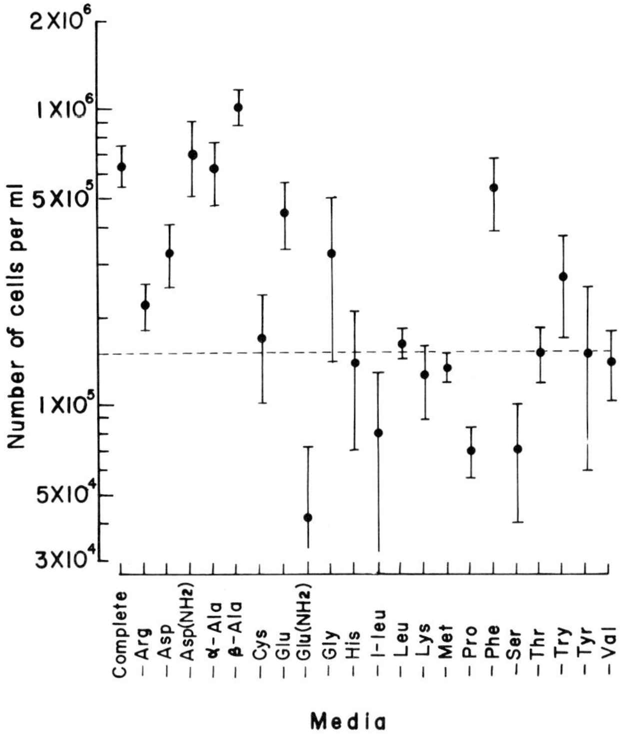

Fig. 3. Number of cells after 8 days cyltivation of Px 58 cells in MGM-431 media lacking oneamino acids. (After Mitsuhashi, 1976).

growth of the cells. In addition to the basic amino acids, threonine, serine, glycine, valine, phenylalanine, β-alanine, tryptophan, and glutamic acid decreased only slightly. Among these amino acids, threonine, serine, valine and tryptophan were found to be essential while glycine, phenylalanine, β-alanine, and glutamic acid were found to be non-essential.

Glutamine, asparagine, and cystine decreased considerably in the growing culture. Glutamine and cystine were found to be essential, while asparagine was non-essential. Following this group, aspartic acid, methionine, tyrosine, isoleucine, leucine, and proline were consumed. All of these amino acids were found to be essential.

The common patterns of amino acid changes in growing cell cultures, which can be summarized from the data on the four cell lines hitherto studied, are the increase of α-alanine, and the marked decrease of aspartic acid, cystine, methionine, proline, and tyrosine. (Grace and Brzostowski, 1966; Landureau and Jollès, 1969; Hink *et al*, 1973).

On the other hand a characteristic difference in the Px 58 cell line was found in regard to glutamic acid. In the cultures of Ae, EPa and CP cell lines, glutamic acid decreased markedly as the cultures increased in age. In the Px 58 cell culture, however, glutamic acid increased at the early stage of the cultivation, and then decreased to some extent still keeping an increased level at the end of 8 days cultivation. Deletion of glutamic acid from MGM-431 medium did not impair the growth of the cells. Therefore, glutamic acid may be non-essential for Px 58 cells or it may be required in a very small amount. When the glutamic acid was deleted from the medium it nevertheless contained a very small amount of glutamic acid derived from the fetal bovine serum (Table 1). On the other hand, glutamine was the most used amino acid, and was found to be essential for the growth of Px 58 cells. Deletion of glutamine stopped cell growth, which suggests that glutamic acid was not converted to glutamine in the cultures.

α-alanine accumulated considerably over the period of cultivation. This accumulation might be the product of degradation of other amino acids or transamination of pyruvic acid which was produced by the metabolism of the cells.

Glutamic acid and tryptophan increased during the first few days, and then decreased. The significance of this increase is obscure. In CP cell cultures, most of the amino acids were reported to increase during the first few days of cultivation (Hink *et al*, 1973).

It is evident that different cell lines have different patterns of amino acid utilization and requirements. Hink *et al* (1973) showed that the patterns of amino acid utilization were different even between two cell lines, CP 1268 and CP 169, derived from the same tissue of the same species.

The different nutritional requirements of the cell lines suggest that the culture medium should be formulated individually for each cell line in order to get the best cell growth. This may be true in primary cultures also. At present, some media are used widely even for insects of different orders. To get successful outgrowth in primary cultures, the media may have to be formulated especially for the tissues concerned.

V. References

Clements, A.N. and Grace, T.D.C. (1967). *J. Insect Physiol.* *13*, 1327.

Grace, T.D.C. and Brzostowski, H. (1966). *J. Insect Physiol.* *12*, 625.

Hink, W.F., Richardson, B.L., Schenk, D.K. and Ellis, B.J. (1973) *Proc. 3re Intern. Colloq. Invert. Tissue Culture, Smolenice 1971*, 195.

Landureau, J.C. (1969). *Exp. Cell Res.* *54*, 399.

Landureau, J.C. and Jollès, P. (1969). *Exp. Cell Res.* *54*, 391.

Mitsuhashi, J. (1972). *Appl. Ent. Zool.* *7*, 39.

Mitsuhashi, J. (1973). *Appl. Ent. Zool.* *8*, 64.

Mitsuhashi, J. (1976). In: Invertebrate Tissue Culture. (K. Maramorosch, ed.) Academic Press, New York, In press.

Chapter 23

A COMPARISON OF AMINO ACID UTILIZATION BY CELL LINES OF *CULEX TARSALIS* AND OF *CULEX PIPIENS*

J. Chao and G.H. Ball

I. Introduction

An analysis of the amino acids in a culture medium during cell growth may offer an insight to the requirement of cells for these compounds. Igarashi *et al.*, (1973) found serine and, to a lesser degree, proline were required by Singh's *Aedes albopictus* cells growing in Eagle's minimal essential medium supplemented with 10% calf serum. They also found that "non-essential" amino acids stimulated growth over a 7-day period of culture. Grace and Brzostowski (1966) demonstrated that cells of the moth, *Antheraea eucalypti* used 14 of 21 amino acids over 7 days, while the concentration of 7 others either increased or remained unchanged. Hayashi and Sohi (1970) showed that Grace's *Aedes aegypti* cell line, now considered to be *Antheraea eucalypti* cells (Greene *et al.*, 1972), incorporated leucine, the only amino acid analyzed, over a 5-hr period.

Landureau and Jollès (1969) showed that embryonic *Periplaneta americana* cells required 15 amino acids for optimum growth. Řeháček and Brzostowski (1969) reported the utilization of amino acids by cells of the tick *Rhipicephalus sanguineus* based on an analysis of the culture medium by means of an amino acid analyzer. Stanley (1972) reviewed growth, nutrition, and metabolism of arthropod cells in culture including conflicting results of various investigators on the utilization of amino acids.

Our *Culex pipiens* and *Culex tarsalis* cell lines (Chao and Ball, 1973, 1975) have been adapted to grow in Hsu's, Schneider's, or Singh's medium. Both lines could also be grown in these media without the addition of FBS for at least two serial transfers. We thought that an analysis of amino acids at a few chosen points of cell growth would offer a first step to the understanding of the utilization of amino acids by the two mosquito cell lines. Amino acid metabolism, however, has to be studied by other techniques such as the use of radioisotopes.

II. Materials and methods

C. pipiens and *C. tarsalis* cell lines were isolated from embryonic eggs and were routinely grown in Hsu's, Singh's, or Schneider's medium containing 10% FBS (Chao and Ball, 1973, 1973, 1975). Antibiotics were not used. For amino acid analyses, the concentration of the FBS was reduced to 5%. The analyses were done by a Beckman Model 120 amino acid analyzer. Three samples of 50 μl each were withdrawn from a culture and analyzed at 24 hr, 72 hr, and 7 days after the inoculation of cells. The cell number in an inoculum was 2 x 10^6 and 4 ml of medium was used for each culture. Cultures were grown in small Falcon flasks at 25°C and remained healthy for as long as two weeks. The pH of the media was adjusted to 6.8 and showed very little change during the 7 day period of animo acid analysis.

Amino acid determinations of the stock medium were made before and after each serial run. The average of the two was used for comparison with the readings shown by the cell incubation medium. Analysis of Schneider's medium without FBS was done in the following manner. A culture in Schneider's medium containing FBS was washed three times with serum-free Schneider's medium. An inoculum of 2 x 10^6 cells were transferred to a flask containing 4 ml Schneider's medium without FBS. Three consecutive samples, 50 μl each, were withdrawn from the culture and analyzed for amino acids. The readings were compared with the average of those from the FBS-free Schneider's medium.

III. Results

The amino acids showing changes above 20% in 7-day culture, whether decreasing or increasing, are listed in Table 1 and Table 2. The figure 20% is arbitrarily chosen to exclude lower values which may be due to experimental error and error in estimating the quantities of the amino acids from the graph plotted by the analyzer. Such errors are especially great with amino acids having low readings (small quantities). The difference of each amino acid in the two readings of a stock medium is within 5%. A sweeping conclusion is difficult to draw on the amino acid utilization of the two mosquito cell lines. However, a few general statements can be made with qualifications.

The analyses run at 24 hr, 72 hr, and 7-day culture showed that, with few exceptions, the percentage of change increased with the incubation time, the greatest being at the 7th day. Fewer amino acids showed increase in *C. tarsalis* culture than in *C. pipiens* culture except in Hsu's medium. More amino acids increased in Hsu's or Singh's medium than in Schneider's medium (with or without FBS). Aspartic acid and methionine were used in great amount by both cell lines in all the three media, followed by cystine with the exception of Hsu's medium, in which case, cystine did not appear on the graph and methionine decreased only slightly in *C. tarsalis* culture. The decrease of cystine in *C. pipiens* culture growing in Schneider's medium without FBS was, however, of a low order. Aspagarine and glutamine (not separable, AGN in the Tables) were used by both cell line growing in Schneider's medium but greatly accumulated in Hsu's medium and also in Singh's medium greatly accumulated in Hsu's medium and also in Singh's medium with *C. pipiens* cell.

All media in which *C. pipiens* cells were growing showed increase of glycine. The increase was especially high in Hsu's medium. Proline increased in Hsu's or Singh's medium and the increase appeared high in *C. tarsalis* culture growing in Singh's medium. The total amount of amino acids after 7 days culture increased in

TABLE 1.

Amino Acids Decreasing 20% or more after 7-day Culture

Singh's medium plus 5% FBS

C. pipiens: ASP (87%) MET (57%) CYS (50%)

C. tarsalis: ASP (100%) MET (86%) CYS (70%) GLU (50%) AGN (27%) PHE (22%)

Hsu's medium plus 5% FBS

C. pipiens: THR (91%) MET (85%) ASP (81%) GLU (32%) LEU (21%) PHE (20%)

C. tarsalis: GLU (87%) ASP (85%)

Schneider's medium plus 5% FBS

C. pipiens: CYS (99%) ASP (80%) MET (69%) AGN (32%) GLU (25%) TRY (21%)

C. tarsalis: ASP (100%) CYS (99%) MET (90%) PHE (56%) AGN (29%) βALA (27%) TRY (21%)

Schneider's medium without FBS

C. pipiens: MET (98%) ASP (46%) THR (57%) AGN (29%) PRO (29%) CYS (23%)

C. tarsalis: MET (99%) ASP (97%) CYS (88%) PHE (72%) AGN (35%) SER (22%) GLY (21%)

TABLE 2.

Amino Acids Increased 20% of More after 7-day Culture

Singh's medium plus 5% FBS

C. pipiens: AGN (198%) ALA (123%) GLY (82%) PRO (53%) GLU (48%) SER (48% HIS (42%)

C. tarsalis: PRO (84%) TYR (68%) THR (47%) SER (39%) ALA (27%)

Hsu's medium plus 5% FBS

C. pipiens: GLY (232%) AGN (200%) ALA (113%) PRO (64%)

C. tarsalis: ALA (202%) AGN (89%) GLY (70%) PRO (60%) THR (25%)

Schneider's medium plus 5% FBS

C. pipiens: ALA (202%) GLY (32%) VAL (31%)

C. tarsalis: ALA (173%) TYR (28%)

Schneider's medium without FBS

C. pipiens: ALA (1-35%) GLY (34%)

C. tarsalis: ALA (1650%)

TABLE 3.

Ammonia Concentration Change after 7-day Culture

	Hsu's medium 5% FBS	Singh's medium 5% FBS	Schneider's medium 5% FBS	Schneider's medium without FBS
C. pipiens	-66%	-94%	-63%	+700%
C. tarsalis	-83%	-94%	-91%	+230%

media containing FBS except in *C. tarsalis* culture growing in Singh's medium, in which case the amount remained about the same as in the control medium. In Schneider's medium without FBS, the total amount of amino acids decreased.

The change of ammonia concentrations in media after 7 days culture is shown in Table 3. It decreased in media containing FBS but increased in Schneider's medium without FBS.

IV. Discussion

Our results showed that the pattern of amino acid utilization by cells of *C. pipiens* and *C. tarsalis* depends on: 1, the kind of cell culture tested; 2, media in which the cells were grown; and 3, the exclusion of a source of non-defined amino acids such as FBS. A 4th factor which is easy to realize but was not entirely demonstrated in our experiments would be the "well being" of the culture, its age, and cell number at the time of analysis. The difference in these and other factors would undoubtedly explain the conflicting results of the various investigators on the amino acid utilization of arthropod cells in culture as reviewed by Stanley (1972).

The media used in this study differ not only qualitatively and quantitatively in respect to amino acids but differ also in the inorganic ion concentrations. For example, the concentrations of NaCl and KCl 210 mg, KCl 160 mg in Schneider's medium; and NaCl 700 mg, KCl 20 mg in Singh's medium. It is conceivable that these differences would greatly alter amino acid transport and consequently metabolism.

The results from this study may serve as a guide for the improvement of the media in growing these two mosquito cell lines.

Acknowledgements

The authors wish to express their appreciation to Miss June Baumer for her skill in the use of the amino acid analyzer. The research was supported by Grant AI-00087 from NIAID, U.S. Public Health service, by Research Grant 254, Zoology, University of California, and by a Bio-medical Science Support Grant to University of California. Minor support was provided by NIH Contract 72-2527 to Dr. C. J. Bayne. Thanks are also due to Dr. J. R. Allen for his critical reading of the manuscript.

V. References

Chao, J., and Ball, G.H. (1973). *Int. Colloq. Invertebr. T.C., 3rd,* pp. 93-104, *Slovak Acad. Sci.,* Czechoslovakia.

Chao, J., and Ball, G.H. (1973). *In Vitro, 8,* 406.

Chao, J., and Ball, G.H. (1975). *Inv. Conf. Invertebr. T.C., 4th,* Mont Gabriel, Québec, Canada.

Grace, T.D.C., and Brzostowski, H.W. (1966). *J. Insect Physiol. 12,* 625.

Greene, A.E., Charney, J., Nichols, W.W., and Coriell, L.L. (1972). *In Vitro 7,* 313.

Hayashi, Y., and Sohi, S.S. (1970). *In Vitro 6,* 148.

Igarashi, F., Sasao, R., and Fakai, K. (1973). *Biken J. 16,* 95.

Landureau, J.D., and Jollès, P. (1969). *Exp. Cell Research 54,* 391.

Rehacek, J., Brzostowski, H.W. (1969). *J. Insect Physiol. 15,* 1683.

Stanley, M.S.M. (1972). In: Growth, Nutrition, and Metabolism of Cells in Culture (G.H., Rathblat and B.J. Cristofalo, eds.), Vol. 2, 327-370, Academic Press, N.Y. and London.

Chapter 24

UTILIZATION OF SOME SUGARS BY A LINE OF *TRICHOPLUSIA NI* CELLS

H. Stockdale and G.R. Gardiner

I. Introduction

At present the nutrition of invertebrate cells *in vitro* is poorly understood. While a study of nutrition for its own sake seems esoteric, the fact that insect cells support replication of arboviruses and baculoviruses provides some justification for such a study. A knowledge of those nutrients which are actually essential for cell growth, coupled with quantitative data on cell yields, could lead to the formulation of cheap and efficient media for virus production. Our knowledge of nutrients essential for growth of insect cells *in vitro* has been clouded to some extent by the complexity of the media employed. Some studies, notably those of Hink *et al* (1972) and of Clements & Grace (1967) have demonstrated the removal of amino acids and sugars during growth. However, little is yet known as to whether a particular nutrient is essential for growth, whether it is metabolised gratuitously, or whether cells can be adapted to grow in its absence.

The carbon and energy sources of a cell line appears to be a convenient point at which to start a survey of this nature.

II. Materials and methods

For our model we took Hink's *Trichoplusia ni* (TN-368) cell line, which was grown in a protein hydrolysate-based medium designated BML-TC/7A (Table 1). In preliminary studies we used a balanced salt solution (BSS) for washing and resuspension of the cells. This was based on the inorganic constituents of TC/7A with 7.5 g. Potassium chloride per liter added to adjust the osmotic pressure to 320 milliosmoles (Table 2). In later studies a medium was produced as far as possible devoid of usable carbohydrate. This "CF-TC/7A" was made as shown in Table 3.

Cells were grown in 30 ml "Falcon" tissue culture flasks with 5 ml of medium at 27°C. Growth was measured by cell counting using a Model B Coulter counter. Carbohydrate levels in the medium were determined by the anthrone method of Trevelyan and Harrison (1952).

TABLE 1.

Formula of growth medium BML - TC/7A

Component	*Amount per liter*
$NaH_2PO_4.2H_2O$	1.145 g
K Cl	4.61 g
$CaCl_2.2H_2O$	1.32 g
$MgCl_2.6H_2O$	2.28 g
$MgSO_4.7H_2O$	2.78 g
Na HCO_3	0.35 g
Lactalbumin Hydrolysate	10.00 g
Tryptose Broth	5.00 g
Glucose	2.00 g
Grace's Vitamins (1,000x)	1.0 ml
Foetal Calf Serum	50.00 ml

TABLE 2.

Formula of balanced salts solution for cell washing

Component	*Amount per litre*
$NaH_2PO_4.2H_2O$	1.145 g
K Cl	10.37 g
$CaCl_2.2H_2O$	1.32 g
$MgCl_2.6H_2O$	2.28 g
$MgSO_4.7H_2O$	2.78 g
Na HCO_3	0.35 g

TABLE 3.

Formula of carbohydrate deficient BML - TC/7A medium for growth studies

Constituent	*Amount per litre*
$NaH_2PO_4.2H_2O$	1.145 g
K Cl	4.61 g
$CaCl_2.2H_2O$	1.32 g
$MgCl_2.6H_2O$	2.28 g
$MgSO_4.7H_2O$	2.78 g
$NaHCO_3$	0.35 g
Lactalbumin Hydrolysate	10.00 g
Bacto Tryptose	4.00 g
Grace's Vitamins (1,000x)	1.0 ml
Dialysed foetal calf serum	50 ml

Uptake of oxygen by cells was initially measured at 27°C using a Warburg manometer assembly, in later experiments by means of a Clark-type oxygen cell (Rank Brothers Instruments, Bottisham, Cambridgeshire, U.K.).

III. Results

The groups of compounds which are most likely to provide carbon and energy sources for these cells are carbohydrates, carboxylic acids ans some amino acids. The standard method for detecting a potential energy source of micro-organism is to incubate a washed cell suspension in a respirometer such as a Warburg Manometer or an oxygen cell. When an oxidizable compound is added to the suspension, an elevated rate of respiration above the basal or endogenous rate of respiration is observed. Such data alone do not prove that a compound supports growth but provides a rapid screen to eliminate compounds which are not oxidizable.

However, preliminary tests showed this technique to be unsuitable in the case of *T. ni* cells. When a cell suspension in balanced salts solution was put into an oxygen cell, a steady endogenous rate was observed which was not detectably increased by the addition of glucose.

Table 4 shows the various manipulations performed in an attempt to demonstrate a difference between the glucose and endogenous respiration rates or Q_{O_2} (Expressed as μl. O_2 per mg. dry weight of cells per hr.). Centrifugation at 55 xg for 10 mins did not appear to damage the cells. A build-up of metabolites in the growth medium does not appear to be the cause of any suppression of glucose Q_{O_2} as cells in fresh and used medium show virtually the same Q_{O_2}. When cells are incubated with BSS and glucose there is a marked suppression of respiratory rate and this suppression is reversed to a certain extent by the addition of a colloid (0.1% 1,500 centipoise methyl cellulose (MC)). In a second set of experiments cells were suspended in various media with and without glucose and in no case were we able to detect a difference between the endogenous and the glucose Q_{O_2}.

The addition of increasing levels of glucose over a range of 0.1 to 1.0% did not increase the glucose Q_{O_2}. Starvation of the cells in BSS with MC over a period of three days did not lower the endogenous rate and no increase in respiration was noted when glucose was added after this period.

Therefore, the only method available left was the addition of a test compound to otherwise carbohydrate-free medium, and observation as to whether or not it supported cell growth over a prolonged period. An inoculum of 1.0 x 10^5 cells per ml was used and test compounds were added to a level of 0.2% w/v. Cells in carbohydrate-free medium maintained themselves without evident multiplication for a week; but then showed degeneration, which was not prevented by addition of fresh CF medium. The same pattern was observed when compounds which did not support growth were added. Vigorous growth occurred in the presence of a utilizable substrate and was maintained over 5 passages of the cells, which suggests that cells can be grown continuously when these compounds provide the sole energy source.

Table 5 shows compounds tested and those which supported growth. Among the test compounds were several mono and di-saccharides, two polyols, four Krebs cycle intermediates, pyruvic acid, glycerol and β-glycerophosphate. Only glucose, fructose, mannose and the two glucose dimers, maltose and trehalose supported growth. The anthrone test showed that those sugars which did not support growth remained in the

TABLE 4.

Respiratory rates of T.ni cells under varying conditions and in different media.

Condition	$\mu l\ O_2\ 10^6$ cells/h	$\mu l\ O_2$ mg D.W./h
Suspended, not spun	20.2	41.6
Spun in growth medium	20.2	41.6
Fresh medium	21.8	44.9
BSS + 0.1% glucose	7.17	14.8
BSS + 0.1% glucose + MC	10.6	21.8
CF TC/7A	20.0	41.2
TC/7A + 0.1% glucose	19.4	39.9
BSS	12.4	25.5
BSS + 0.1% glucose	13.8	28.4
BSS + MC	16.3	33.6
BSS + MC + 0.1% glucose	16.0	32.9

TABLE 5.

List of compounds tested for support of growth of T.ni cells and those compounds found to support growth.

Glucose	Sucrose	Succinic acid
Fructose	Lactose	Malic acid
Galactose	Cellobiose	Fumaric acid
Mannose	Melibiose	2–Oxo–Glutaric acids
Sorbose	Turanose	Pyruvic acid
Maltose	Sorbitol	Glycerol
Trehalose	Mannitol	Na–β–glycerophosphate

Compounds supporting growth

Glucose	Mannose	Trehalose
Fructose	Maltose	

TABLE 6.

Yields of T.ni cells from different carbohydrate substrates after 8 days growth in monolayer culture in BML - TC/7A

Substrate	% utilization	Cell yield/ML	Dry wt/ML	µg dry wt / µM monomer
0.1% glucose	100	8.99×10^5	436 µg	79.3
0.1% fructose	76.3	10.03×10^5	486 µg	116
0.1% maltose	94.8	10.09×10^5	489 µg	88.9
0.1% trehalose	99.9	9.60×10^5	466 µg	80.3
0.1% mannose	77.5	2.09×10^5	101 µg	23.7

medium at the same level at the end of the experiment. No data on the fate of the polyols and carboxylic acids were obtained.

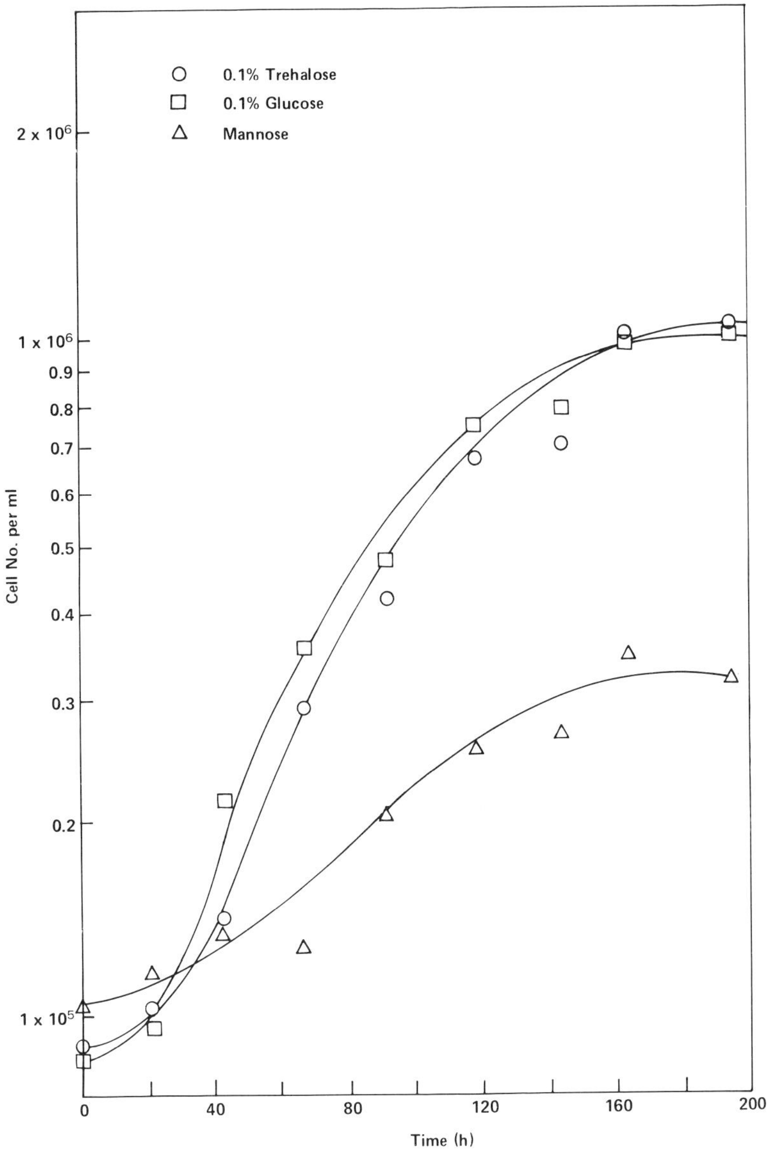

Fig. 1. Growth of *T.ni* cells in monolayer culture in BML - TC/7A with different carbohydrate substrates.

Finally, an attempt was made to measure the efficiency of those substrates which would support growth. This required cells to be grown under conditions in which the energy source was limiting. In monolayers with 0.2% added glucose, 2×10^6 cells/ml are produced and 1×10^6 cells/ml in the presence of 0.1% glucose. Consequently, 0.1% (w/v) of the various substrates were added to CF-TC/7A. A series of monolayers were set up and sampled over an eight day period (three monolayers per determination). The results are as shown in Figure 1. All substrates produced yields comparable to that of glucose except in the case of mannose where growth is slow and the yield low. The yields of cells obtained from a given quantity of a utilizable carbohydrate are shown in Table 6. All the carbohydrates utilized gave roughly comparable cell yields, ranging from 79.3 to 116 µg dry wt of cells per µ mole of sugar monomer utilized except in the case of mannose, which yielded 23.7 µg dry wt of cells per µ mole metabolized.

IV. Discussion and conclusions

It is interesting to note that galactose did not serve as a growth substrate, as the ability to utilize this sugar is a fairly general characteristic of mammalian cells in culture. The utilization of trehalose by an insect cell line is hardly surprising, but it is a property which is shared by several mammalian cell lines. The finding that sucrose neither supports growth of *T.ni* cells nor is metabolised by them appears to be at variance with the work of Clements & Grace (1967) who noted the disappearance of a substantial amount of this sugar during growth of *Antheraea eucalypti* cells. However, these workers used a different cell line which was growing in the presence of glucose, fructose, and *Bombyx mori* hemolymph.

It appears that *T.ni* (TN-368) cells are strictly limited in the number of monosaccharide isomers they can metabolise and also in the types of disaccharide bridges they can hydrolyse. Galactose is an epimer of glucose at C_4 and is not utilized, while mannose, an epimer of glucose at C_2 is, possibly through a conversion to glucose by inversion of the uridine diphosphate derivative. Fructose is metabolised, but sorbose is not (Figure 2). In a similar manner the disaccharidases or the permeases

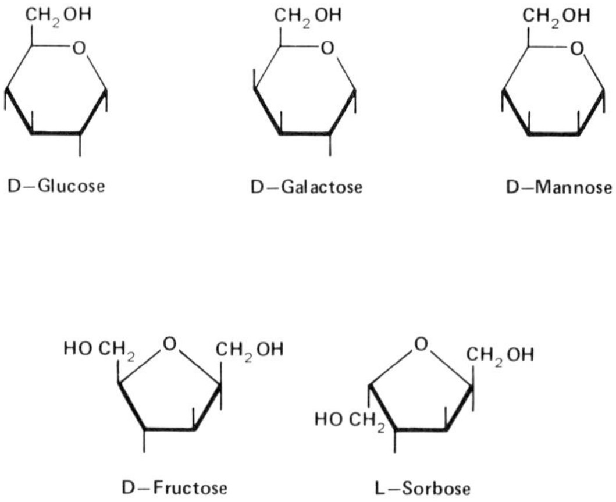

Fig. 2. Structure of monosaccharides tested as growth substrates for *T.ni* cells.

involved in carbohydrate metabolism appear to be highly specific and take into account the whole molecule rather than a particular monomer and its linkage. Maltose with a 1-α-glucoside link is used, but turanose and sucrose are not. The linkage between glucose units is critical: those in maltose and trehalose are acceptable while that in cellobiose is not (Figure 3).

In conclusion, it appears that the number of energy sources which support growth of *T.ni* cells is limited, but at the same time subsidiary sources such as Krebs' cycle intermediates are not required. One might explain the low yield from mannose on the basis that its conversion to glucose is slow and consequently a greater proportion of the derived energy is required for maintenance. Under certain circumstances, for instance where a high cell yield is required, it may be advantageous to use a disaccharide in place of a monosaccharide as one would obtain a higher cell yield for a relatively smaller increase in osomotic pressure.

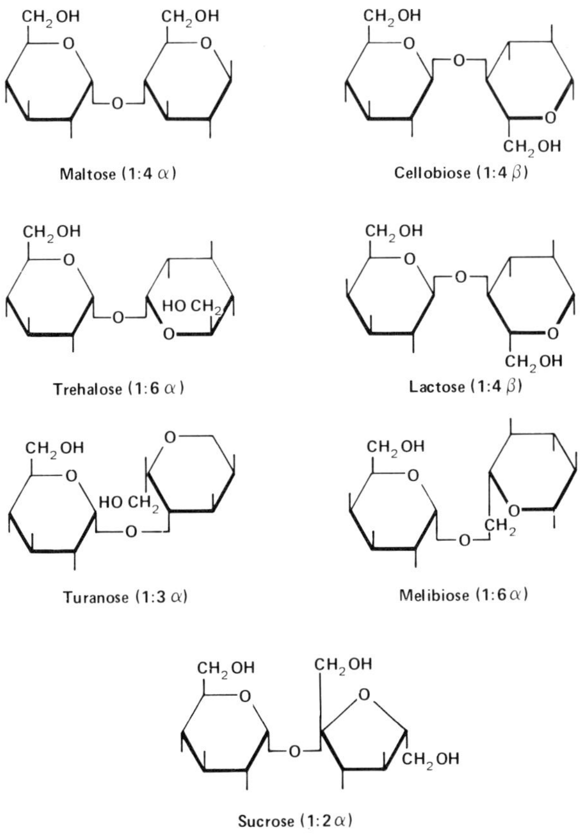

Fig. 3. Structure of disaccharides tested as growth substrates for *T.ni* cells.

V. References

Clements, A.N. and Grace, T.D.C. (1967). *J. Insect. Physiol.* *13,* 1327.

Hink, W.F., Richardson, B.L., Schenk, D.K. and Ellis, B.J. (1973). *Proc. 3rd Intern. Coll. Invert. Tissue Culture,* 1971, Smolenice, Czechoslovakia.

Trevelyan, W.D. and Harrison, J.S. (1952). *Biochemical J.* *50,* 298.

Chapter 25

INFLUENCE OF POLYPHENOL OXIDASE ON HEMOCYTE CULTURES OF THE GYPSY MOTH

H.M. Mazzone

I. Introduction

The establishment of insect cell cultures is generally preceeded by a very long inactive period prior to cell multiplication and subsequent subculturing (Grace, 1962, Sohi, 1971). This phenomenon has been studied in our laboratory and it appears that the polyphenol oxidase (PPO) reaction, which occurs in many insect cells and produces toxic melanin, is a principal factor for the delay of cell development *in vitro.* The usual practice of heating hemolymph plasma in order to nullify the action of PPO is not sufficient, since as will be demonstrated in this study, the cells can also contain the enzyme.

II. PPO in the Gypsy moth

PPO is very active in the Gypsy Moth *(Porthetria (Lymantria) dispar, Linnaeus),* the major forest insect pest of the Northeastern United States. In this insect catechol is a substrate for the enzyme (Mazzone and Brown, 1975) and in the presence of copper ions and oxygen, is converted to melanin. Prior to this study blockage of melanin formation was reported by utilization of competitive substrate analogs and by chelation of the copper ions, notably with cysteine (Mazzone, 1968, 1970). Phenylthiourea (PTU) chelates copper ions and is an effective inhibitor of the enzyme, but has not been commonly employed in insect tissue cultures because of its reported adverse effects (Wyatt, 1956). The utilization of PTU was reinvestigated in this study and seemed to serve a useful purpose in the cell cultures investigated.

The PPO reaction Gypsy Moth hemolymph and the effect of PTU inhibition were studied by polyacrylamide gel electrophoresis (PAGE), Figure 1. Hemolymph from a 5th larval instar was obtained and divided into three aliquots. From the first aliquot a sample was applied quickly to a PAGE tube and electrophoresis commenced. Melanization did not occur under these conditions and this sample served as a control for the experiment. Protein staining revealed 13 bands. Aliquot 2 contained a crystal of PTU and was exposed to air for the same time as aliquot 3, which was allowed to melanize. PAGE samples of aliquots 2 and 3 are represented by the middle and bottom gels, respectively. The PTU inhibited sample (middle gel) closely resembled the non-melanized sample (top gel). By comparison, the melanized sample (bottom gel) showed a decrease in concentration (ban width) of the two broad bands near the origin, as did the next band immediately following. In addition, there was a buildup of stained material at the origin, representing protein complexes too large to penetrate the pores of the gel. These results on melanization vs. non-melanization (top and bottom gels) are essentially in agreement with results reported by McKinstry and Steinhauer (1970).

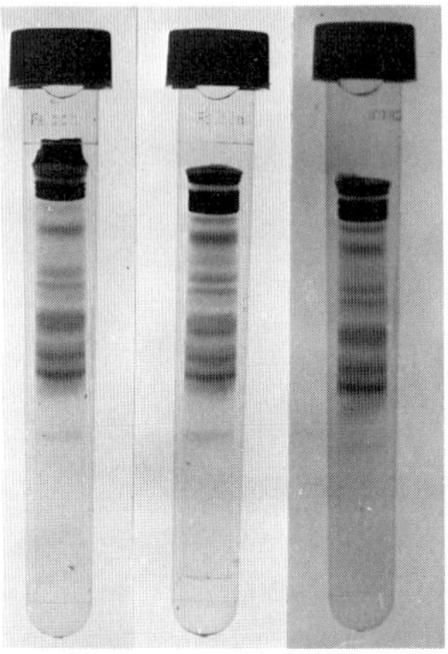

Fig. 1. Polyacrylamide Gel Electrophoresis (PAGE) of Larval Hemolymph. 7.5% PAG, 2.5 mA/tube; 4°C, 0.4M Tris-Glycine, pH 8.3. Proteins stained with naphthalene black. (refer to text).

As demonstrated in our laboratory the egg proteins and the hemolymph proteins of larva, pupa and adult forms contain PPO. The enzyme is maximal in the 5th larval instar as indicated by PAGE - catechol incubation (Brown and Mazzone, 1975). In the present study hemocytes were also observed to contain PPO by the pyrogallol staining method of Vercauteren and Aerts (1958), Figure 2a and by PAGE - catechol incubation of hemocyte extracts, Figure 2b.

III. B. Hemocyte cultures

Our laboratory has been concerned with the control of populations of the Gypsy Moth. A potential insecticide is the nucleopolyhedrosis virus (NPV) of the insect. In order to study host cell-virus relationships, and to consider an alternate means of producing the virus over *in vivo* procedures, tissue cultures of the Gypsy Moth are necessary. Hemocytes of the insect show a high sensitivity of NPV infection as shown in Figure 2c. This infection occurs naturally in virus epizootics or can be obtained by addition of inclusion bodies of NPV to the diet of laboratory reared insects.

Hemocyte cultures were set up in the following manner. A drop volume of whole hemolymph was added to Leighton tubes or T-flasks which contained a crystal of PTU (1 mg/Leighton tube; 4 mgs/T-flask). Grace's medium, containing 10% fetal bovine serum (FBS) and penicillin-streptomycin, was added to the culture vessel (1 ml/Leighton tube; 4 ml/T-flask). The cell number in the Leighton tube averaged 1000-2000, while the cell number in the T-flask averaged 4000-8000. The cell types

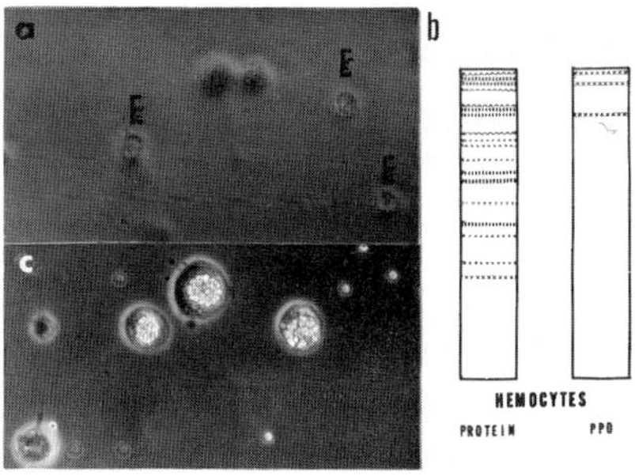

Fig. 2. Observations from *in vivo* hemocytes. a) Hemocytes indicating the presence of PPO(E). A drop of larval hemolymph was smeared on a slide, dried, fixed in formaldehyde fumes, and incubated with pyrogallol (Vercauteren and Aerts, 1958). x400. b) PAGE of protein extract of larval hemocytes with PPO bands obtained by catechol incubation of the corresponding gel (experimental conditions as noted in Fig. 1). c) Larval hemocytes infected with NPV. x400.

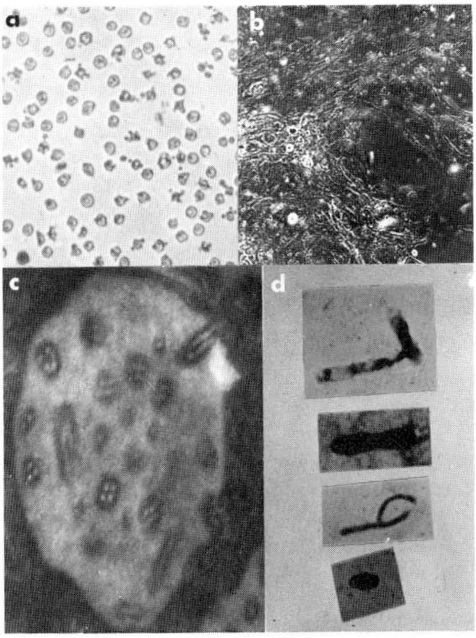

Fig. 3. Observations from *in vitro* hemocyte cultures. a) Cells floating in medium. x400. b) Attached cell culture. x100. c) Thin section of an inclusion body of the type obtained from infected cell cultures. x57,000. d) Chromosomes obtained from attached cell culture (see text).

which developed *in vitro* were the following: free floating or cells loosely attached, Figure 3a. This cell type morphologically resembled the hemocytes as they appeared when first added to culture vessels. Subculturing was accomplished by transfer of these cells through pipetting. The other cell type was that which attached tenaciously to the bottom of the culture vessel, Figure 3b, and in 48-72 hours occupied most of the vessel area. Subculturing was achieved by enzymatic dissociation with trypsin or with pronase. The two cell types responded to NPV infection as shown in Figure 3c. Infection was achieved by the addition to the cultures of viral infectious hemolymph plasma (cells and inclusion bodies centrifuged out) containing PTU. The *in vitro* hemocyte cell types could serve as excellent source material for studies involving viral invasion of cells. In addition, the cell cultures would afford a highly efficient method for the *in vitro* production of the viral insecticide.

Preliminary observations have been made on the chromosomes of these cell cultures. The attached cells were cochicnized (2.5 x 10^{-5}M), fixed in ethanol-acetic acid (3:1), scraped from the vessel surface, and gently homogenized. After low speed centrifugation a drop of the remaining suspension was spread on a Langmuir trough. A grid was allowed to touch the surface. After air drying, the grid-sample was examined in a transmission electron microscope. Some of the individual chromosomes noted are shown in Fig. 3d, and reveal a significant range in size.

Each type of cell culture had been subcultured 10 times. At this stage of investigation the following points are being considered: a) attempting to substitute other compounds for native hemolymph, an essential ingredient in these cultures (0.01 ml/ Leighton tube; 0.04 ml/T-flask). A suitable alternative is FBS (30%). b) testing the cell cultures for continued presence of PPO. In early subcultures (passage numbers 3-5) the hemocytes which float in the medium stained positively (Vercauteren and Aerts, 1958) for the presence of the enzyme, whereas the PAGE-catechol incubation system as utilized, did not indicate the presence of PPO. In later subcultures (passage number 7-9), although PPO may be present in the floating hemocytes (staining procedure), its activity is considerably reduced. These observations tend to support the contention that PPO can persist for a period of time in cell cultures. c) while conducting the current study it was noted that certain hemocytes in cultures, where no PTU was added, survived the effects of melanization and for many weeks continually darkened the medium, presumbably producing melanin. These cell types are being investigated.

IV. References

Grace, T.D.C. (1962). *Nature 195,* 788.

Sohi, S.S. (1971). *Canadian J. of Zoology 49,* 1355.

Mazzone, H.M. and Brown, S. (1975). In preparation.

Mazzone, H.M. (1968). *Second International Colloquium on Invertebrate Tissue Culture,* Milan, Italy, 14-21.

Mazzone, H.M. (1970). In: Current Topics In Microbiol. and Immunol. *55,* 196-200.

Wyatt, S.S. (1956). *J. Gen. Physiology 39,* 841.

McKinstry, D.M. and Steinhauer, A.L. (1970). *J. Invert. Pathology 16,* 123.

Brown, S. and Mazzone, H.M. (1975). In preparation.

Vercauteren, R.E. and Aerts, F. (1958). *Enzymologia 20,* 167.

Chapter 26

EFFECTIVE COLONY FORMATION IN DROSOPHILA CELL LINES USING CONDITIONED MEDIUM

S. Nakajima and T. Miyake

I. Introduction

In vitro tissue culture has offered a new tool for genetic studies in animal cells. *Drosophila* has well-analyzed genetic backgrounds and simple chromosome sets. Thus it seems suitable for genetic studies at the cellular level. In order to rescue a few mutant cells surviving under restricted conditions, it is very important to create conditions favoring cell colony formation from small numbers of cells. Unfortunately, *Drosophila* cells have usually low plating efficiencies (Echalier, 1971). On the other hand, in mammalian cell cultures, "conditioning" of the culture medium often raised the plating efficiencies of cells and was used to obtain colonies from small numbers of cells (Fisher and Puck, 1956, Ichikawa *et al.*, 1969, Pluznik and Sachs, 1966, Puck and Marcus 1955, Sanford 1948, Worton *et al.*, 1969).

In this report we show that the use of a conditioned medium and glass dishes remarkably increases the colony forming efficiencies of GM_1 and GM_2 cell lines derived from embryos of *Drosophila melanogaster*.

II. Materials and Methods

A. *Cell culture*

Drosophila melanogaster cell lines, GM_1 and GM_2 (Mosna *et al.* (Mosna and Dolfini, 1972)), were kindly supplied by Drs. G. Mosna, and S. Dolfini and originally cultured in D-225 medium (Echalier and Ohanessian, 1970). These cell lines were adapted to Schneider's medium (Schneider, 1972) in our laboratory. Schneider's medium supplemented with 7% not-heat inactivated fetal bovine serum (FBS) (Microbiological Associates, Inc.) was used as a culture medium in all experiments. The cultures were incubated at 27°C.

B. *Used medium*

Unless otherwise stated, the used medium was prepared as follows; GM_1 and GM_2 cells were seeded in MA-140 flasks (Miharu Seisakusho, Co. Ltd., Tokyo) with 30 ml of culture medium. When cells reached confluency after 3-5 days, the culture fluids were harvested and centrifuged at 3000 rpm for 10 min. The supernates were freeze-thawed with dry ice-aceton and water bath and used as a used medium. Culture fluids harvested at different times after seeding cells were expressed as "1 day-old medium" and so on. One day-old medium means a culture fluid harvested on the second day after seeding cells.

C. *MC medium*

MC medium was of the used type and prepared as follows unless otherwise stated; when the cells were grown to confluency in MA-140 flasks, the culture fluids were replaced by equal volumes of fresh culture medium and harvested 24 hrs later by the same procedure described in "used medium".

D. *Colony formation*

Cells were suspended in a fresh culture medium and diluted with media containing the conditioned medium in various proportions. The culture vessels were 60x15 mm glass or plastic (Falcon Plastics, No.3030) dishes. The cells were incubated at 27°C for 10 days in a humid atmosphere of 5% CO_2 and 95% air. In order to examine contaminations of surviving cells in the used media, control experiments without inoculum cells were always performed.

E. *Colony count*

Colony numbers were counted 10 days after the inoculation. Colonies were stained with Giemsa staining solution and counted under a dissecting microscope. Only colonies consisting of more than 30 cells were counted. The number of colonies of 4 dishes were averaged in every experiment.

F. *Cell count*

One ml of cell suspension was mixed with 1 ml of 0.2% trypan blue solution and kept at room temperature for 5 min. Viable cells were counted in a hemocytometer.

G. *Mycoplasmas*

The media described by Hayflick (Hayflick, L., 1965) were used for the detection of mycoplasmas. Tubes containing 7 parts of mycoplasma broth (Baltimore Biological Laboratories, BBL), 2 parts of horse serum (Grand Island Biological Company), and 1 part of 25% yeast extract (Difco Laboratories) were inoculated with cells or supernates from GM_1 and GM_2 cell cultures. The incubation was performed at 37°C for 7 days in a humid atmosphere of 5% CO_2 95% air. These samples were further subcultured on mycoplasma agar (BBL) plates supplemented with 20% horse serum and 10% yeast extract at 37°C for 2 weeks in 5% CO_2 and 95% air.

III. Results

A. Adaptation of GM_1 and GM_2 cell lines to Schneider's medium

GM_1 and GM_2 cell lines originally cultured in D-225 medium have been successfully adapted to Schneider's medium supplemented with 7% FBS for more than 6 months. Compared to the original cultures, these adapted cells have reduced tendency to float in the medium. Fig. 1 represents growth curves of these adapted GM_1 and GM_2 cell lines and Fig. 2 and Fig. 3 show phase-contrast photomicrographs of typical colonies of GM_1 and GM_2 cells. Under these conditions, the population doubling times of GM_1 and GM_2 cell lines were 26 hrs and 28 hrs, respectively.

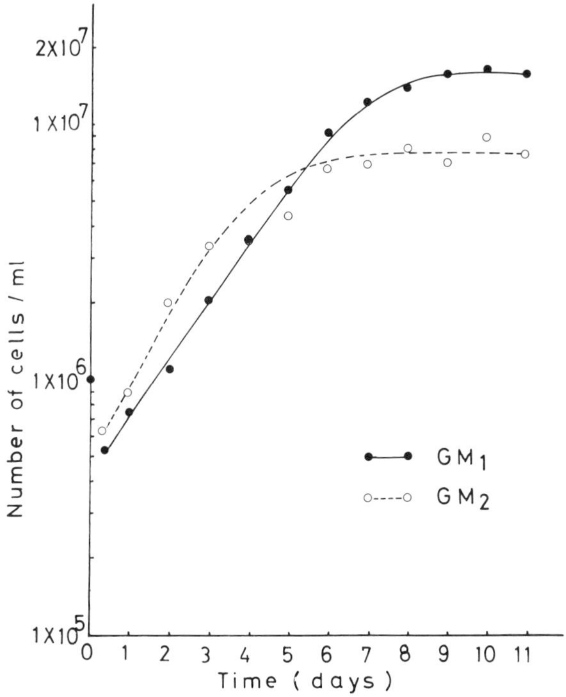

Fig. 1. Growth curves of GM_1 cell line and GM_2 cell line at 27°C. GM_1 and GM_2 cells were plated in 60x15 mm glass dishes in 5 ml of fresh culture medium (Schneider's medium supplemented with 7% FBS). They were incubated at 27°C in a humid atmosphere of 5% CO_2 and 95% air. Cells were periodically detached in the culture medium by a rubber policeman and viable cells were counted in a hemocytometer.

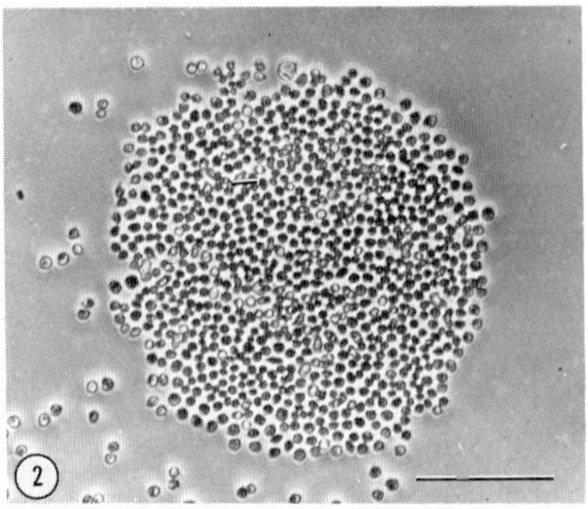

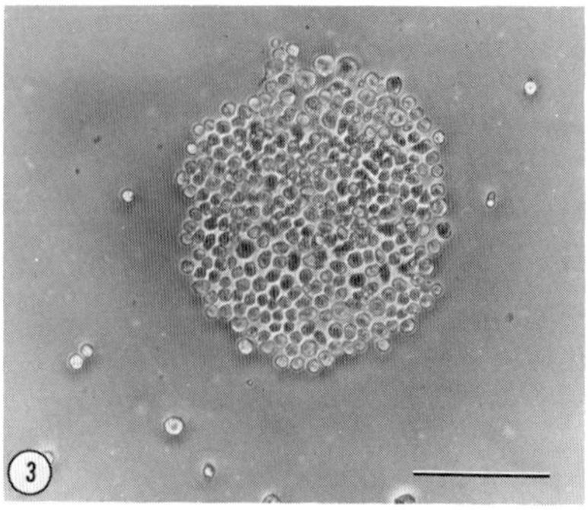

Fig. 2. A typical GM_1 cell colony at 10 days, Phase-contrast, Bar: 100 μm.

Fig. 3. A typical GM_2 cell colony at 10 days, Phase-contrast, Bar: 100 μm.

B. Colony formation in the conditioned medium

GM_1 cells at various cell concentrations were inoculated into 60x15 mm glass or plastic dishes in 5 ml of fresh culture medium or conditioned medium. As shown in TABLE I, in the fresh culture medium, GM_1 cells were unable to grow to form colonies at inoculum cell sizes of less than $5x10^4$/dish. On the other hand, in the conditioned medium prepared by mixing 1 part of the fresh culture medium and 1 part of the used medium obtained from GM_1 cell culture, colonies were formed more effectively even at smaller inoculum cell sizes. In glass dishes, the plating efficiencies at inoculum cell sizes of $5x10^4$-5x10/dish were 14.3%-2.6%. In plastic dishes, the efficiency was 1.7% at an inoculum cell size of $5x10^4$/dish. These results indicated that the use of the conditioned medium and glass dishes was effective for increasing the colony forming efficiency of GM_1 cells. In GM_2 cell line, colonies were also unable to grow from cells of less than $5x10^4$/dish (data not shown).

Culture fluids of GM_1 cells harvested at different times were mixed with fresh culture medium in different proportions and these conditioned media were then tested for the colony forming activity. Five thousand cells of GM_1 cell line were inoculated into 60x15 mm glass dishes. As shown in TABLE II, when 3 day-old modium was used in 1 : 4 or 1 : 1 proportion, colonies were formed with efficiencies of about 20% or more, wheras 1 day-, 2 day-, or 5 day-old mediums had less effect on the colony formation. In this series of experiments, cells reaches confluency on 4th day and when MC medium prepared from these monolayer cultures (4MC medium) was used in 0 : 1 proportion, that is 100% MC medium, the efficiency of the colony formation was 33.6% and highest. The efficiency of the colony formation of 5MC medium harvested on 6th day was higher than that of 5 day-old medium. From these experiments, the use of 100% MC medium prepared at the time when cells reached confluency and the use of glass dishes were found to be most effective to increase the colony forming efficiency of GM_1 cells.

C. Colony formation with small number of cells in MC medium

Two to 1000 cells were inoculated into 60x15 mm glass dishes in 5 ml of 100%

TABLE I.

Colony formation of GM_1 cells in the conditioned medium[a]

Vessel	No. of cells inoculated per dish	Fresh medium: No. of colonies per dish	Fresh medium: % Plating efficiency	Conditioned medium: No. of colonies per dish	Conditioned medium: % Plating efficiency
glass	500000	1608	0.3	–	–
	50000	0	0	7155	14.3
	5000	0	0	265	5.3
	500	0	0	22	4.4
	50	–	–	1.3	2.6
	0	–	–	0	0
plastic	500000	9900	2.0	–	–
	50000	0	0	841	1.7
	5000	0	0	0.8	0
	500	0	0	0	0
	50	–	–	0	0

[a]Cells at various concentrations were inoculated into 60x15 mm glass or plastic dishes in 5 ml of fresh culture medium or conditioned medium. The conditioned medium consisted of 1 part of the fresh culture medium and 1 part of the used medium of GM_1 cell culture. The number of colonies at the inoculum cell size of $5x10^5$/dish in the fresh culture medium was calculated from numbers of 10 optional grid areas.

TABLE II.

Colony formation in various conditioned media which were harvested at different times during cell cultures and mixed in various proportions with fresh culture medium[a]

Time of harvest (days)	No. of colonies per dish (% Plating efficiency) — Proportion of fresh culture medium to used medium: 1 : 1	1 : 2	1 : 4	0 : 1
1	0 (0)	0 (0)	0 (0)	1 (0)
2	0 (0)	19 (0.4)	80 (1.6)	137 (2.4)
3	37 (0)	323 (6.5)	816 (16.3)	1146 (22.9)
4[b]	0 (0)	531 (10.6)	1378 (27.6)	1348 (27.0)
4MC[c]	2 (0)	391 (7.8)	1335 (26.7)	1679 (33.6)
5[d]	1 (0)	306 (6.1)	621 (12.4)	126 (2.5)
5MC[e]	8 (0.2)	380 (7.6)	893 (17.9)	1031 (26.2)

[a] 5000 cells of GM_1 cell line were inoculated into 60x15 mm glass dish in 5 ml of medium. Used media were prepared as described in the test.

[b]Cells reached confluency.

[c]On 4th day, medium was replaced with fresh culture medium and harvested on 5th day.

[d]Cell drops were seen in part of the cell population.

[e]On 5th day, medium was replaced with fresh culture medium and harvested on 6th day.

MC medium. In the experiment using GM_1 cells and the MC medium of GM_1 cell cultures (GM_1-MC medium), colonies were formed with plating efficiencies of 60.8%-68.5% at all inoculum cell sizes. In 5 day-old medium, however, colonies were not formed in these inoculum cell ranges (TABLE III). In the experiment with GM_2 cells and the MC medium of GM_2 cell cultures (GM_2-MC medium), colonies were also formed with efficiencies of 3.2%-5.8% at inoculum cell sizes of 100-1000/dish (TABLE IV).

TABLE III.

Colony formation with small number of GM_1 cells in MC medium

Used medium	No. of cells inoculated per dish [b]	No. of colonies per dish	% Plating efficiency
	1000	642	64.2
	500	304	60.8
	4	2.75	68.5
GM1-4MC [a]	2	1.25	62.5
	0	0	0
	1000	0	0
	500	0	0
4 [a]	100	0	0
	0	0	0

[a] GM_1-4MC medium and 4 day-old medium were prepared by freeze-thawing.

[b] GM_1 cells were inoculated into 60x15 mm glass dishes.

TABLE IV.

Colony formation with small number of GM_2 cells in MC medium

Used medium	No. of cells inoculated per dish [b]	No. of colonies per dish	% Plating efficiency
	1000	40	4.0
	500	16	3.2
GM_2-4MC [a]	100	5.8	5.8
	0	0	0
	1000	0	0
	500	0	0
4 [a]	100	0	0
	0	0	0

[a] GM_2-4MC medium and 4 day-old medium were prepared by freeze-thawing.

[b] GM_2 cells were inoculated into 60x15 mm glass dishes.

D. Effect of preparation methods of MC medium on the colony forming activity

GM_1-MC medium was prepared by 5 different ways: (Echalier and Ohanessian, 1970) culture fluid was centrifuged at 3000 rpm for 10 min and the supernate was freeze-thawed (Echalier, 1971) culture fluid was centrifuged at 3000 rpm for 10 min and the supernate was treated at 56°C for 30 min (Fisher and Puck, 1956) culture fluid was centrifuged at 3000 rpm for 10 min and the supernate was filtered through a filter (Sartorius membrane filter, pore size 0.2 μm) (Grace 1968) culture fluid was centrifuged at 15000 rpm for 20 min and the supernate was used (Hayflick, 1965) culture fluid was centrifuged at 30000 rpm for 30 min (Spinco model L-3, No. 30 rotor) and the supernate was used. One thousand cells of GM_1 cell line were inoculated in 60x15 mm glass dish in 5 ml of medium. Control experiments without inoculum cells were paralleled for all MC media. As shown in TABLE V, colonies formed effectively in all cases. The colony forming efficiency of the MC medium prepared by very high speed centrigufation (30000 rpm, 30 min) was a little lower than others. In three MC media prepared by freeze-thawing, heat treatment and filtration colonies did not develope in control dishes. In two MC media prepared by high speed centrifugations, however, colonies developed in control dishes.

E. Effect of preservation at -20°C of MC medium on the colony forming activity

The used media or the MC media were usually stored at -20°C. The effect of preservation at -20°C on the colony forming activity was tested using GM_1-MC media preserved for various periods. One thousand cells of GM_1 cell line were inoculated into 60x15 mm glass dish in 5 ml of medium. In all samples, colonies formed with high efficiencies (45.2%-66.4%) (TABLE VI). These results showed that the used media or the MC media could be preserved at -20°C without decreasing the activity; for the colony formation for at least 10 days.

IV. Discussion

In insect cell cultures, cell clones have already been isolated in several species. Suitor *et al.* (1966) isolated one clone from *Aedes aegypti* cell line and Grace (1968) isolated 10 clones from *Antheraea eucalypti* cell lines. In the former experiment, dilution and capillary method and in the latter, dilution method was employed. Isolated cells could grow in the fresh culture medium in the latter case. In *Drosophila* cell lines, 7 clones from *Drosophila melanogaster* cell line were isolated using coverslip fragments method (Echalier, 1971). Recently, a method of cloning in agar medium was developed in insect cell lines of *Trichoplusia ni* (McIntosh and Rechtoris, 1974). The plating efficiencies of the insect cells, however, were usually very low and the procedures used by the cloning methods seem not to be suitable to rescue a few mutant cells surviving in restricted conditions. By using the system shown in this report, the plating efficiencies of *Drosophila melanogaster* cells can be remarkably increased and also colonies easily isolated from single cells of these cell lines.

By using the MC medium and glass dishes, GM_1 cells could grow to form colonies at an efficiency of 62.5% even when only two cells per 60x15 mm dish were inoculated (TABLE III). GM_2 cells could also grow to form colonies under the same condition, but the efficiency was lower than that of GM_1 cells and was 5.8% at the inoculum cell size of 100 (TABLE IV). The MC medium or the used medium could prepared by several methods. In each case, the possibility that the cells surviving in the used media developed to colonies should be excluded. In control experiments

TABLE V.

Effect of preparation methods of MC medium[a] on the colony forming activity

Methods	No. of cells inoculated per dish[b]	No. of colonies per dish	% Plating efficiency
Freeze-thawing	1000	469	46.9
	0	0	0
Heating 56°C, 30 min	1000	572	57.2
	0	0	0
Filtration	1000	485	48.5
	0	0	0
Centrifugation	1000	784	78.4
15000 rpm, 20 min	0	63	6.3
Centrifugation	1000	238	23.8
30000 rpm, 30 min	0	33	3.3

[a]MC medium was obtained from GM_1 cell culture.

[b]GM_1 cells were inoculated into 60x15 mm glass dishes.

TABLE VI.

Effect of preservation at -20°C of MC medium[a] on the colony forming activity

Preservation period (days)	No. of colonies per dish[b]	% Plating efficiency
0	664	66.4
1	628	62.8
2	452	45.2
8	557	55.7
10	471	47.1

[a]GM_1-MC medium was prepared by freezing with dry ice-acetone and stored at -20°C.

[b]1000 cells of GM_1 cell line were inoculated into 60x15 mm glass dishes.

without inoculum cells, colonies did not develope in three media prepared by freeze-thawing, heat treatment and filtration methods. In two MC media prepared by high speed centrifugation, however, colonies developed in control dishes (TABLE V). Among these methods, freeze-thawing was simplest and the MC medium prepared by this method retained good activity, so it could be employed routinely. The used medium or the MC medium could be stored at -20°C without decreasing the activity for the colony formation for at least 10 days (TABLE VI).

The culture fluid of a confluent monolayer could not sustain the colony formation starting from small number of cells, while the MC medium could (TABLE III), and also constant good results for the colony formation were obtained by treating the MC medium at 56°C for 30 min (TABLE V). This suggested that several active factors for the colony formation were included in the used culture medium.

There have been many reports on growth stimulating factors in the conditioned medium in mammalian cell cultures. There exist two types of such factors. Chick cells at high density were able to produce nondialyzable heat-sensitive substances (Rubin, 1966). These "conditioning factor" were capable to counteract the inhibitory material present in the calf serum (Shodell *et al.*, 1972). The second type of factors is represented by small dialyzable growth factors in the culture media of certain normal and tumor mouse cells (Ichikawa *et al.*, 1969), or human fibroblastic cells (Pohjanpelto and Raina, 1972), or rabbit cells (Melbye and Karasek, 1973). The analysis of the active factors for the colony formation in our system is now in progress.

V. Conclusions

Various conditions for the effective colony formation in *Drosophila melanogaster* cell lines, GM_1 and GM_2, were examined. The use of the MC medium (a type of the conditioned medium) and glass dishes remarkably increased the colony forming efficiencies of these cell lines. When these cells were inoculated into glass dishes with 100% MC medium colonies were formed effectively starting from a small number of cells (2-1000 cells per 60x15 mm dish). The plating efficiencies of the cell lines GM_1 and GM_2 under these conditions were about 50% and 4%, respectively. In the fresh culture medium, however, GM_1 and GM_2 cells were unable to form colonies at an inoculum cell size of less than 5×10^4 cells per dish.

Acknowledgment

The authors are deeply grateful to Drs. G. Mosna and S. Dolfini for the supply of cells, and Miss Ritsuko Aihara for her help in the preparation of media and in cell culturing.

VI. References

Echalier, G. and Ohanessian, A., (1970). *In Vitro 6,* 162.
Echalier, G., (1971). *Curr. Top. Microbiol. Immunol. 55,* 220.
Fisher, H.W. and Puck, T.T., (1956). *Proc. Natl. Acad. Sci. 42,* 900.
Grace, T.D.C., (1968). *Exptl. Cell Res. 52,* 451.
Hayflick, L., (1965). *Tex. Rep. Biol. Med. 23 (Suppl. 1),* 285.
Ichikawa, Y., Paran, M. and Sachs, L., (1969). *J. Cell. Physiol. 73,* 43.
McIntosh, A.H. and Rechtoris, C., (1974). *In Vitro 10,* 1.
Melbye, S.W. and Karasek, M.A., (1973). *Exptl. Cell Res. 79,* 279.
Mosna, G. and Dolfini, S., (1972). *Chromosoma 38,* 1.
Pluznik, D.H. and Sachs, L., (1966). *Exptl. Cell Res. 43,* 553.
Pohjanpelto, P. and Raina, A., (1972). *Nature 235,* 247.
Puck, T.T. and Marcus, P.I., (1955). *Proc. Natl. Acad. Sci. 41,* 432.
Rubin, H., (1966). *Exptl. Cell Res. 41,* 149.
Sanford, K.K., Earle, W.R. and Likely, G.D., (1948). *J. Natl. Cancer Inst. 9,* 229.
Schneider, I., (1972). *J. Embryol. Exp. Morph. 27,* 353.
Shodell, M., Rubin, H. and Gerhart, J., (1972). *Exptl. Cell Res. 74,* 375.
Suitor, E.C., Jr., Chang, L.L. and Liu, H.H., (1966). *Exptl. Cell Res. 44,* 572.
Worton, R.G., McCulloch, E.A. and Till, J.E., (1969). *J. Cell. Physiol. 74,* 171.

Chapter 27

COMPARATIVE STUDIES WITH CLONES DERIVED FROM A CABBAGE LOOPER OVARIAN CELL LINE, TN-368

L.E. Volkman and M.D. Summers

I. Introduction

One of the most widely used Lepidopteran cell lines for producing and studying nuclear polyhedrosis virus replication is the cabbage looper ovarian cell line TN-368 (Hink, 1970). This line has been used routinely as the indicator cell line in a plaque assay developed for these viruses (Hink and Vail, 1973). As this line is uncloned, we thought that perhaps we could reduce the variability of the results (Habel, 1969) and increase the efficiency of the assay by employing cloned lines derived from TN-368 as plaque assay indicators.

In this paper we describe the method by which we successfully cloned several daughter lines and discuss their morphological differences as well as their different capabilities to serve as plaque assay indicator lines.

II. Materials and methods

Cells. The parent cell line, TN-368 was originally established in continuous culture from the ovaries of newly emerged virgin female moths (Hink, 1970). The cells are subcultured routinely every three days in TNM-FH (Hink, 1970) by merely pipetting the loosely adhering cells several times to make a uniform suspension then transferring them to a new flask. Plastic Falcon tissue culture flasks (25 ml) are seeded with 1-2 x 10^5 cells/ml in 6 ml of media.

Development of cloned lines. The clones were developed from single cells which were selected microscopically and isolated using hand drawn capillary micropipettes and a micromanipulator. The single cells were drawn up into the capillary pipettes from flasks which contained very few isolated cells attached to the plastic, and from which moments before, all media had been removed. The capillary pipettes, which contained several microliters of media were positioned just above but very close to the selected cells. The media was slowly expressed and then taken in again, until, after usually one to five cycles, the cells would yield to the pressure and would accompany the media into the pipette. They then were transferred to single chambers of a Lab-Tek chamber/slide along with .2 to .3 ml of media. Following this, the chamber slides were sealed off from the external environment with melted dental wax.

Each of the clones' progress was monitored visually, and from time to time the dental seal was broken and additional media was added. Even the single cells could be viewed microscopically with relative ease because of the size of the fairly good optical properties of the glass chamber/slides. When the clones outgrew the chamber slides, they were transferred to small glass T flasks; and from there to 25 ml plastic Falcon tissue culture flasks. We isolated 10 cloned lines in this manner.

Autographa californica inoculum. The inoculum, kindly supplied by Dr. Pat Vail, was the supernatant fluid of a 1000 x g for 15 minutes centrifugation of five day old infected TN-368 cells. The cells were infected with *Trichoplusia ni* hemolymph containing *Autographa californica* nuclear polyhedrosis virus.

Plaque method for the assay of infectivity. We did the assay essentially as it was first described (Hink and Vail, 1973), with a few minor changes (Volkman and Summers, 1975). The cells we used as indicators were log phase cells taken when they were between the concentrations of 2 and 6 x 10^5 cells per ml. 3.5 x 10^5 cells were seeded per 35 millimeter Petri dish and all assays were routinely done in triplicate. The plates were scored after 72 hours of incubation at 28°C.

III. Results

Characteristics of the clones. Morphologically, the clones appeared to be slightly different from one another, although all of them were still somewhat heterogeneous as was the parent TN-368. (Fig. 1A, B, C, D, E, F).

Clone 3: The cells are relatively large and spindle-shaped, with plump centers and long, graceful termini. Nuclei are large and round and contain numerous nucleoli. Protoplasmic extensions are present and fairly prominent. Clone 3 cells are relatively homogeneous and the average of 70 random cells measured with a micrometer is 103 microns in body lenght, 200 microns from tip to tip of the protoplasmic extensions and 35 microns in width.

Clone 8: The cells are relatively small, only 60 microns in body length, and commonly occur with exocentric nuclei. Protoplasmic extensions may be present but are not prominent.

Clone 10: This clone is relatively heterogeneous and contains many large cells, wider than the clone 3 cells and with correspondingly larger nuclei. The larger cells are not necessarily spindle-shaped, but rather can be somewhat circular or rectangular, and possess multiple, though not very long protoplasmic extensions. Others retain the spindle-shape and the long protoplasmic extensions but are still very wide through the center.

Clone 13: These cells are heterogeneous and morphologically resemble a cross between clones 10 and 3.

Clone 5: These cells are spindle-shaped with very prominent protoplasmic extensions.

If the cells are passed once every 3 days at an initial concentration of between 1 and 2 x 10^5 cells/ml and are thus maintained in log phase, their growth curves are all very similar and their generation times are between 14 and 18 hours.

If the cells are passed at a concentration lower than 10^5/ml or if they are passed from late stationary phase, then a lag period in their growth ensues.

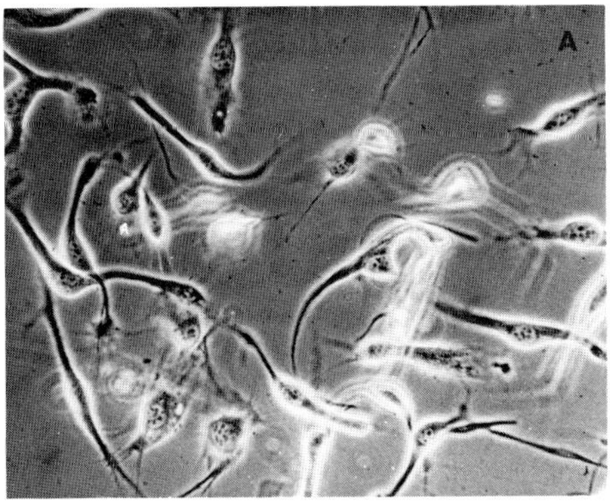

Fig. 1 A.

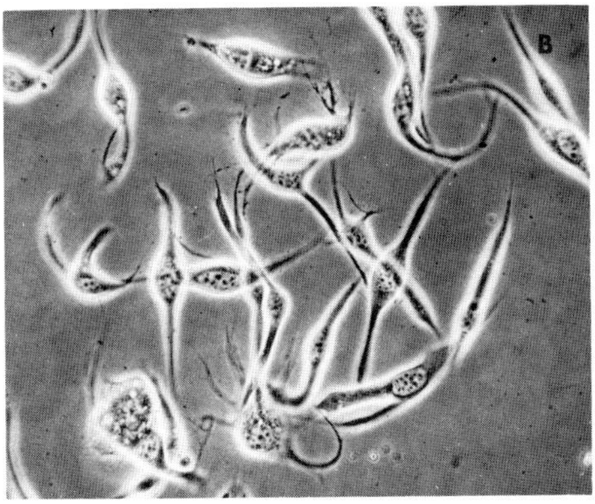

Fig. 1 B.

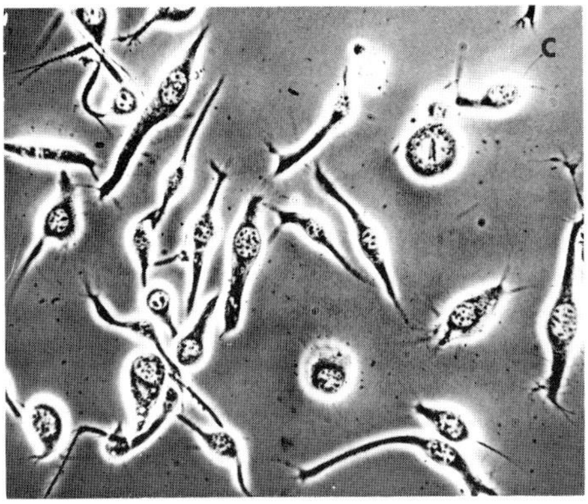

Fig. 1 C.

Fig. 1. Photomicrographs of daughter cell lines developed from single cells of *Trichoplusia ni* ovarian cell line TN-368. Log phase cells were transferred to glass coverslips to which they attached in a 2 hour incubation period at 28°C. x 2,000. (A) parent cell line, TN-368 (B) clone 3 (C) clone 8 (D) clone 10 (E) clone 13 (F) clone 5

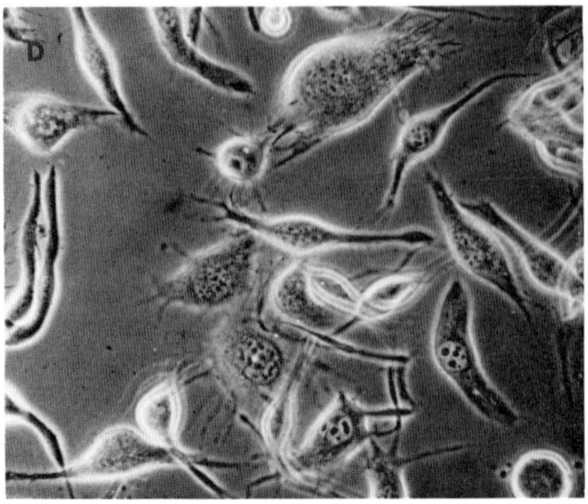

Fig. 1 D.

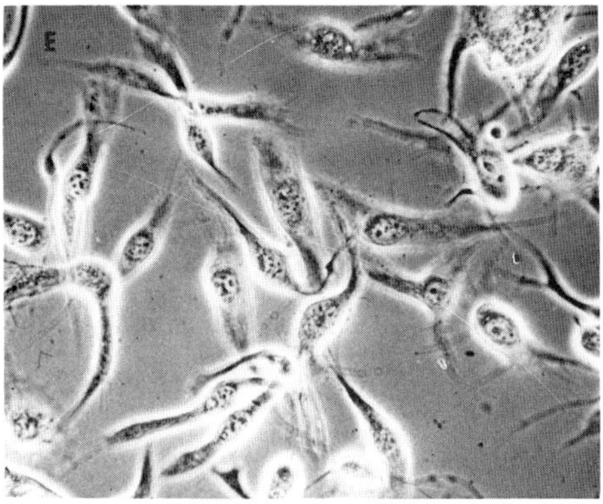

Fig. 1 E.

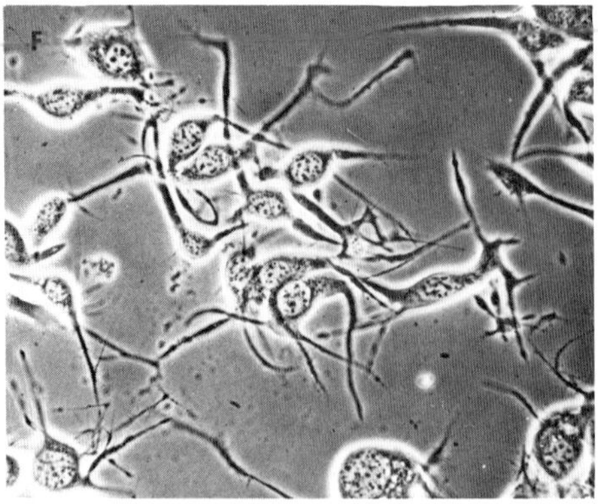

Fig. 1 F.

Effects of phase of growth on plaque of growth on plaque assay indicator cells. Cells taken from the lag phase, logarithmic phase and the stationary phase were compared as indicator cells in the plaque assay (Figure 2). In the case of both TN-368 and clone 10, 2 to 3 times more plaques were apparent when using log phase cells than when using the lag phase cells and approximately 2 times more appeared with log phase than with stationary phase.

In view of these results all plaque assays were done with logarithmic phase cells.

Comparison of the cloned lines and TN-368 as indicator cells in the plaque assay. The cloned lines were compared to TN-368 with regard to their capability to serve as indicator cells in the plaque assay.

Table 1 shows the relative number of plaques that appear on the various indica-cator lines when they are infected with an identical inoculum. The parent line and clone 10 are the best indicators giving the highest number of plaques, followed by clones 3 and 8 and then by 5 and 13. It is of interest to note that a line's relative capability of serving as an indicator seems to be independent of the generation time. That is, for example, even though clone 13's population doubling time was as short or shorter than either TN-368's or clone 10's, fewer plaques appeared. These results indicate that either the cells of some clones are less receptive to infection than cells of other clones, or that a larger proportion of infected cells in some clones produce polyhedra less readily; or both. We tried to shed some light on the situation by setting up comparative synchronous infection experiments, being careful to use cells from the same populations that we used as indicator cells in coincident plaque assays.

For each clone we removed 2 million (2×10^6) cells and transferred them to new flasks where they were allowed to attach to the plastic. After an attachment time of about 2 hours we removed all the media and added 1 ml of inoculum containing 10 million plaque-forming units (when assayed on TN-368 cells) which was a multiplicity of infection of 5. This is the amount of virus predicted by the Poisson distribution needed to insure infection of more than 99% of the cells. We allowed the virus to adsorb 1 hour while the flasks were slowly being rocked on a rocker platform, then added 4 ml of fresh media to each flask and allowed them to incubate at $28^{\circ}C$. Periodically, samples were withdrawn and 300 to 400 cells were inspected with a phase microscope for signs of infection.

The data obtained revealed that both the parent line and clone 10, the highest capable indicators in the plaque assay began to show prepolyhedral CPE at about 10.5 hours post infection while the other lines start from 1.5 to 3.5 hours later. The rates of development of total CPE, i.e., the sum of both pre-polyhedral and polyhedral CPE, once it began to appear, were all about the same with maybe those of clones 13 and 5 being slightly less. By 36 hours 100% of the cells in all the lines showed some sort of CPE.

It was also observed that in all cases the relative rates of appearance of polyhedra were less than the rates of appearance of total CPE. This may indicate that a rate limiting step occurs for crystal maturation. As might be expected, clones 10 and the parent line began polyhedra production before the other clones did.

For both clone 10 and TN-368 there was a 7 to 8 hour lag between the first appearance of pre-polyhedral CPE and the first appearance of polyhedra. For clones 8, 5 and 13, this period was 9 to 10 hours.

PFU per plate

	lag	log	stationary
TN 368	38	95	67
Clone 10	41	101	47

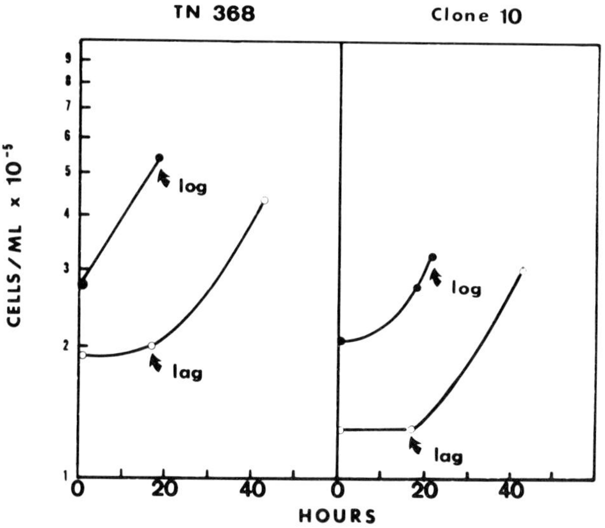

Fig. 2. Dependence of plaque assay results on stage of indicator cell growth. Cells from both clone 10 and TN-368 in lag, logarithmic and stationary phase of growth were compared as indicator cells in the plaque assay. (Arrows indicate the time when the cells were taken for use in the plaque assay). The conditions for the assay are described in the text.

Table 2 shows the percentage of polyhedra containing cells relative to clone 10 at 24 and 48 hours. These values were not determined at 72 hours because by 72 hours there is considerable cell lysis in the cultures.

At 24 hours the percentage of polyhedra containing cells in each of the clones roughly correlates with the plaque assay values in that clone 10 and TN-368 contain the highest percent polyhedra and are about equal, with clones 8 and 3, 5 and 13 containing less. By 48 hours, however, these latter values have increased considerably, suggesting that perhaps those clones that do relatively poorly as indicator cells in the plaque assay would reveal more plaques with a lengthened incubation period.

TABLE I.

Relative number of PFU revealed by TN-368 and daughter cell lines

Cell Population	PFU per plate[a]	Generation Time(hours)[b]
10	280	17
	269	14
TN-368	264	18
	264	16
8	156	16
	213	15
3	190	16
	196	16
5	64	16
	52	17
13	46	14
	85	15

[a]Results of two experiments, each assayed in triplicate. An identical inoculum was used on all the plates in each of the assays. Assay conditions are given in the text.

[b]Refers to the population doubling times of the cells used as indicators in the assays.

TABLE 2.

Percentage of polyhedra-containing cells relative to clone 10 at 24 and 48 hours

Cell Population	24 hour	48 hour
TN-368	.92[a]	1.05[a]
	1.36	1.03
8	.49	.83
	.35	.89
3	.46	.84
	.45	.89
5	.31	.98
	.38	.85
13	.44	.81
	.16	.82

[a]Results expressed as the ratio of the percent of polyhedra-containing cells in each of the cloned lines and TN-368 relative to the percent of polyhedra-containing cells in clone 10.

Plaque assay with extended incubation. In view of the above results it seemed that perhaps those clones that did relatively poorly as indicator cells in the plaque assay would reveal more plaques with a lengthened incubation period. Table 3 shows the results of a comparative plaque assay incubated for 96 hours. It can be seen that the extended incubation period makes no difference, and that just as in the 72 hour assay, clones 8 and 3 still reveal only 60 to 70 percent of the PFU that TN-368 and clone 10 do, and that 5 and 13 reveal only about 25 to 30 percent.

TABLE 3.

Effect of extended incubation on plaque assay results

Indicator cell line	Incubation period	
	72 hours	96 hours
Clone 10	1.03[a]	1.04[a]
Clone 8	.70	.69
Clone 3	.72	.60
Clone 5	.24	.27
Clone 13	.25	.33

[a]Results expressed as the ratio of the average PFU/plate obtained when using each of the cloned lines as the indicator line relative to the average PFU/plate obtained with TN-368 cells.

IV. Conclusions

We have cloned several daughter lines from TN-368 which differ from the parent cell line as well as from each other both morphologically and in their capability to serve as indicator cells for the plaque assay. These latter differences cannot be attributed to rates of population growth and cell division as the generation times for the clones are all very similar. The time required for polyhedra production may be involved as the lines that are the best indicators produce polyhedra before the others do. This notion is not strengthened, however, by the observation that the clones that indicate second best produce polyhedra in about the same time as the worst indicators do. Still, the kinetics are not exactly the same for these clones and this possibility cannot be ruled out. Evidence was obtained that all the cells in each of the clones will support infection when exposed to a relatively high MOI, but it is not known whether every random cell in each of the clones will support infection upon coming into contact with a single potentially infectious particle. The numbers obtained from the plaque assays indicate that the answer is no. Nevertheless, it is not certain that fewer plaques means in fact that fewer cells have been infected. The difference might be that in the poor indicator clones the infected cells more often fail to transmit the infection to adjacent cells, or, that once infected, the cells fail to produce polyhedra.

Acknowledgements

We gratefully acknowledge the excellent experimental assistance of Ms. Ching-Hsiu Hsieh and photographic assistance of Mr. Gale Smith. This investigation was supported by Public Health Service grant AI-09765 and Environmental Protection Agency grant R803666.

V. References

Hink, W.F. (1970). *Nature* (London). 226:466.

Hink, W.F. and Vail, P.V. (1973). *J. Invertebrate Pathol.* 22:168.

Habel, K. (1969). In: Fundamental Techniques in Virology. (Habel, K. and Salzman, N.P., eds.) Academic Press. pp. 288-296.

Volkman, L.E. and Summers, M.D. (1975). In preparation.

Chapter 28

GROWTH OF THE *TRICHOPLUSIA NI* (TN-368) CELL LINE IN SUSPENSION CULTURE

W.F. Hink and E. Strauss

I. Introduction

This paper reports successful attempts at culturing an insect cell line (TN-368) in suspension cultures. This system is being developed for large scale production of cells and will be evaluated as a method for production of pathogenic insect viruses.

II. Methods and materials

The *Trichoplusia ni* (TN-368) cell line established by Hink (1970) was used in this study. At the start of this investigation they had been in continuous culture for 5 1/2 yrs and were in the 700th subculture.

The medium used for suspension culture of cells was a modification of TNM-FH medium (Hink, 1970) and contained 90.0 ml Grace (1962) medium, 8.0 ml fetal bovine serum (FBS), 0.25 gm bovine albumin, 0.3 gm lactalbumin hydrolysate, and 0.3 gm TC yeastolate. In experiments designed to test effects of omission, combination, or addition of ingredients this formula was altered and the new formulas are noted in the text.

Four kinds of vessels were used for suspension culture of cells. Water jacketed spin flasks (Wheaton Scientific) with 100 ml volumes were used in all experiments where not noted otherwise. A Vibromixer-Glass fermentor (Chemapec, Inc.), Bioflow (New Brunswick Scientific Co.), and MF-205 fermentor (New Brunswick Scientific Co.), were also evaluated for growth of cells in suspension.

All experimental cultures had initial populations of 1-2x10^5 cells/ml and the inocula were from 2-3 day old parent cultures. Cells were grown at 28-29°C.

III. Results and discussion

Initial efforts to propagate cells in suspension produced growth curves as in the open circles in Figure 1. This growth rate was reduced with a lower final cells density in comparison to stationary cultures, where maximum cell densities of

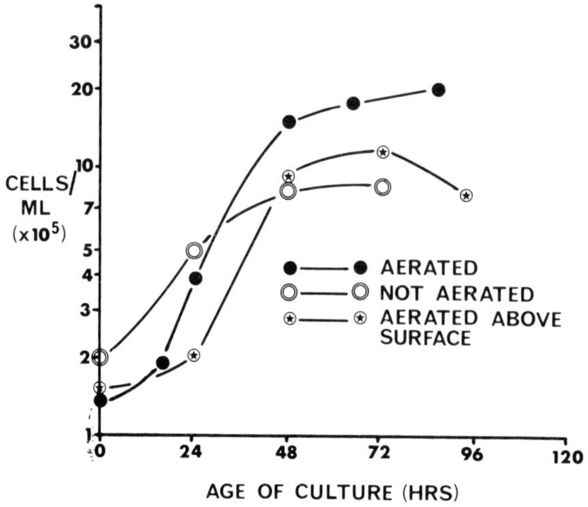

Fig. 1. Effect of aeration on cell growth.

2×10^6 cells/ml are reached in 48 to 72 hrs. Another problem in our early experiments was cell clumping. This was alleviated by adding 0.1% methylcellulose, 50 cps, (Methocel® 65HG, Dow Chemical Co.), to the suspension culture medium.

Growth was increased by aeration. Passing a stream of air above the surface of the medium resulted in higher cell densities and sparging air through the medium gave even higher densities (Fig. 1). The air flow rate was not determined but the flow was adjusted so that an air bubble emerged from the end of a 1.2 mm inside diameter glass tube about every second. If rate of aeration was increased it was clearly detrimental to the cells.

As cells grew in aerated suspension flasks, the pH of the cultures increased. The pH of fresh medium was 6.3 and it rose to 7.0 within 96 hrs (Fig. 2). Figure 3 illustrates the effect of adjusting pH. When the cultures were 48 hrs old the pH was approximately 6.7. At this point cell growth rate slowed down. When 0.1N HCl was used at 48 hrs to adjust pH to 6.4, cell growth ceased. However, adjustment of pH to 6.4 at 48 hrs with 0.1N H_3PO_4 resulted in a maximum cell density of 3×10^6 cells/ml. This was the highest density we have obtained and was greator than we usually observed in stationary cultures.

Results with three other suspension culture systems were as follows. Cells grown in the Bioflow in 400 ml volumes or the MF-205 fermentor in 2 liter volumes also reached a density of 3×10^6 cells/ml. However, the growth period is approximately twice as long as in spin flask cultures, i.e. it takes 6-7 days in the Bioflow or fermentor and 3-4 days in the spin flasks to reach 3×10^6 cells/ml. Cells cultured in the vibromixer had slower population doubling times and appeared to be damaged by this method of keeping cells in suspension.

In an effort to reduce cost of medium, various constituents were eliminated or reduced and cell growth evaluated in these modified formulations (Table 1). Stationary 5.0 ml cultures, grown in Falcon polystyrene 25 cm^2 flasks, were used for initial evaluation of media. These required smaller amounts of media than spin flasks and cells could be conveniently carried through passages to determine any long term effects on growth. The elimination of egg ultrafiltrate and bovine albumin did not effect cell growth. In medium without these two ingredients, the FBS was reduced from the previously used amount of 8% to 4% and cells continued to grow normally through 16 subcultures. In egg ultrafiltrate and bovine albumin-free medium with 1% FBS and 2% FBS the cells stopped replicating at the 5th and 6th subculture respectively.

TABLE 1.

Maximum densities of cells grown in various media formulations in stationary 5.0 ml cultures

Medium	Maximum Population after 2 days cells/ml	Passage
Complete TNM-FH medium	1.6 to 1.3×10^6	-
Egg ultrafiltrate-free	1.53×10^6	25
Egg ultrafiltrate (EU) + bovine albumin (BA)-free	1.53×10^6	23
EU + BA-free with 6% FBS	1.44×10^6	4
EU + BA-free with 4% FBS	1.56×10^6	16
EU + BA-free with 2% FBS	7.1×10^5	4
EU + BA-free with 1% FBS	3.0×10^4	4

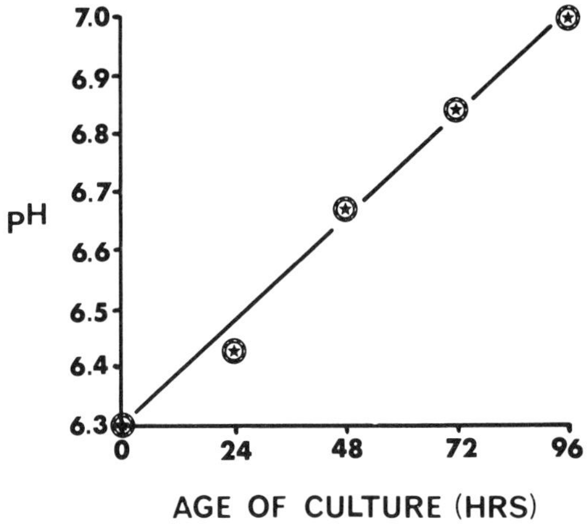

Fig. 2. Change in pH of culture as cells replicate.

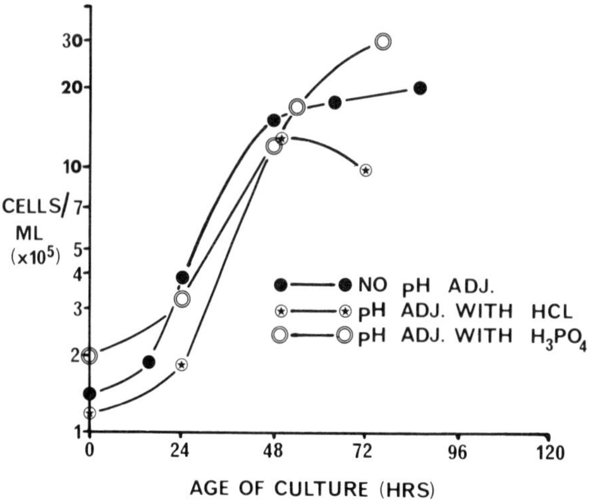

Fig. 3. Effect of pH adjustment on cell growth. pH adjusted to 6.4 at 48 hrs.

In spin flasks, the cells also replicated as well in medium without egg ultrafiltrate and bovine albumin as in medium containing these two ingredients. However, unlike the stationary cultures, in the absence of these two ingredients the FBS level could not be reduced without effecting cell growth. Media with 4% or 6% FBS did not support as high cell densities as medium with 8% FBS. By elimination of two medium constituents the cost has been reduced by 50%.

Acknowledgements

This research was supported, in part, by Grant R-802516 from the Environmental Protection Agency and Grant GB-38154 from the National Science Foundation.

V. References

Grace, T.D.C. (1962). *Nature, 195,* 788-789.

Hink, W.F. (1970). *Nature, 226,* 466-467.

Chapter 29

EFFECTS OF EXTRACTS FROM ECHINODERMS ON CELL CULTURES FROM MOLLUSKS AND ECHINODERMS

J.T. Cecil, G.D. Ruggieri, and R.F. Nigrelli

I. Introduction

A biologically active compound isolated from the Cuvierian tubules of the Bahamian sea cucumber *Actinopyga agassizi* was shown by Nigrelli and Jakowska (1960) to display an extensive variety of effects on various biological systems. This substance was the first steroid saponin of animal origin to be isolated, and was given the name holothyrin by Nigrelli (1952). Holothurin has been shown to suppress the growth of sarcoma 180, and Krebs-2 ascites tumors in Swiss mice (Nigrelli and Zahl, 1952; Sullivan *et al.*, 1955; Sullivan and Nigrelli, 1956; Nigrelli and Jakowska, 1960). One fraction of holothyrin, referred to as holothurin A, has been chemically characterized by Chanley *et al.*, (1966), and Fries *et al.*, (1965). Recent studies have shown that holothurin and other steroid saponins suppress or completely inhibit human epidermal carcinoma (KB) cell lines in monolayer cultures (Nigrelli *et al.*, 1967).

In continuing studies on the effects or extracts derived from various echinoderms in biological systems, our previous results have directed attention to specific cellular abnormalities associated with echinoderm extracts on animal cells cultured *in vitro*. This report considers some of these effects on a fresh water invertebrate cell line and several marine invertebrate cell lines in culture.

Cardiac tissue from the surf clam, *Spisula solidissima* has been grown in monolayer cell culture for extended periods. These cells show mitoses and have continued to grow and form monolayers, for a period of six or more months (Cecil, 1969). Recently we have been able to transfer these cells, and they are now in the fifth generation passage. Cells are fibroblast-like and frequently aggregate.

Utilizing the incorporation of the chemical carcinogen 3-methyl-cholanthrene in the growth medium, we have been able to initiate the primary aspects of transformation in surf clam cardiac cells. The transformation induced consists of loss of contact inhibition of movement, resulting in criss-crossing of cells, juxtaposition of cells, invasion of the initial predominate cell type by another, and morphological transformation from a fibroblast-like cell culture to that having an essentially epithelial-like cell characteristic (Cecil *et al.*, 1974). We have not yet been able to induce unrestrained cell proliferation nor tumor formation in host clams of the same species.

Sea star explants from *Asterias forbesi* and *Asterias vulgaris* axial organs grow in monolayer cell cultures having a characteristic fibroblast-like morphology. We have now established an eight generation transfer culture of these cells. Axial organ explants from the sea urchin, *Arbacia punctulata* grow in a similar manner to the sea stars, having a typical fibroblast-like appearance in the cell monolayers. The axial organ explant may pulsate spontaneously with the cells contracting and relaxing.

The cell line designated *Bge*, *Biomphalrai glabrata*, the host for schistosomiasis, was established by Dr. Eder Hansen of the Clinical Pharmacology Research Institute, Berkeley, California (Hansen, 1974). These cells are fibroblast-like, showing cytoplasmic extensions between cells.

II. Materials and methods

A. Preparation of medium.

Medium for *Bge* cells was published by Hansen (1974). Medium for surf clam cardiac cultures and echinoderm axial organ cell cultures follow:

1. Salt water base:

Ingredient	*Grams per 500 ml.*
NaCl	23.50
$MgCl_2 . 6H_2O$	7.10
KCl	0.75
$MgSO_4 . 7H_2O$	4.19
K_2HPO_4	0.19
Glucose	0.50
Trehalose	0.50
Galactose	0.50

The above ingredients are dissolved in 400 ml. of distilled water by using a magnetic stirrer.

$CaCl_2 . 2H_2O$ 1.48 grams

Dissolve in 40 ml. of distilled water; then, stirring slowly, add to the above. The resulting solution is then brought up to 500 ml. with distilled water. The solution is millipored through a 0.45 micron filter.

2. Final medium

Ingredient	*Milliliters*
Salt water base	3.40
Sterile distilled water	3.00
Fetal Bovine Serum	1.00
Wolf's Amphibian Medium	2.00
Taurine (0.45 gm. per 100 ml. H_2O)	0.25
Antibiotic - Antimycotic solution (GIBCo)	0.15
Phenol red (0.01% solution)	0.01
$NaHCO_3$ (7.5% solution)	0.55

(The final pH of this medium, after equilibration in a T-45 flask for 15 minutes, should be 7.2. The milliosmolarity should be about 695.)

B. Source of cell cultures:

The heart of surf clam is a two chambered vessel which loops around the intestine. To prevent unwanted contamination by protozoa, bacteria, and mycoses, the cardiac tissue must be dissected carefully away from the intestine.

The axial organ of echinoderms is conspicuously located alongside the stone canal, directly below the madreporite. The axial organ of *Asterias forbesi* is white, while that of *Asterias vulgaris* is purple. *Arbacia punctulata* contains an axial organ near Aristotle's lantern which is immersed in a purplish dye-like substance. This organ must be washed very thoroughly with 2.6% NaCl. The organs are removed from the animals with comparative ease.

Dr. Eder Hansen kindly furnished us with the *Bge* cell line in the 40th passage. It has been passed at least ten times in our laboratory in the ensuing period.

C. Processing of specimens for cell culture:

1. Cardiac tissue from the surf clam was carefully dissected from the intestine. The hearts were pooled in groups of 10 and washed for 30 minutes in running sea water, having a milliosmolarity of 750. The sea water is pumped directly from the ocean and filtered through sand and gravel before reaching the laboratory.

The cardiac tissue was minced finely with iris scissors, collected in centrifuge tubes, and washed five times in a solution of 2.6% NaCl containing streptomycin (600 micrograms per ml.), penicillin (600 units per ml.), and fungizone (25 units per ml.). The tissue was collected and placed in 10 ml. of 0.25% trypsin in a 100 ml. trypsinizing flask containing a teflon coated stir-bar. The flask was placed on a slowly spinning magna-stir of 20°C and trypsinized for six hours.

At the end of the digestion period, the cells were filtered three times through four layers of surgical gauze into a centrifuge tube. The cells were washed three times in 2.6% NaCl containing the above antibiotics, sedimented at 800 rpm for 10 minutes and counted with a haemacytometer. A total of 10° cells/ml. of growth medium was inoculated into Leighton tubes or T-45 plastic flasks. Tubes and flasks were incubated at 20°C.

2. Axial organ cultures from the sea stars *Asterias forbesi*, *Asterias vulgaris* and the sea urchin *Arbacia punctulata* were prepared as follows:

The axial organ was carefully dissected away from the stone canal. Organs were washed in six changes of 2.6% NaCl containing the same antibiotics above. Generally, the entire axial organ was placed on the bottom of a flask or on a cover slip in a Leighton tube containing the same growth medium for clams. The tubes and flasks were incubated at 20°C. (Cecil and Nigrelli, 1972; Cecil *et al.*, 1974).

D. Echinoderm extracts:

Echinoderms (Table 1) were collected from various sources during the past several years. Holothurin was obtained from the Cuvarian tubules of the Bahamian sea cucumber *Actinopyga agassizi. Patiria miniata* (red web sea star) and *Pisaster brevispinus* (short spine sea star) were obtained from the North Pacific while *Asterias forbesi* was obtained locally from Long Island Sound. Extracts from the echinoderms were prepared by the method of Rio *et al.* (1965).

E. Visualization of activity:

With the exception of holothurin, all the extracts may be lyophilized and are water soluble. Holothurin is only partially soluble in water. Since holothurin is heat stable, Nigrelli and Zahl (1952), have shown that the active fraction may be recovered by evaporation. Dilutions of the extracts were prepared w/v, in the media described above, and sterilized through a swinney millipore filter. Cells were removed from stock and either passed at 10^4 cells per ml., allowed to monolayer and extracts incorporated, or cells passed and grown in dilutions of these extracts. Leighton tube cultures of cells and extracts were incubated at 20°C.

At intervals after treatment with echinoderm extracts, coverslip preparations were stained with acridine orange fluorochrome and observed by fluorescent microscopy. Other cover slip preparations were observed by phase microscopy or stained with hematoxylin and eosin.

F. Acridine orange staining:

This staining procedure offers the property of cytochemical differentiation (Armstrong 1956). DNA-containing cellular components fluoresce yellow green and RNA-components fluoresce rusted.

G. Fluorescent microscopy:

An American Optical Fluorolume microscope and illumination system were used. An Osram 200 watt bulb in conjunction with a BG-12 exciter filter and an OG-6 barrier filter was the illumination source. A Zeiss 35 mm camera was used, with a Zeiss automatic timer. Either type B Kodak Ektachrome, Panatomic-X, or Tri-X film was used.

H. Phase microscopy:

Phase was utilized either by cover slips on slides observed by American Optical phase equipment, or by the use of a Wild #60 inverted microscope and phase optics.

III. Results

A. Relative activities of echinoderm extracts:

Samples of echinoderm extracts were tested under assays utilizing the medium above for marine invertebrates, and Hansen's medium for the fresh water snail. Medium, containing the extracts, was placed on the monolayered cell cultures; or extracts were incorporated in the growth medium when cells were passed and incubated. Medium on controls was changed when extracts were assayed. The additional milliosmolarity between control and extract medium was negligible.

Depending upon the extent of morphological abnormalities and cytotoxicity produced, activity was recorded on a comparative basis as + = 25%, ++ = 50%, +++ = 75%, +++ = 100%. Results are recorded in Table 1.

TABLE 1

Extract (25 µgm/ml)

	Bge	Spisula	Echinoderms
Actinopyga agassizi (crude)	++++	++++	+++
Patiria miniata (extract)	+++	+++	++
Pisaster brevispinus (extract)	+++	+++	++
Asteria forbesi (extract)	+++	+++	+

B. Normal cell cultures:

Illustrations from normal cell cultures or mollusks and echinoderms are shown in figures 1, 2, 3, 4 and 5.

C. Cytoplasmic vacuolization:

When cell cultures from invertebrates are exposed to marine invertebrate extracts, lysis and vacuolization may occur in 24 hrs. Figure 6 and 7.

Fig. 1

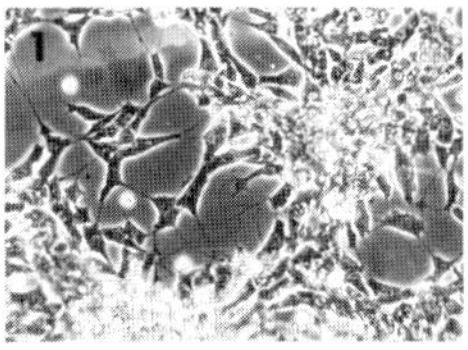

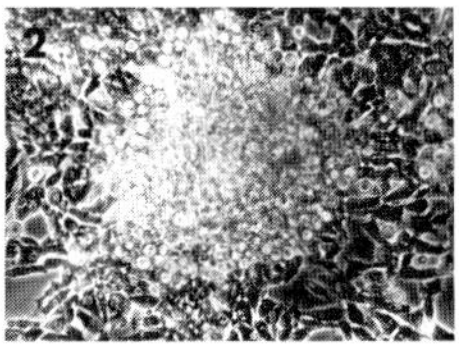

Fig. 2

Fig. 3

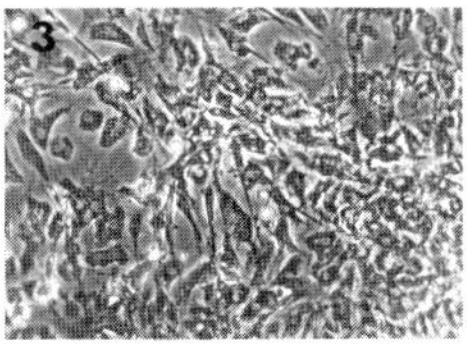

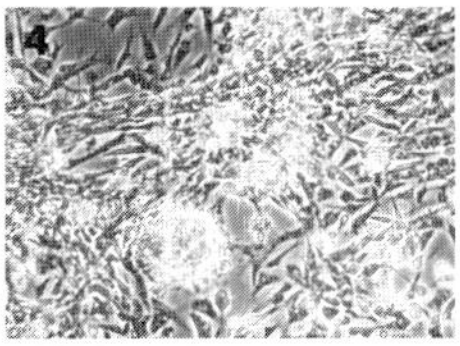

Fig. 4

Fig. 5

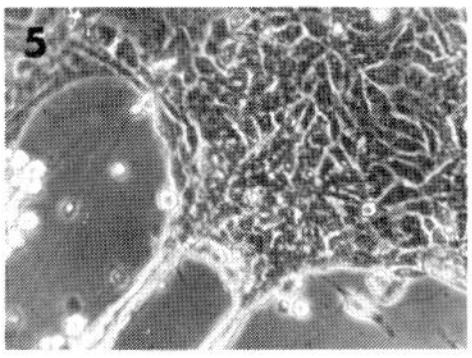

Fig. 1. Normal cells from the fresh water snail cell line *Biomphalaria glabrata, Bge.*

Fig. 2. Older cell culture of *Bge.* Cellular aggragation.

Fig. 3. Normal cardiac cell culture from the Surf clam *Spisula solidissima.*

Fig. 4. Normal axial organ cell culture from echinoderms.

Fig. 5. Normal "gland" cultures from echinoderms.

D. Karyokinesis without cytokinesis and cellular RNA accumulation.
Several nuclei may form without cell division, resulting in giant cell formation. Bright-white granules in cytoplasm are rust red in color. Figure 8 and 9.

Fig. 6 Fig. 7

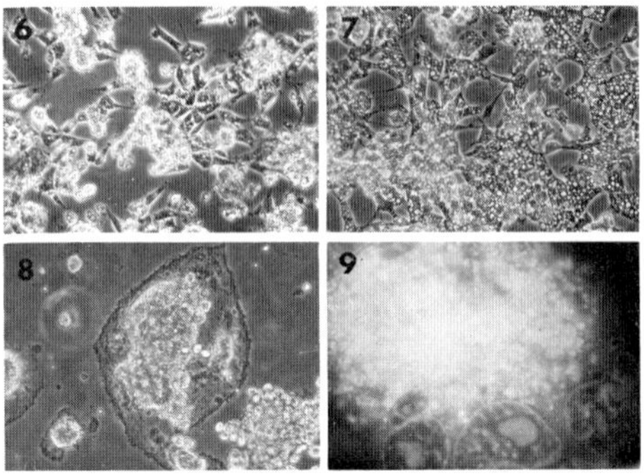

Fig. 8 Fig. 9

Fig. 6 and 7. Lysis and vacuolization of invertebrate cell cultures exposed to marine invertebrate extracts.

Fig. 8 and 9. Giant all formation.

IV. Discussion

In this paper we describe some of the cytological abnormalities which result after exposure of cell cultures from the axial organs of the Echinoderms, *Asterias forbesi*, *Asterias vulgaris* and *Arbacia punctulata*, to extracts from echinoderms. We have also exposed molluscan cells from the fresh water snail *Biomphalaria glabrata* and the marine clam *Spisula solidissima* to these extracts *in vitro*.

Cytological abnormalities include cytoplasmic vacuole formation, morphological transformation, massive globular RNA production, karyokinesis without cytokinesis-sometimes resulting in giant cell formation.

The occurrence of all these conditions may imply the presence of several biologically active compounds, a concentration effect of the extracts, or multiple activity of a single compound. These findings could also indicate the biological activity of the extracts occurs on one or several stages in the cell replicative cycle.

Other possibilities include cellular membrane activity. Apparently, the extracts do not penetrate the nuclear membrane, or they may prevent the formation of the cleavage furrow. The very large accumulation of cytoplasmic RNA suggests the possibility of tRNA inhibition, preventing the synthesis of protein. The excessive RNA accumulation also suggests that mRNA may be activated by these extracts.

Nigrelli *et al.*, (1967) have shown toxic effects on human KB (oral carcinoma, Eagle, 1955) cells in culture, utilizing extracts from several sea stars, holothurin, and bonellin. Lytic effects, usually resulting in cellular death, are observed. Ruggieri (1965, 1966), has described the animalization action of echinoderm toxins on sea urchin larval development. These results show the diversity of action of some of the echinoderm extracts on different biological systems.

Acknowledgment

Supported by a grant from The Scaife Family Charitable Trusts.

V. References

Armstrong, J.A. (1956). *Exp. Cell. Res., 11,* 640.

Cecil, J.T. (1969). *J. Invert. Path.,* 407-410.

Cecil, J.T. and Nigrelli, R.F. (1972). *Society for Invertebrate Pathology Newsletter,* Vol. IV, No. 3, 13.

Cecil, J.T., Ruggieri, G.D. and Nigrelli, R.F. (1974). *VIIth Annual Meeting; Society for Invertebrate Pathology,* Arizona State University. Tempe, Arizona.

Cecil, J.T., and Nigrelli, R.F. (1974). *In Vitro, 9,* 388.

Chanley, J.D., Mezzetti, T., and Sobotka, H. (1966). *Tetrahedron, 22,* 1957.

Eagle, H. (1955). *Proc. Soc. Exp. Biol. (N.Y.), 89,* 362.

Hansen, E.A. (1974). *Biomphalaria glabrata. IRCS, 2,* 1703.

Fries, S.L., Durant, R.C., Chanley, J.D., and Mezzetti, T. (1965). *Biochem. Pharmacol., 14,* 1237.

Nigrelli, R.F. (1952). *Zoologica, 39,* 89.

Nigrelli, R.F., and Zahl, P.A. (1952). *Proc. Soc. Expl. Biol. Med., 81,* 379.

Nigrelli, R.F., and Jakowska, S. (1960). *Ann. N.Y. Acad. Sci., 90,* 884.

Nigrelli, R.F., Stempien, M.F. Jr., Ruggieri, G.D., Liguori, V.R., and Cecil, J.T. (1967). *Fed. Proc., 26,* No. 4, 1197.

Rio, G.J., Stempien, M.F. Jr., Nigrelli, R.F., and Ruggieri, G.D., S.J. (1965). *Toxicon, 3,* 147.

Ruggieri, G.D., S.J. (1965). *Toxicon, 3,* 157.

Ruggieri, G.D., S.J. (1966). *Am. Zoologist, 6,* No. 593.

Sullivan, T.D., Ladue, K.T., and Nigrelli, R.F. (1955). *Zoologica, 40,* 49.

Sullivan, T.D., and Nigrelli, R.F. (1966). *Proc. Am. Assoc. Cancer Res., 2,* 151.

Chapter 30

CYTOTOXIC AND ANTIPROLIFERATIVE SUBSTANCES IN INVERTEBRATES AND POIKILOTHERMIC VERTEBRATES

M.M. Sigel, W. Lichter, L.L. Wellham and D.M. Lopez

I. Introduction

This conference has been concerned with problems relating to the cultivation and proliferation of invertebrate cells, their social behavior, interactions and susceptibilities to viruses. Problems facing the investigators working with cells from invertebrate organisms are in part like the problems that faced (and are still facing) tissue culturists and virologists dealing with vertebrate cells. We have heard a good deal about nutritional requirements, about mutations and about hormonal effects. What interests me most is the issue of determination and regulation of cellular behavior in mixed populations. On the one hand there appears to be a dependence of one cell upon its fellows for the initiation of growth or for the sustainment of replication or function. On the other hand cells may interact in an antagonistic fashion with the result of loss of one subset of cells or the dominance of another subpopulation. One may view these interactions as an expression of diversity of regulatory controls. It is therefore appropriate to speak about certain aspects of regulation exerted by cytotoxic and antiproliferative substances derived from vertebrate and invertebrate sources. Because of the limitation of time and space I will limit my remarks to brief descriptions of relevant findings obtained from our researchers on serum proteins of the shark and a biologically active extract from the tunicate *Ecteinascidia turbinata*.

II. Shark serum proteins

The nurse shark (*Ginglymostoma cirratum*) which represents a living form with an origin over 100 million years ago has received major attention in our phylogenetic studies. The serum proteins of this animal present an interesting array ranging from what one might call primitive protein analogous to sugar binding lectins, which are ordinarily present in plants and invertebrates to highly specific antibodies usually associated with higher vertebrates. Between the two extremes are antibody-like proteins with unusually reactive combining sites which differ from conventional antibodies.

A. Sugar binding proteins

The nurse shark contains at least three sugar binding proteins in concentrations which differ among individual animals. One protein is specific for rye grass levan, while the second is specific for dextran B 1355-S-4 which contains a high percentage of alpha (Harisdangkul, *et al.,* 1972; Sigel, 1974) linked glucopyranosyl residues. The third is a fructosan specific protein (FSP) which was studied in considerable detail (Harisdangkul, *et al.*, 1972). The molecular weight of FSP was calculated to be 280,000 and the macromolecule was shown to consist of four subunits of equal weight (70,000 daltons) held together by noncovalent bonds. FSP was found to be antigenically distinct from the shark immunoglobulin (IgM) by immunoelectrophoretic and double diffusion studies (Harisdangkul, *et al.,* 1972). Although similar to concanavalin A (Con A) in its high content of aspartic and glutamic acids and by virtue of its subunit structure with noncovalent bonds, FSP differs from Con A by having a high content of cystine which is not found in this plant lectin (Harisdangkul, *et al.*, 1972). Moreover, unlike Con A FSP appeared to lack hemagglutinating and blastogenic activities (Sigel, 1974).

B. Immune antibodies

Sharks are capable of responding to a variety of antigens with the production of specific antibodies (Sigel and Clem, 1963; Clem and Sigel, 1965; Sigel and Clem, 1965). These occur both as 7S and 19S molecules, but unlike in other vertebrate species both forms of antibodies belong to the IgM class (Clem, *et al.,* 1967). IgG and IgA antibodies have not been detected in fishes except for the lungfishes. The existence of 7S monomeric IgM antibodies is characteristic of many fishes. The polymeric IgM of fish may occur either as a pentamer or as a tretramer (Bradshaw, *et al.,* 1971). The 7S molecule is not a precursor or a breakdown product of the polymer but is an independent substance differing physiologically, chemically and immunologically from the polymeric IgM (Small, *et al.*, 1970; Sigel, 1974; Voss and Sigel, 1972).

Primary immune responses can be engendered without too much difficulty although some antigens appear to be more efficient than others. However, it is the secondary immune responses which are defective in the shark (Sigel and Clem, 1975). The shark is essentially devoid of immunologio momory and in order to achieve the significant response to secondary immunization it is usually essential for the primary immunization to be of low magnitude (Sigel and Clem, 1966). This apparent deficit is not entirely due to the inability to make IgG because other fishes whose immunoglobulins are also restricted to the IgM class are nevertheless capable of mounting a strong secondary immune response (active immunologic memory) (Bradshaw, *et al.,* 1967). We have succeeded in immunizing sharks against bacterial and viral antigens and these antibodies are capable of causing agglutination, precipitation or neutralization (Clem and Sigel, 1963).

C. Natural antibodies

The sera of normal unimmunized nurse sharks possess immunoglobulins of the IgM class which react with a large assortment of antigens. Thus, they can agglutinate and lyse (in the presence of shark complement) a variety of red blood cells; they can neutralize several kinds of viruses and kill bacteria (Sigel, *et al.,* 1970). In point of fact, the presence of these substances was first discovered serendipitously when it was observed that sharks were resistant to infectious agents present in the estuarine environment. In experiments in which it was essential to perform surgery on sharks in order to remove the spleen, there were numerous occasions where the sutures had become corroded or dissolved after the sharks were returned to their pens in the sea and the wounds had opened releasing the viscera. In spite of such severe trauma and exposure to bacteria in the environment, we have never seen peritonitis or any other kind of infection in sharks, which led us to search for antibacterial substances in sharks body fluids and the demonstration of these natural antibodies. We refer to them as antibodies because they are IgM molecules (sedimentation coefficients of 19S) with binding sites in the Fab region of the molecule which is characteristic of true antibodies (Rudikoff, *et al.,* 1970; Voss, *et al.,* 1971).

In addition to viruses and bacteria these substances also react with small molecules such as dinitrophenyl (Rudikoff, *et al.,* 1970) (DNP) and nucleotides. What is more immediately relevant to our subject is that they exert profound cytotoxic activity against tumor cells and cells in culture. One can readily demonstrate activity of high potency against tumors transplanted to experimental animals (Sigel and Fugmann, 1968) or against cells *in vitro* (Sigel, *et al.,* 1970). The antiproliferative and cytotoxic effects (at least *in vitro*) require the presence of shark complement. Natural antibodies can be adsorbed to and eluted from immunoadsorbents and thereby purified. Stuart Rudikof, while working in our laboratory, accomplished this by using DNP protein complexes (Rudikoff, *et al.,* 1970). The isolated antibody possessed a reactivity for DNP; that is as it should be, but surprisingly this antibody also reacted with red blood cells. This indicated that the molecules possessed combining sites with an unusually broad reactivity, a property which sets off this substance from the conventional type of antibody endowed with combining sites restricted to one antigenic specificity.

Other experiments led to the same conclusion. For example, it was shown that the natural antibodies could cause mixed agglutination. This type of reaction involves the following steps: natural antibodies of the shark are added to chicken red blood cells (CRBC) at a concentration which is insufficient to permit agglutination of the cells. After thorough washing, the cells are mixed with sheep red blood cells (SRBC) (which apparently do not share common antigens with CRBC). The two kinds of cells interact forming rosettes with SRBC adhering to the CFBC. Since CRBC and SRBC surface antigens are different, the ability of antibody to combine with each surface and to cause this mixed agglutination is an indication that the antibody has a multiple specificity. A third kind of proof for the presence of polyspecific binding sites come from experiments on competitive binding. These experiments indicated that the binding of *different* antigens occurred in or close to the site. For example, if DNP was first allowed to react with the natural antibody there occurred a decrease in binding of another ligand, e.g., GMP, compared to this antibody binding of GMP alone (Sigel, 1974). Thus, the natural antibodies have some properties of specific antibodies of the IgM class but possess the additional quality of a broadly reactive binding site. In this respect these antibody resemble myeloma proteins with polyspecific binding capabilities (Rosenstein, *et al.,* 1972).

III. Biologically active extracts from ecteinascidia turbinata

This marine tunicate which is a member of the Ascidiaceae class of the phylum Chordata was studied for several biological activities by our group. Most of the work has been done with water-alcohol extracts designated Ete (*Ecteinascidia turbinata* extract) in tests measuring antitumor and immunosuppressive activities.

A. Antitumor activity in vivo

We chose one of the standard procedures to determine the antitumor activity of Ete which employs BDF_1 mice implanted with lymphocytic leukemia (P-388) cells one day before treatment. The tumor bearing animals were treated daily for 10 days with Ete and the mean survival of the group was compared to the control group which was treated with saline. The mean survival of the control group was 10-11 days while the Ete treated groups showed a marked prolongation of the mean survival time with some animals surviving the 30 day observation period (Sigel, *et al.,* 1970).

B. Antitumor activity in vitro

This assay was performed using KB cells which is an established cancer cell culture line originating from a human nasopharyngeal carcinoma. These cells attach to glass or plastic surfaces where they divide with a generation time of 18-24 hours when incubated (37°C) in an optimal medium. On day 0 the cultures were seeded with a desired cell number (usually 6 x 10^4 in 2 ml per culture) and subdivided into groups supplemented with various concentrations of Ete or with buffered saline; (controls). After 3 days of incubation the extent of cell growth was measured and the difference between cultures treated with Ete and the controls was ascertained. The measurement of cellular proliferation was based on determinations of protein content at day 0 versus 3 days. It was found that Ete had a significant growth inhibiting effect: 50% endpoints of inhibition were obtained with Ete concentrations ranging from 5 to 24 mcg/ml (Sigel, *et al.,* 1970).

C. Immunosuppressive activity in vivo

To test the effect of Ete on the humoral immune system Swiss mice were immunized with sheep red blood cells (SRBC) with and without accompanying treatment with Ete. The immune response was determined by measuring the hemagglutinating antibody titer of the aminals. Ete exerted a marked immunosuppressive effect when given to animals prior to and after immunization. Treatment prior to immunization was essential while treatment after immunization was mainly ancillary (Lichter, *et al.,* 1973).

To test the effect of Ete on the cellular immune systems, skin was transplanted from C57Bl/6J X A/J to A/J mice. A marked prolongation of graft survival was noted in animals which were treated with Ete (Lichter, *et al.,* 1973).

D. Immunosuppressive activity in vitro

Mammalian cells are being categorized as thymus derived (processed) T cells and thymus independent B cells (Greaves, *et al.,* 1974). The latter are concerned with antibody production while the former participate in a variety of so called cells-mediated reactions i.e., delayed hypersensitivity and graft rejection. The experiment described above indicates that Ete may have inhibited the function of the T cells as it inhibited host versus graft reactions. To test directly the effect on T cells *in vitro,* tests were performed based on inhibition of selective stimulation of T cells. It has been known for a long time that T cells can be stimulated by plant lectins i.e., phytohemagglutinin (PHA) and concanavalin A (Con A). The stimulatory effect can be recognized in a variety of ways i.e., cell enlargement (blast

TABLE 1.

Effect of Ecteinascidia turbinata extract on human peripheral blood lymphocytes stimulated with PHA and Con A

	PHA		*Con A*	
Cultivated in:	*Counts per minute*	*S.I.*	*Counts per minute*	*S.I.*
Growth medium (control)	410	1.00	2564	1.00
Mitogen	123794	302.00	21299	8.30
Mitogen + Ete at time 0	1002	2.44	2557	0.99
Mitogen at time 0; 24 hours later Ete	8904	21.70	3645	1.42

Ete = 100 mcg/ml final concentration
PHA = 1:200 final concentration
Con A = 5 mcg/ml final concentration
S.I. = Stimulation index

formation, hence the reaction is referred to as blast transformation), increased synthesis of macromolecules and cell division. The increased rate of macromolecular synthesis lends itself to quantitative measurement. By means of radioactive precursors it is possible to quantify the rate of synthesis of protein, RNA or DNA. In our work we measured the rate of DNA synthesis as an indicator of blast transformation utilizing tritiated thymidine as a precursor of DNA. In table 1 are shown results of stimulation of human peripheral blood lymphocytes by PHA and Con A. Also shown is the inhibitory effect of Ete in concentrations of 100 mcg/ml. In this and other experiments the extract introduced into the culture at the time of stimulation by mitogen caused a total inhibition of reaction (Lichter, *et al.,* 1975). This inhibition was not due to killing of lymphocytes (T cells) as demonstrated by tests for cell viability.

Since Ete was given at the same time as mitogen it was considered possible that the inhibition was due to Ete preventing the binding of mitogen to lymphocyte membranes. The fact that PHA and Con A bind to different receptors argues against this notion as Ete would have to react with two different receptors in order to prevent stimulation by PHA and Con A. In fact, subsequent experiments revealed that receptor blockade was not the explanation for the inhibitory action as the substance could inhibit the lymphocyte response even when administered 24 hours after mitogen at a time when maximal commitment of the cell to the stimulus has already taken place It is as if the extract could reverse the signal which normally initiates transformation. This appears to be a unique function of the extract.

Acknowledgement

This study was supported by Sea Grant No. NOAA04-3-158-27 administered by the University of Miami and National Cancer Institute Contract No. C-70-2225.

IV. References

Bradshaw, C.M., Clem, L.W. and Sigel, M.M. (1971). *J. Immunol., 106,* 1480.
Bradshaw, C.M., Clem, L.W. and Sigel, M.M. (1967). *Bacteriol. Proc.,* 61.

Clem, L.W. and Sigel, M.M. (1965). *Fed. Proc.*, *24*, 504.

Clem, L.W., De Boutaud, F. and Sigel, M.M. (1967). *J. Immunol.* *99*, 1226.

Clem, L.W. and Sigel, M.M. (1963). *Fed. Proc. (Symposium)*, *22*, 1338.

Greaves, M.F., Owen, J.J. and Raff, M.C. (1974). American Elsevier Pub., Co., Inc. New York.

Harisdangkul, V., Kabat, E.A., McDonough, R.J. and Sigel, M.M. (1972). *J. Immunol.*, *108*, 1259.

Harisdangkul, V., Kabat, E.A., McDonough, R.J. and Sigel, M.M. (1972). *J. Immunol.*, *108*, 1244.

Lichter, W., Welham, L.L., VanderWerf, B.A., Middlebrook, R.E. and Sigel, M.M. (1973). In: 1972 Food-Drugs from the Sea Proceedings (L.E. Worthen, ed) MTS, Washington.

Lichter, W., Lopez, D.M., Welham, L.L. and Sigel, M.M. (1975). *Proc. Soc. Exptl. Biol. Med.* (In press).

Rosenstein, R.W., Musson, R.A., Armstrong, M.K., Konigsberg, W.H. and Richards, F.F. (1972). *Proc. Nat. Acad. Sci.*, *69*, 877.

Rudikoff, S., Voss, E.W., Jr., and Sigel, M.M. (1970). *J. Immunol.*, *105*, 1344.

Rudikoff, S., Sigel, M.M. and Voss, E.W., Jr., (1970). *Fed. Proc.*, *29*, 771.

Sigel, M.M. (1974). *Ann. N.Y. Acad. Sci.*, *234*, 198.

Sigel, M.M. and Clem, L.W. (1963). *Nature*, *197*, 315.

Sigel, M.M. and Clem, L.W. (1965). *Ann. N.Y. Acad. Sci.*, *126*, 662.

Sigel, M.M. and Clem, L.W. (1966). In: Phylogeny of Immunity (R.I. Smith and R.A. Good, eds). U. Fla. Press.

Sigel, M.M., Voss, E.W., Jr., Rudikoff, S., Lichter, W. and Jensen, J.A., (1970). In: Homologies in Enzymes and Metabolic Alterations in Cancer (W. Whelan and J. Shultz, eds). *North-Holland Pub. Co., Amsterdam*, *1*, 409.

Sigel, M.M. and Fugmann, R.A. (1968). *Cancer Res.*, *28*, 1457.

Sigel, M.M., Wellham, L.L., Lichter, W., Dudeck, L.E., Gargus, J.L. and Lucas, A.H. (1970). *Food-Drugs from the Sea Proc.*, *281*.

Small, P.A., Jr., Klapper, D.G. and Clem, L.W. (1970). *J. Immunol.*, *105*, 29.

Voss, E.W., Jr., and Sigel, M.M. (1972). *J. Immunol.*, *109*, 665.

Voss, E.W., Jr., Rudikoff, S. and Sigel, M.M. (1971). *J. Immunol.*, *107*, 12.

IV

STUDY OF VIRUSES AND PROTOZOA OF AGRICULTURAL AND FOREST IMPORTANCE

Chapter 31

IMMUNOCHEMICAL CHARACTERIZATION OF THE BACULOVIRUSES PRESENT STATUS

R. A. DiCapua and P. W. Norton

I. Introduction

Increasing debate on the necessity for more appropriate use of insecticides has focussed attention on the natural predators of insects. The characterization of insect viruses is therefore an area of growing research, and the nuclear polyhedrosis and granulosis viruses (Baculoviruses), as natural enemies, will likely continue to be agents under investigation for control of noxious pests. However, the information required for developing a comprehensive picture of the Baculovirus antigenic complex has not yet been attained.

Present knowledge of the mode of entry, multiplication, epizootiology, ecology, nomenclature, biochemical, biophysical, and morphological characteristics of the Baculoviruses allowed W.A.L. David (1975) to state, "this genus is the best known of insect viruses". Nevertheless, our increasing knowledge, if devoid of serology, will not satisfy the growing mandate for rigorous characterization required for the safe and effective implementation of biological pesticides. Classifications almost certainly will be subject to change upon re-evaluation which includes serological characterization (Norton and DiCapua, 1974 1975; DiCapua and Norton, 1974).

Efforts directed at the classification and interrelationships of Baculoviruses will remain incomplete until the number of assumed distinct viruses is verified. Defining their serological interrelationships is a means to this end. At minimum, rigorous characterization must include a determination of the serological relationship between insect and non-insect viruses which can multiply in both insect and non-insect hosts, in particular man. The necessity and proven value of serological characterization is quickly established by recalling the advent of commercial vaccines (rubella, polio, and influenza, for instance), and the difficulties encountered in the various stages of their development and application.

Basic serological investigations of insect viruses have been conducted. Aoki and Chigasaki (1921), Gratia and Paillot (1938, 1939), Tanada (1954), and Krywienczyk and Bergold (1960a, 1960b, 1960c, 1960d, 1961) have contributed to the initiation of a serological characterization of the Baculoviruses. Considering that such investigations began over 50 years ago, our ability to follow the lead of these pioneers appears unstimulated when compared, for example, to the serological data of the Leuko-, Myxo-, Paramyxo, and Arbovirues now at hand, Recent general descriptions of the Baculoviruses have been provided by Smith (1967), Bellet, *et al* (1973), and David (1975), and need not be included.

II. Physico-chemical characteristics of Baculoviruses: immunochemical considerations.

Serological characterization of antigens and the generation of an antibody response is dependent primarily upon the molecular conformation, display, and accessibility of antigenic determinants. Proteins possess a finite number of antigenic determinants, possibly in restricted regions which are most exposed to the external environment. It has been recognized for many years that changes in conformation, due to denaturation or chemical alteration, will result in changes in antigenicity. For instance, cleavage of intramolecular dissulfide bonds of proteins, which results in unfolding of polypeptide chains, will alter antigenicity to such an extent that antibodies to the native protein will react only weakly, if at all, with the unfolded protein. It is equally likely that antibody to the unfolded protein will react only weakly with the native molecule. (For a review of protein conformation and antigenic structure, see Crumpton, 1974).

The nuclear polyhedrosis viruses contain complex antigenic structures which can be altered or modified by chemical extraction techniques routinely used to study these viruses. The polyhedra themselves represent very complex antigenic determinants, which may be degraded *in vivo* by experimental animals to expose internal antigenic components.

Matrix protein, a major structural protein of polyhedra or granules (capsules), forms a crystalline lattice surrounding and occluding virions. Such polymerized proteins are formed by the interaction of identical (or non-identical) protein monomers. The polymerized protein has antigenic determinants (metatopes) at the surface of both monomer and polymer. The interaction of monomers to form polymers may result in the masking of surface determinants between molecules (cryptotopes), or may result in the creation of new antigenic determinants (neotopes) by 1) conformational or allosteric transitions in the monomer, or 2) contribution of amino acid residues from different monomers juxtaposed to form new antigenic determinants (see Neurath and Rubin, 1971). Therefore, the degree of aggregation or disaggregation of matrix protein may have significant impact upon the antigenicity of proteins or polypeptides isolated for serological study. The inhibition of proteolytic activity described in matrix protein fractions (Kozlov, 1975) may therefore preserve neotopes for antibody stimulation. If proteolysis is allowed to take place, criptotopes may be exposed.

Virions, like the polyhedra, are antigenic complexes containing membrane structures, nucleic acids, and associated proteins. In their natural state, virions which are enveloped present or display specific antigenic structures. Under different isolation conditions, such as variations in salt concentrations, the antigenic complex of the virion may be modified, by selective removal of proteins from envelopes or by altering the conformation of proteins in envelopes. Capside or core proteins may also be aggregated or polymerized, and would be subject to the same considerations of antigenic determinants as matrix protein. Internal antigenic determinants of protein monomers may also be exposed by treatments designed to isolate antigens, such as mercaptoethanol treatment of proteins in preparation for sodium dodecyl sulfate-polyacrylamide gel electrophoresis (SDS-PAGE). Obviously, with the true number of antigenic determinant possibilities unknown, combined with increasing modification of a given dissolution procedure (see section III), the present methods of antigen preparation for serological characterization of the Baculoviruses are left in a tenuous state.

III. Preparation of Baculovirus antigens: criticism.

The serological characterization of the nuclear polyhedrosis (NPV) and granulosis (GV) viruses has and continues to involve the harvesting of matrix protein and virions by the method of Bergold (1947, 1953, 1958, 1963). The procedure consists of the dissolution of inclusion bodies (polyhedra or capsules) at an alkaline concentration (0.005-0.02 M Na_2CO_3 in 0.05 M NaCl) suitable for the production of homogeneous solutions of polymerized protein with minimum disintegration of component parts. This is generally followed by centrifugation (at approximately 4000g) and subsequent separation of matrix protein, virions, and polyhedral (or capsular) "membrane fractions".

Krywienczyk (1962, 1963, 1967), Krywienczyk, MacGregor, and Bergold (1958), and Krywienczyk and Bergold (1960a, 1960b, 1960c, 1960d, 1961) based their serological, immunoelectrophoretic, and immunofluorescent studies on Bergold's method for procurement or antigen, providing a standard or comparison in their work. Krywienczyk defined (1960b) Bergold's weak alkali (1953) to be at a suitable pH *below* 9, a level not followed by all investigators who cite Bergold's method.

Hukuhara and Hashimoto (1966), Shapiro and Ignoffo (1970), Longworth, *et al* (1972), Harrap and Longworth (1974), Scott and Young (1973), and Norton and Di Capua (1975) have generally applied the Bergold method with variations occurring in the duration of dissolution, ionic strength, and pH which is generally in excess of 9. Cunningham (1967), for example, utilized a high concentration of Na_2CO_3 (0.10 M), pH 11.1, and a 10 minute dissolution period.

The presence of an alkaline enzyme of either endogenous or exogenous origin (Eppstein and Thoma, 1975; Kozlov, 1975) in matrix proteins, which have a high buffering capacity, fosters the extremes to which variations in the Bergold procedure will progress. However, such variations necessarily alter the native state of the matrix and virion proteins, and may change the antigenicity of the by-products (see section II), thereby reducing or eliminating the validity of comparing serological data concerning homologous and heterologous matrix and virion components from the same or different laboratories. We now are forced not just to ask, but to determine to what extent prior serological investigations of a given nuclear polyhedrosis or granulosis virus incorporated the same or pure antigens. These previous serological comparisons of the Baculoviruses (cited earlier) have generally utilized antigen preparations separated by differential centrifugation, and consequently the matrix and virion fractions may contain mixtures of both. Harrap and Longworth (1974) point to this problem, providing procedures for a more critical comparison of both biophysical, biochemical, and immunological properties of these viruses. Although the answer is not readily apparent, differences in dissolution conditions, compounded by other variations, i.e. interval between solubilization and inoculation, or usage as test antigen, the manner of storage, presence of bacterial and enzyme inhibitors, haptens, host proteins, and so forth, are factors which are known to effect the antigenicity of any given agent (see section II).

Experiments conducted in our laboratories indicate, as expected, that comparisons of short (3-5 minute) and long (30-60 minute) dissolutions of inclusion bodies from *Porthetria (Lymantria) dispar* and *Autographa californica* NPV demonstrate increased degradation of the matrix protein with increased time (pH and ionic strength constant), as shown by the SDS-PAGE technique of Maizel (1971). There is 1) no detectable difference, via immunodiffusion, between degraded and undegraded matrix protein when it is isolated from the NPV of infected larvae, and 2) a distinct difference (see section VIII) in the precipitates formed (number and type of reaction) when isolated from the NPV of ovarian cells grown in culture, depending upon the method used to procure homologous antisera (total protein inoculated, injection route and interval, species utilized, etc.).

Obviously, rigorous characterization for the purpose of defining the antigenic structure of Baculoviruses literally demands standardized procedures for antigen procurement. Kawanishi and Paschke (1970) compared the relationship of buffer, pH, and ionic strength on the yield and infectivity of virions obtained upon dissolution of *Rachiplusia ou* NPV. Their protocol could easily be adopted for measuring the effect on antigenicity of native versus degraded matrix protein or virion proteins, allowing for more properly controlled manipulation of the variables.

IV. Preparation of Baculovirus antibody: immunochemical considerations.

The production of specific antibodies in laboratory animals is a problem of practical and academic importance in many investigations. As discussed (see section III), the dissolution procedure for obtaining matrix protein and virions to be used as antigens is not standardized. ''Antigen'' in the context is simply a substance with the potential of stimulating an immune response, its identity or purity usually unknown at the time of inoculation. The number of variations in the Bergold procedure with which Baculovirus antigens (more properly termed immunogens) have been obtained is often matched by the methods reported for the production of antibody. Descriptions such as ''antisera were made in rabbits and collected in the usual manner'' (especially when unreferenced) are meaningless. Injection schedules and routes of inoculation for NPV antigens, which allow for duplication, have been provided (Norton and Di Capua, 1975).

The investigator who wishes to produce specific antisera for such purposes as identifying 1) serological relationships between viruses, 2) virus associated products in infected cells, 3) their site of production, or 4) level of host response to an infectious process, amongst others, must have at least a working knowledge of basic immunological concepts. Only then is he in a position to either recognize references which are appropriate to the experimental demands or empirically devise an immunization procedure.

Experimental requirements are most often unique to each investigation. Consequently, strict guidelines can never be provided, and general guidelines appropriate for all types of antibody likely to be required would be exhaustive. Methods employed for antiserum production and collection, the factors which effect the same, the choicc of assay in conjunction with the type of antibody obtained (precipitating, agglutinating, neutralizing, etc.), are matters of "the state of the art". Detailed procedures and rationals have been provided by numerous investigators including Ouchterlony (1962), Campbell, *et al* (1964), Moreland (1965), Pike (1967), Hyde (1967), and Harper and Martos (1973). Such procedures are absolutely necessary if we wish to attain an accurate serological characterization of the Baculoviruses.

V. Serological characteristics of polyhedrin and granulin (matrix protein).

Serological relationships between polyhedra of different hosts were first observed by Gratia and Paillot (1939) who determined that no relationship existed between the polyhedrins of *Bombyx mori* and *Euxoa segetum* Schiffermuller. Friedrich-Freksa (1943) demonstrated that the polyhedrins of *Lymantria monacha* and *Porthetria dispar,* although unrelated to *B. mori,* are interrelated.

Norton and Di Capua (1975) have shown, using multiple injections of immunogen in excess of 3 mg of protein in rabbits, that the polyhedrins (matrix protein) and possibly virions (see section VI) of *P. dispar* and *Neodiprion sertifer* are serologically related (see plate 1a). This is contradictory to the data of Krywienczyk and Bergold (1960a) who, using a single injection of 3 mg of immunogen in guinea pigs, found no relationship between the NPV's of these phylogenetically distinct hosts. In their study, Krywienczyk and Bergold examined the interrelationship of the polyhedrins of 13 insect species from widely separated geographical areas. Employing complement fixation assays, the polyhedrins from Lepidopterous and Hymenopterous NPV's did not cross-react. These findings were later confirmed by immunodiffusion assays (1961). As can be seen from plate 1a, a reaction of partial identity (type III relationship) occurs with either anti- *P. dispar* or (anti- *N. sertifer*) polyhedrins. The spur formation indicates, in addition to the presence of a group specific polyhedrin antigen (common to all but one polyhedrin tested to date), that sub-group specific polyhedrin antigens also exist (see section VII). The contradiction is directly related to the total amount of matrix protein inoculated and the route of administration, that is, 3 mg of immunogen given by the intraperitoneal route does not stimulate the production of sub-group specific antibody. This fact has been substantiated in our experiments (unpublished) in which less than 50% of guinea pigs inoculated produced demonstrable antibody. Of the remaining animals, one of 24 produced antibody (ies) which reacted with both matrix proteins. By comparison, 1 mg of *P. dispar* matrix protein in Complete Freund's Adjuvant (CFA) inoculated via multiple routes (intradermal, intramuscular, and intraperitoneal) stimulated antibody in 4 of 6 guinea pigs which cross-reacted with *N. sertifer* matrix protein, and in 6 of 6 which reacted with homologous matrix protein. Failure of low doses of immunogen to stimulate group or sub-group specific antisera may explain the failure of the 20,000 molecular weight split component of *B. mori* matrix protein to stimulate antibody in guinea pigs or rabbits, as reported by Krywienczyk and Bergold (1961).

Using degraded or undegraded (see section III) *P. dispar* matrix protein as the immunogen in either rabbits or guinea pigs (dose range 1-280 mg) we have observed type I relationships, i.e. reactions between antibody and the matrix protein group specific antigen, with *N. sertifer, N. taedae, Trichoplusia ni, A. californica, Heliothis zea,* and *Pseudoplusia includens,* and type III relationships, i.e. reactions between antibody and the matrix protein sub-group specific antigen, with *N. sertifer* and *N. taedae,* (plate 1a). Anti-*N. sertifer* matrix protein produced the reciprocal reactions. The matrix protein of *Spodoptera frugiperda,* at equivalent concentration, does not react with *P. dispar* or *N. sertifer* anti-matrix protein, suggesting that its group antigen is at a concentration unsuitable for precipitate fomation, sterically hindered, or absent. The phenomenon is currently under investigation, the outcome of which may effect the phylogenetic relationship of *S. frugiperda* NPV.

As indicated, we have observed cross-reactivity between homologous and heterologous matrix protein and virion fractions, and antisera to either from Lepidopterous and Hymenopterous NPV's. Similar cross-reactions have been observed by Long-

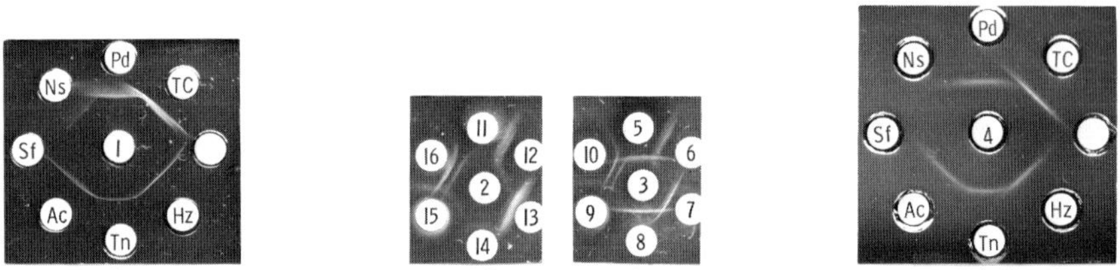

1a **1b** **1c** **1d**

PLATE 1

Ac = *Autographa californica* matrix protein (polyhedrin)
Hz = *Heliothis zea* matrix protein
Ns = *Neodiprion sertifer* matrix protein
Pd = *Porthetria dispar* matrix protein
Sf = *Spodoptera frugiperda* matrix protein
TC = *Porthetria dispar* matrix protein isolated from tissue culture NPV
Tn = *Trichoplusia ni* matrix protein

1, 2 = anti-*P. dispar* matrix protein (rabbit)
3 = anti-*P. dispar* virion (rabbit)
4 = anti-*P. dispar* matrix protein isolated from tissue culture NPV (guinea pig)
5 = Intragel absorbtion (IGA) with anti-*N. sertifer* virion (rabbit) followed by *P. dispar* virion
6, 12 = *N. sertifer* matrix protein
7, 13 = *P. dispar* matrix protein
8, 14 = *P. dispar* virion
9 = IGA with anti-*N. sertifer* virion (rabbit) followed by *N. sertifer* matrix protein
10 = IGA with anti-*N. sertifer* virion (rabbit) followed by *P. dispar* matrix protein
11 = IGA with anti-*N. sertifer* matrix protein (rabbit) followed by *P. dispar* virion
15 = IGA with anti-*N. sertifer* matrix protein (rabbit) followed by *N. sertifer* matrix protein
16 = IGA with anti-*N. sertifer* matrix protein (rabbit) followed by *P. dispar* matrix protein

worth, *et al* (1972) who detected the presence of two matrix proteins, A and B, of the granulosis virus (GV) of *Pieris brassicae.* Protein B was found on the surface of the enveloped virus particle, and in the inclusion body ''envelope''. Antiserum to purified virions detected the B protein only in matrix protein.

Scott and Young (1973), base on their investigation of the antigens of *T. ni* NPV, observed 5 precipitates, via immunodiffusion, between the virion fraction and homologous antiserum. One of these precipitates is also found in the matrix protein, and they suggest its presence could be due to degradation during dissolution of the polyhedra. Although we are in agreement with this explanation, which may account for the cross-reactivity reported by Longworth, *et al* (1972), the *T. ni* matrix protein and virion fractions were in alternate rather than adjacent immunodiffusion wells, preventing the possible formation of a type I reaction, thereby establishing common antigenicity. With anti- *P. dispar* or *N. sertifer* matrix protein sera, we have observed the type I reaction between virion and matrix protein fractions of homologous and heterologous origin (see plate 1b), which can also be explained by the presence of a common virion-matrix protein antigen(s). If the question of cross-reactivity between matrix protein and virions can be resolved as contamination or true shared antigenicity, then the nature of matrix protein as an intrinsic by-product of the viral genome or as a host protein may be further reflected upon. To answer this question, the necessity of appropriate standard dissolution procedures is re-emphasized.

The first investigation of the serological relationship between the granulosis and nuclear polyhedrosis viruses was conducted by Tanada (1954). He found the two inclusion body types of *P. rapae* to be related, which Bergold (1963) considered to be more closely associated with the host species than a serological relationship between polyhedra and capsule. Bergold did not rule out the possibility of contamination of either viral protein by host protein. Krywienczyk and Bergold (1960b) in their studies of the matrix protein (granulin) of *Recuvaria milleri* reported a strong serological relationship with *B. mori, Malacosoma disstria, Choristoneura fumiferana* (Clem.), *Colias philodice eurytheme, Malacosoma americanum,* and *P. dispar.* The serological cross-reactivity indicated is misleading, as discussed in section VI, and the actual cross-reaction may be due to the presence of a common (C) antigen found by Croizier and Meynadier (1973b) in polyhedrin and granulin via carbonate or thioglycolate dissolution procedures. Their study also revealed the presence of a T antigen in the granulin of *P. brassicae* and a B antigen in the polyhedrin of *B. mori* through thioglycolate dissolution only, indicating again the effect of dissolution conditions on antigen variation.

Tanada and Watanabe (1971) have observed the presence of common antigens in the matrix proteins of two strains of a granulosis virus of *Plodia unipuncta,* which are distinguishable by synergy and pathogenicity. They indicate the effect host synthesized proteins incorporated into the inclusion bodies would have, that is, establishment of a false serological relationship, stressing that ''further studies should be conducted comparing the reactions of proteins from the inclusion body, enveloping membranes and virus particle.''

Glaser and Stanley (1943) detected cross-reactivity between anti-polyhedra and ''healthy'' hemolymph of *B. mori.* One implication, based on the experiments of Aizawa (1954) and Krywienczyk and Bergold (1961), who could not repeat the observation, is that the reaction was due to the presence of anti-hemolymph activity. The results of these investigations were thought to prove clearly that no serological relationship existed between insect hemolymph and matrix proteins, a finding our experiments clearly substantiate. However, Young and Johnson (1972) have detected virus specific soluble antigens in the fat body of infected larvae of *T. ni.* The presence of virus-specific antigen, directly in the hemolymph or through a fat body-hemocoel access, would therefore provide an alternative explantation of Glaser and Stanley's result, implying that the criteria for determination of healthy hemolymph were inappropriate. This should serve to remind us that the need for infectious free controls is infinite.

Serological relationships of matrix protein and virion (see section VI) have been examined by Krywienczyk and Bergold (1960d) who presented evidence that 1) insect viruses and their matrix protein were only slightly, if at all, related serologically, and that 2) cross-reactivity occurred only within the Lepidopterous NPV's. We have demonstrated 1) cross-reactivity of homologous antisera to viral components of *N. sertifer,* 2) reciprocal cross-reactivity of antisera to viral components of *P. dispar* and *N. sertifer,* and 3) through hemagglutination inhibition, the cross-reactivity of antisera to polyhedra, matrix protein, and virions of *N. sertifer* with the hemagglutinin of *P. dispar* matrix protein (Norton and Di Capua, 1975). As previously indicated, the cross-reactivity between homologous matrix protein and virions of either source may be due to immunogen contamination; nevertheless, the cross-reactivity between antisera to each of the polyhedral components of *N. sertifer* and the matrix protein of *P. dispar* clearly demonstrates that such reactions are not exclusive to the Lepidopterous viruses, and that a serological relationship exists between two viruses which have widely separated hosts.

Hemagglutinin activity has also been found in the matrix protein of *S. frugiperda* NPV (Reichelderfer, 1974), virion fractions of the cytoplasmic polyhedrosis virus (CPV) and NPV of *B. mori* (Miyajima and Kawasi, 1969), and NPV of *H. zea* (Shapiro and Ignoffo, 1970a, 1970b). Since the matrix proteins of *P. dispar* and *S. frugiperda* can be distinguished by immunodiffusion, the hemagglutinin activity of these two NPV's has significant diagnostic value, (see plate 1a).

VI. Serological characterization of Baculovirus virions.

The serological analysis of Baculovirus virions is intuitively more complex, because of the presence of a larger number of protein moieties found in virions. Young, *et al* (1973) demonstrated the presence of 12 polypeptides in *T. ni* virions, and Padhi, *et al* (1975) demonstrated the presence of 14 polypeptides in *P. dispar* virions, when analyzed by SDS-PAGE. However antisera to *T. ni* virions does not form 12 precipitates with homologous antigen in immunodiffusion studies, nor does antisera to *P. dispar* virions form 14 precipitates with its homologous antigen. Scott and Young (1973) demonstrated the presence of 5 precipitates in *T. ni* virion-antivirion systems, one band being in common with matrix protein. We have demonstrated (unpublished) the presence of at least 5 precipitates with *P. dispar* virions, perhaps 2 bands being in common with matrix protein (see plate 1c). More precipitates may be present, but the strength of reactions and superimposition of bands make specific band identification difficult. Although the number of protein bands seen on acrylamide gels cannot be directly correlated with the number of distinct, antigenically unrelated proteins to be found in virions, it may be possible that antisera to virions contain specific antibodies to more than 5 virion antigens. These specific antibodies, being in low concentration, will not precipitate with homologous antigen in a state of antigen excess.

Krywienczyk, *et al* (1958, 1960a) studied the interrelationships of virions from NPV's and GV's infecting Lepidopterous and Hymenopterous insects by the complement fixation technique. Based on these studies, the following serological classification of viruses was established, paralleling the classification of matrix proteins from these viruses: group 1, the granulosis viruses of Lepidoptera; group 2, the nuclear polyhedrosis viruses of Lepidoptera; and group 3, the nuclear polyhedrosis viruses of Hymenoptera. The data presented in these studies may be immunologically invalid. For instance, the data presented to interrelate granulosis and nuclear polyhedrosis viruses are based upon the comparison of antisera to several viruses as antigens. Not only is it invalid to compare two antisera made to different antigens with a third antigen, but the results can also be misleading, because any one antiserum from an individual animal may be the result of a poor response to antigenic stimulation. In these studies for instance, the homologous titers range from 768 to 6400 to the same level of inoculated antigen, which reflects the individual responses of animals. Inconsistencies in the amount of antigen used in the complement fixation test also make comparisons untenable. An objective serological analysis must be based upon comparison of a given antisera and its homologous antigen tested against equal concentrations of several heterologous antigens. Subjective serological analysis occurs when one compares several antisera to one antigen, because of the individual response variation.

We have attempted to analyze reciprocal cross-reactions between antisera and virion antigens of *P. dispar* and *N. sertifer* NPV's by immunodiffusion. Heterologous precipitation was present but very difficult to detect. To clarify the antigenic relationship between these two viruses, we assayed heterologous antigens by the intragel absorbtion technique (Feinbert, 1957; Schmidt, *et al*, 1965). In immunodiffusion plates, specific wells were preloaded with antisera, and, after complete diffusion, antigen was loaded into the same wells. It was found, for instance, that the number of precipitates between *P. dispar* virion and its homologous antisera was reduced if the antigen well was preloaded with *N. sertifer* virion antisera (see plate 1c). The absence of bands between homologous antigen and antibody due to absorbtion by heterologous antibody implies a serological relationship between the virus reagents.

Krywienczyk (1960d) also reported "slight" cross-reaction of virion antisera with matrix protein antigen, but considered this to be the possible result of contamination of virus inoculum with matrix protein. Scott and Young (1973) reported that there was a common antigen between virions and matrix protein from *T. ni* NPV, and Shapiro and Ignoffo (1970b) reported similar results with *Heliothis* NPV. Longworth, *et al* (1972) reported a common antigen (B) between virions and polyhedra. We have also found significant reciprocal cross-reactivity between virions and matrix proteins from NPV's which have been regorously cleaned, by physical and chemical manipulations designed to minimize the possibility of contamination (see section III). Even with such precautions taken, it is possible to absorb *P. dispar* matrix protein antisera with *P. dispar* virions to the point at which the matrix antisera no longer reacts with its homologous antigen. In immunodiffusion tests of virions and matrix protein from *P. dispar* and *N. sertifer* NPV (see plate 1c), cross-reactivity of virion and matrix antisera with heterologous antigens demonstrates the serological relationship between these antigens. Hemagglutination inhibition (Norton and Di Capua, 1975) also demonstrated that virion antisera could very effectively inhibit hemagglutination of chicken erythrocytes by matrix protein, nearly as efficiently as matrix protein antisera.

The question of cross-reactivity between virions and their matrix proteins has not been resolved. Matrix protein antisera, made to antigens from polyhedra rigirously cleaned and dissolved under restrictive conditions of pH, temperature, and time, is still cross-reactive with virions isolated from sucrose gradients after ultracentrifugation. Contamination being minimized, we believe their is significant evidence to support the concept of a common or shared antigen between virions and matrix protein. Such a common antigen may serve as a receptor site for matrix protein monomers and as an initiator of polymerization of matrix protein for the formation of polyhedra.

VII. Group antigens in matrix proteins.

Serological cross-reactions between matrix protein fractions from a large number of nuclear polyhedrosis and granulosis viruses were investigated by Krywienczyk, *et al* (1958, 1960a, 1960b, 1960c, 1960d, 1961) and Tanada (1954). Tanada concluded that the serological relationship between viruses is not necessarily associated with the infecting virus, but more likely with the species of the host insect. Krywienczyk, *et al*, concluded that there are serological groups among insect viruses, namely the NPV's of Lepidoptera, the NPV's of Hymenoptera, and the GV's of Lepidoptera.

In studies of matrix protein fractions from several NPV's, employing several different sources of antisera to matrix protein, we believe the serological cross-reactivity among matrix proteins from NPV's of Lepidopterans and Hymenopterans is due to the presence of a groupe antigen(s) in matrix protein.

Porthetria dispar matrix protein antisera, whether made in rabbits or guinea pigs to high or low doses of degraded matrix antigen or to low doses of undegraded matrix antigen from tissue culture grown polyhedra, precipitates with matrix protein fractions of *T. ni, H. zea, A. californica,* and *P. includens* in a type I relationship (reaction of identity) as described by Ouchterlony (1962) in double (radial) immunodiffusion tests. *P. dispar* matrix protein antisera precipitates with *N. sertifer* and *N. taedae* matrix protein in a type 3 relationship (reaction of partial identity). There are slight variations in antisera from different sources resulting in formation of spurs between various fractions, but the recognition of heterologous matrix protein fractions by antisera to *P. dispar* matrix protein suggests the presence of group specific antigenic determinants. *N. sertifer* matrix protein antisera shows the same relationships with heterologous matrix proteins, previously mentioned in relation to *P. dispar* antisera. The type reactions are reciprocal, i.e. type 1 with *N. sertifer* and *N. taedae* and type 3 with *T. ni, H. zea, A. californica* and *P. includens*. Spur formation or type 3 reactions may indicate the presence of sub-group antigenic determinants on some molecules of matrix protein which are not found on other molecules. In addition to group specific and sub-group specific antigenic determinants, there may also be species specific antigenic determinants on matrix protein molecules, a relationship which is found to exist in antigens of other viruses (Davis, *et al* 1975; Barron, 1971). Chemical manipulation of antigens, such as proteolytic digestion, may expose antigenic determinants which cross-react as a type 1 reaction. An example may be the differences in antigenicity between matrix protein from tissue culture grown polyhedra when compared with matrix protein from polyhedra grown in the natural host (see section VIII).

Amino acid analyses of matrix proteins from NPV's of several insect species demonstrate significant differences in content (Wellington, 1951, 1954). However, sequence studies of matrix proteins have not been accomplished to determine the extent of homology between matrix proteins from different NPV's. Small regions of homology are sufficient for manifestation of group specific or sub-group specific antigenicity, provided that they are also sufficiently immunogenic.

Common antigenic determinants on virions (see section VI) demonstrated by intra-gel absorbtion, may also be a manifestation of group antigenicity distinct from matrix protein.

Croizier and Meynadier (1973a, 1973b) found a common F antigen in all strains of GV tested, a common C antigen in 5 species tested, and species specific antigens for some of the viruses. They also determined that GV and NPV of Lepidoptera shared common antigens, ''to show the homogeneity of the Lepidopter Baculovirus group.''

VIII. Serological relationship between matrix protein isolated from tissue culture and host nuclear polyhedrosis viruses.

Employing tissue culture to grow virus has for many years been one of the most important tools for the virologist and immunologist in their efforts to characterize viruses. The recent rapid development of invertebrate tissue culture, especially insect tissue culture, has similiarly given virologists, pathologists and others interested in characterizing insect viruses a new opportunity to study particular strains of virus, isolate mutants, study infection pathways or plaque morphology, and so forth. These methods also give the immunologist an opportunity to study strains of virus and perhaps develop a better serological classification of insect viruses.

Nuclear polyhedrosis viruses grown in tissue cultures have not been characterized serologically. Attention is now being focussed in this area, primarily to ascertain whether virus produced in cell culture is identical to that grown in larvae. Limited physico-chemical and serological work has been undertaken with *P. dispar* NPV grown in cultures of ovarian epithelial cells (IPL 49, Goodwin) and *A. californica* NPV grown in culture of *S. frugiperda* pupal ovarian cells (IPLB-21, Vaughn).

Matrix protein from these two NPV's has been isolated by modifications of the alkaline dissolution procedure of Bergold. SDS-PAGE of matrix protein fractions revealed no apparent endo- or exogenous alkaline protease activity associated with matrix protein, which has been found in matrix protein of NPV's produced *in vivo*. Consequently, there is no degradation of matrix protein over extended periods of time in alkaline conditions, as determined by SDS-PAGE of matrix protein fractions assayed at various time intervals up to 7 days.

Antisera made in rabbits and guinea pigs to degraded matrix protein from *P. dispar* NPV's does not discriminate antigenically between matrix protein from *in vivo* and *in vitro* produced polyhedra, when assayed by the immunodiffusion technique. However, employing antisera made in guinea pigs to one intraperitoneal, intramuscular, and intradermal inoculation of matrix protein (1 mg) in Complete Freund's Adjuvant, demonstration of an antigenic difference between matrix proteins of both sources was accomplished (see figure 1d). One major precipitate from tissue culture matrix protein shows a type 3 relationship (reaction of partial identity) with matrix protein from *in vivo* produced polyhedra. This is an indication that there is an antigenic determinant(s) found in tissue culture matrix protein, which perhaps is not found in native *(in vivo)* matrix protein. This antigenic difference may be due to the enzymatic degradation of native matrix protein, resulting in the loss of an antigenic determinant(s), or the cleavage of an antigenic determinant(s) which is monovalent and not precipitable.

Whether this antigenic difference is a reflection of the loss of enzyme activity in matrix protein from tissue culture polyhedra, or selection of aberrant virus strains, or cell types which do not make appropriate enzymes, is an unresolved question. Some answers may be provided by the analysis of proteins from virus grown in tissue culture and then passaged back through larvae. In these same studies, protein profiles (280 mu) or Sephadex G-200 eluants of *P. dispar* matrix protein isolated from tissue culture and host NPV's have been examined for the presence of hemagglutinin. Of significance, the hemagglutinin is consistantly active in the matrix protein of NPV's isolated from infected larvae, is generally inactive or absent from the matrix protein of NPV's isolated from ovarian cell lines, and its activity reobserved when the tissue culture polyhedra are passed through larvae, via diet incorporation.

IX. Other immunological methods of characterization.

Immunofluorescent assays capable of identifying the NPV of *B. mori in vivo* and *in vitro* have been conducted by Krywienczyk (1963, 1967). She has also compared the homogeneity of matrix proteins (polyhedrins and granulins) of Lepidopterous and Hymenopterous origin by immunoelectrophoresis (1962). Young and Scott (1970) were able to detect, by immunoelectrophoresis, differences in the proteins of hemolymph from normal and infected *T. ni* larvae, and that a serological relationship did not exist with proteins present in the virus. Norton and Di Capua (unpublished results) have compared the matrix proteins of *P. dispar* NPV isolated from larvae and tissue culture. Although there is no discernable difference in rate of migration, the matrix protein isolated from host polyhedra lacks a cathodal portion of the precipitate formed with the matrix protein isolated from tissue culture polyhedra, when each is compared to guinea pig anti-tissue culture NPV matrix protein.

Benton, Reichelderfer, and Hetrick (1973) are the first investigators to employ the macrophage migration inhibition assay for the study of insect viruses, successfully differentiating the multiply embedded NPV's of *T. ni* and *A. californica.*

X. Conclusion.

Comparison of previous and current investigations of the serological interrelationships of the Baculoviruses leads to several possible interpretations. The data of Krywienczyk and Bergold (1960c) divides these viruses into three groups: (1) the granuloses of Lepidoptera, (2) the nuclear polyhedroses of Lepidoptera, and (3) the nuclear polyhedroses of Hymenoptera. Bellet (1969) has statistically analyzed the above relationship, and, based on a determination of the molar proportions of guanine and cytosine in DNA, groups the viruses as follows: (1) the granulosis virus of *Choristoneura fumiferana* (Clem.), (2) the nuclear polyhedroses of Hymenoptera, and (3) the nuclear polyhedroses and granuloses of Lepidoptera. Alternatively, when the data of Croizier and Meynadier (1973) is compared with that of Norton and Di Capua (1975), a serological relationship, based on common antigenicity, evolved between the nuclear polyhedroses and granuloses of Lepidoptera *and* Hymenoptera which may allow for discrimination between individual viruses independent of host phylogeny.

Ultimately, empirically devised methods for defining an accurate picture of the antigenic structure of the Baculoviruses must be centered on the development of (1) improved dissolution procedures of polyhedra and capsules for the isolation of native immunogens in pure state and (2) appropriate immunization procedures. Absence of these criteria will prevent a true determination of the overall serological relationship of the polyhedrosis and granulosis viruses of insects or other hosts.

Acknowledgements.

Investigations conducted by the authors were supported by the United States Department of Agriculture, Forest Service Grants, FSNE-14 and -28. We wish to thank Dr. Frank Lewis, USDA, NEFES, Hamden, Connecticut, Dr. Howard Scott, University of Arkansas, and Dr. Charles Reichelderfer, University of Maryland for their contribution of viruses isolated from larvae, and Dr. Ronald Goodwin and Dr. James Vaughn, Insect Pathology Laboratory, USDA, Beltsville, Maryland for their contribution of tissue culture isolated virus.

XI. References

Aizawa, K. (1954). *Virus (Osaka) 4,* 241.

Aoki, K. and Chigasaki, Y. (1921). *Zentr. Bacteriol. Parasitenk. Abt. I 86,* 481.

Barron, A.L. (1971). In: Methods in Virology (K. Maramorosch and H. Koprowski, eds.) vol. V. pp. 347-373. Academic Press, N.Y.

Bellet, A.J.D. (1969). *Virology 37,* 117.

Bellet, A.J.D., Fenner, F. and Gibbs, A.J. (1973). In: Viruses and the Invertebrates (A.J. Gibbs, ed), pp. 43-87. Elsevier, N.Y.

Benton C.V., Reichelderfer, C.F. and Hetrick, F.M. (1973). *J. Inv. Path. 22,* 42.

Bergold, G.H. (1947). *Z. Naturforsch. 26,* 122.

Bergold, G.H. (1953). *Adv. Virus Res. 1,* 91.

Bergold, G.H. (1958). In: Handbuch der Virusforschung (C. Hallauer and K.F. Meyer, eds.), vol. 4, pp. 60-142, Springer, Vienna.

Bergold, G.H. (1963). In: Insect Pathology (E.A. Steinhaus, ed.) vol. 1, pp. 413-456, AcademicPress, N.Y.

Campbell, D.H., Garvey, J.S., Cremer, N.E., and Sussdorf, D.H. (1964). Methods in Immunology, Benjamin, N.Y.

Croizier, G. and Meynadier, G. (1973a). *Entomophaga 18,* 259.

Croizier, G. and Meynadier, G. (1973b). *Entomophaga 18,* 431.

Crumpton, M.J. (1974). In: The Antigens (M. Sela, ed.), vol. 2, pp. 1-72, Academic Press, N.Y.

Cunningham, J.C. (1967). *J. Inv. Path. 11,* 132.

David, W.A.L. (1975). In: Annual Reviews of Entomology (R.F. Smith, T.E. Mittler, and C. S. Smith, eds.), vol. 20, pp. 97-118, Annual Reviews, Inc., Palo Alto, California.

Davis, J., Gilden, R.V. and Oroszlan, S. (1975). *Immunochem. 12,* 67.

Di Cupua, R.A. and Norton, P.W. (1974). *Abstract, VII Annual Soc. Inv. Path. Meeting, Tempe, Arizona*

Eppstein, D.A. and Thoma, J.A. (1975). *Biochem. Biophys. Res. Comm. 62,* 478.

Feinberg, J.G. (1957). *Int. Arch. Allergy 11,* 129.

Friedrich-Freksa, H. (1943). *Biol. Zentr. 1.*

Glaser, R.W. and Stanley, W.M. (1943). *J. Exptl. Med. 77,* 451.

Gratia, A. and Paillot, A. (1938). *Compt. Rend. Soc. Biol. 129,* 507.

Gratia, A. and Paillot, A. (1939). *Arch. ges. Virusforsch. 1,* 130.

Harper, B. and Martos, L.M. (1973). In: The Herpes Viruses (A.S. Kaplan, ed.), pp. 221-252, Academic Press, N.Y.

Harrap, K.A. and Longworth, J.F. (1974). *J. Inv. Path. 24,* 55.

Hukuhara, T. and Hashimoto, Y. (1966). *J. Inv. Path. 8,* 234.

Hyde, R.M. (1967). In: Advances in Applied Microbiology (W.W. Umbreit, ed.), vol. 9, pp. 39-63, Academic Press, N.Y.

Kawanishi, C.Y. and Paschke, J.D. (1970). In: Proc. IV Inter. Colloq. Insect Path. (Soc. Inv. Path.), pp. 127-146, College Park, Md.

Kozlov, E.A., Sidorova, N.M. and Serobrianyi, S.B. (1975), *J. Inv. Path. 25,* 97.

Krywienczyk, J. (1962). *J. Insect Path. 4,* 185.

Krywienczyk, J. (1963). *J. Insect Path. 5,* 309.

Krywienczyk, J. (1967). *J. Inv. Path. 9,* 568.

Krywienczyk, J. and Bergold, G.H. (1960a). *Virology 10,* 308.

Krywienczyk, J. and Bergold, G.H. (1960b). *J. Immunol. 84,* 404.

Krywienczyk, J. and Bergold, G.H. (1960c). *Virology 10,* 549.

Krywienczyk, J. and Bergold, G.H. (1960d). *J. Insect Path. 2,* 118.

Krywienczyk, J. and Bergold, G.H. (1961). *J. Insect Path. 3,* 15.

Krywienczyk, J., MacGregor, D.R. and Bergold, G.H. (1958). *Virology 5,* 476.

Longworth, J.F., Robertson, J.S. and Payne, C.C. (1972). *J. Inv. Path. 19,* 42.

Maizel, J.V. (1971). In: Methods in Virology (K. Maramorosch and H. Koprowski, eds.), vol. 5, pp. 179-246, Academic Press, N.Y.

Miyajima, S. and Kawase, S. (1969). *Virology 39,* 347.

Moreland, A.F. (1965). In: Methods of Animal Experimentation (W.I. Gay, ed.), Academic Press, N.Y.

Neurath, A.R. and Rubin, B.A., (1971). *Monographs in Virology,* vol. 4, S. Karger, Philadelphia, Pa.

Norton, P.W. and Di Capua, R.A. (1974). *Abstract, VII Annual Soc. Inv. Path. Meeting,* Tempe, Arizona.

Norton, P.W. and Di Capua, R.A. (1975). *J. Inv. Path. 25,* 185.

Ouchterlony, O. (1962). *Prog. Allergy 6,* 30.

Padhi, S.B., Eikenberry, E.F. and Chase, T. (1975). Intervirology (In press).

Pike, R.M. (1967). *Bact. Reviews 31,* 157.

Reichelderfer, C.F. (1974). *J. Inv. Path. 23,* 46.

Shapiro, M. and Ignoffo, C.M. (1970a). In: Proc. IV Inter. Colloq. Insect Path. (Soc. Inv. Path.), pp. 147-151, College Park, Md.

Shapiro, M. and Ignoffo, C.M. (1970b). *Virology 41,* 577.

Schmidt, N.J., Lennette, E.H. and Dennis, J. (1965). *J. Immunol. 94,* 482.

Scott, H.A. and Young, S.Y. (1973). *J. Inv. Path. 21,* 315.

Smith, K.M. (1967). Insect Virology Academic Press, N.Y.

Tanada, Y. (1954). *Ann Entomol. Soc. Amer 47,* 553.

Tanada, Y. and Watanabe, H. (1971). *J. Inv. Path. 18,* 307.

Wellington, E.F. (1951). *Biochim. et Biophys. Acta 7,* 238.

Wellington, E.F. (1954). *Biochem. J. 57,* 334.

Young, S.Y. and Johnson, D.R. (1972). *J. Inv. Path. 20,* 114.

Young, S.Y. and Lovell, J.S. (1973). *J. Inv. Path. 22,* 471.

Young, S.Y. and Scott, H.A. (1970). *J. Inv. Path. 16,* 57.

Chapter 32

IN VITRO AND *IN VIVO* COMPARATIVE STUDIES OF SEVERAL NUCLEAR POLYHEDROSIS VIRUSES (NPVs) BY NEUTRALIZATION, IMMUNOFLUORESCENCE AND POLYACRYLAMIDE GEL ELECTROPHORESIS

A.H. McIntosh and S.B. Padhi

I. Introduction

Nuclear polyhedrosis viruses (NPVs) are being isolated in an ever increasing number from various insects. More recently a NPV of pink shrimp has been reported (Couch, 1974). This prevalence together with the potential use of insect NPVs as biological control agents make it mandatory that appropriate tests be established for the identification of these viruses. Many NPVs have proven effective in control of insect pests in the field, with a permit being granted by the United States Department of Agriculture (USDA) for large scale use of *Heliothis zea* NPV. This virus is highly pathogenic for the cotton bollworm or cotton earworm resulting in the death of infected larvae.

Both NPVs and a second group of insect viruses, the granulosis viruses (GVs) are classified under the genus Baculovirus (Wildy, 1971). Whereas it is difficult or impossible to distinguish morphologically between most NPVs, GVs can be readily distinguished from the former because usually only one virion is occluded into an inclusion body called a capsule. On the other hand NPVs are occluded either singly (single embedded virus (SEV)) or in multiples (MEV), into what is termed a polyhedral inclusion body (PIB). Insects feeding on contaminated plants ingest the PIBs which are then dissolved by alkaline and possibly enzymatic action in the gut of the insect thus releasing virions. The latter invade and infect the host cells resulting in eventual death of the insects. Nuclear polyhedrosis viruses are the insect viruses of choice as biological control agents because of the ease with which they can be produced in large numbers *in vivo*, and the protection afforded the virions by the inclusion bodyprotein. Furthermore, they are thought to be species specific. However this last aspect has to be modified in view of the finding that *Autographa californica* NPV will infect insects of several different genera (Vail *et al.*, 1970). Also another NPV, *Trichoplusia ni* (SVE), is capable of infecting an alternate host, the alfalfa looper (Vail *et al.*, 1971). NPVs and GVs are presently named after the host from which they are first isolated. The inadequacy of such a system is readily apparent since some NPVs may infect several different hosts.

The successful production of several NPVs in insect established cell lines (Grace 1967; Goodwin *et al.*, 1970; Faulkner and Henderson, 1972; Vail *et al.*, 1973; Sohi and Bird, 1971; Sohi and Cunningham, 1972) will greatly facilitate the proper identification of these viruses since *in vitro* production of NPVs has decided advantages over the more cumbersome and less controlable *in vivo* system. In this regard it is noteworthy to mention that several immunological procedures have been successfully used to characterize some NPVs before the advent of *in vitro* produced NPVs. These methods included complement fixation, hemagglutination, agar gel diffusion, immunoelectrophoresis and immunofluorescence (Krywienczyk and Bergold, 1960a, b; Shapiro and Ignoffo, 1970; Cunningham *et al.*, 1966; Krywienczyk, 1962; Krywienczyk, 1963). More recently Benton *et al.*, 1973, have employed the macrophage migration inhibition test to differentiate between two NPVs, *A. californica* and *T. ni*. Of equal importance are various physical methods which can be employed for the determination of the number and molecular weights of proteins in NPVs. One such technique, polyacrylamide gel electrophoresis (PAGE) has already been employed for the characterization of inclusion body and viral proteins from several NPVs (Padhi, *et al.*, 1975; Kozlov *et al.*, 1975; Young and Lovell, 1973). The present report described the use of three techniques namely, the neutralization test, immunofluorescence and PAGE for the characterization and differentiation of several NPVs.

II. Materials and methods

A. Cell cultures.

Established cell lines of *Trichoplusia ni* (TN-368; Hink 1970) and *Spodoptera frugiperda* (PRL-21; Vaughn 1970, as reported by Goodwin *et al.*, 1970) were propagated in TC 199-MK (McIntosh *et al.*, 1973) in 25 cm^2 plastic Falcon T flasks or milk dilution bottles at 28°C.

B. Viruses.

Heliothis zea (SEV) was obtained from International Minerals and Chemical Corp., Libertyville, Illinois and *T. ni* (SEV) from Dr. A. Heimpel, United States Department of Agriculture, Beltsville, Maryland. The MEVs, *A. californica* and *T. ni* were obtained from Drs. P. Vail, United States Department of Agriculture, Phoenix, Arizona, and P. Faulkner, Queens University, Kingston, Ontario, respectively. The non occluded Chilo iridescent virus was obtained from Dr. A. Wood of Boyce Thompson Institute, Yonkers, New York. Polyhedral inclusion bodies (PIBs) or the NPVs described and *Porthetria dispar* (MEV) were produced in their respective hosts and purified according to previously described methods (McIntosh and Maramorosch, 1973; Padhi *et al.*, 1975). PIBs of *A. californica* were produced *in vitro* by inoculating several flasks of *T. ni* or *S. frugiperda* cells with low passage virus (5 passages or less). PIBs were harvested 5-7 days following inoculation of cultures by treatment with 2.5% sodium dodecyl sulfate (SDS), and purified on a linear sucrose gradient (65% w/w-50% w/w) at 70,000 xg for 1 h. PIBs isolated from each cell line were analyzed separately. Free virions were obtained from supernatant fluids of cultures infected with *T. ni* or *A. californica* NPVs by low speed centrifugation to remove cells.

C. Antisera.

Antisera against *T. ni* (SEV), *H. zea* (SEV), *A. californica* (MEV) virions and CIV were produced by a previously described method (McIntosh and Maramorosch 1973). Antisera against *T. ni* (MEV) and *P. dispar* (MEV) virions were obtained from Drs. P. Faulkner of Queens University, Kingston, Ontario and R. E. Shope, Yale University, New Haven, Connecticut, respectively.

D. Neutralization test.

0.2 ml of 10^6 TCID50/ml of virus was mixed with 0.2 ml of appropriately diluted antiserum in small sterile disposable plastic tubes (1 x 7.5 cm) and incubated at 28°C for 2 h. Two or three roller tube (1.6 x 15 cm) of *S. frugiperda* (2 x 10^5 cells/ml) were inoculated with 0.1 ml of the various virus-antiserum mixtures. *S. frugiperda* cultures inoculated with virus and virus treated with preantiserum served as controls. A cytopathic effect (CPE) was considered as one in which cell rounding occurred with the presence of PIBs. The titer of the antiserum was that last dilution which was capable of nullifying the CPE of the virus. Titration trays (Linbro, Conn.) were also found useful for running these tests.

E. Immunofluorescence.

Antiserum produced against *A. californica* virions recovered from infectious hemolymph was conjugated with fluorescein isothiocyanate (FITC) by a modified method (Coons, 1958). Leighton tubes seeded with 2 x 10^5 cells/ml of *T. ni* or *S. frugiperda* were inoculated with 0.1 ml of *T. ni* or *A. californica* NPVs ($TCID_{50}$ 10^6/ml). At various time intervals, the supernatant fluid was removed, the cell sheath washed twice with phosphate buffered saline (PBS, pH 7.1), air dried and fixed in acetone for 10 min. The conjugated antiserum (1:16 dilution) was applied to cells on the coverslip and incubated in a moist atmosphere at 37°C for 30 min. Coverslips were then rinsed and counterstained for 10 min in PBS containing Evans blue (0.05 ml of 1% Evans blue in 250 cc PBS), rinsed briefly in distilled water, dried, and mounted in buffered glycerol. Controls consisted of infected cultures treated with postantiserum prior to the application of conjugated antiserum and uninoculated cultures treated with the conjugated antiserum. Specimens were examined with a Zeiss "Aus Jena" microscope by epi-immunofluorescence with BC-38 and FITC filters. Slides were kept for future reference by storage at -20°C.

F. Polyacrylamide gel electrophoresis.

Inclusion body protein was dissolved and separated from virus rods by treatment with 0.02 M Na_2CO_3-0.5 M NaCl. After about 20 min., or as soon as clearing occurred (dissolution of polyhedra) the samples were prepared for electrophoresis. PIBs were also sometimes treated with heat (95°C for 10 min), 2.5% SDS, or 1% sodium deoxycholate (DOC) along with alkali digestion. In other instances PIBs were dissolved at 4°C and kept at that temperature until prepared for electrophoresis. These procedures were used to reduce as much as possible the action of an alkaline protease suspected to be present in the inclusion bodies.

The procedure of Laemmli (1970) was used for preparation and running of polycrylamide gels containing SDS. Gels composed of 7.5%, 8.75%, 10%, and 12.5% acrylamide were used for these analyses. The sample buffer contained 10% glycerol, 5% 2-mercaptoethanol, 3% SDS, and 0.0625 M Tris (pH 6.8). Samples (0.1 µg-4 µg protein/band) were added to 50-300 μl of the buffer and heated to 90°C for 3 min. Care was taken that the final concentration of SDS was always 2% or greater. Bromophenol blue (2 μl of 0.1% solution) was added as the tracking dye. The gels were 16 cm in length having a 14 cm running gel and 2 cm stacking gel. Electrophoresis was carried out at 1 mA/gel until the tracking dye was in the lower portion of the stacking gel, and then the current was increased to 2.5 mA/gel. Staining was done with 0.4% Coomassie Brilliant Blue in 50% w/v trichloroacetic acid and destained 7.5% acetic acid (Eikenberry, personal communications).

A plot of mobility versus log molecular weight was prepared at each acrylamide concentration using, as standards, β galactosidase, ovalbumin, carbonic anhydrase, ribonuclease, bovine serum albumin, β lactoglobulin, and cytochrome C. Molecular weights of the virus and polyhedral proteins were determined form the plot.

III. Results

The results of the neutralization tests are presented in Table I. As can be seen, the antiserum prepared against *T. ni* MEV neutralized *A. californica* MEV at a titer of 1:128. *A. californica* antiserum against its homologous viral antigen had a titer of 1:64, and failed to neutralize the *T. ni* MEV. With regards to other NPVs, none of the antisera against *T. ni* SEV, *H. zea* SEV and *P. dispar* MEV neutralized *A. californica* MEV. Antiserum prepared against the non occluded virus, CIV also failed to neutralize *A. californica* MEV.

The fluorescent antibody results supported the findings of the neutralization test. *S. frugiperda* cells inoculated with *A. californica* MEV showed the presence of antigens when stained with the homologous conjugated antiserum 48 h post inoculation (Fig. 1). However, fluorescent staining was greatly reduced by prior application of unconjugated postantiserum indicating the specificity of the reaction. This blocking reaction was also demonstrated with heterologous *T. ni* postantiserum thus confirming the antigenic relationship between these two NPVs. Uninoculated cultures failed to stain with the conjugated antiserum. Similar results were obtained when TN-368 cells were employed instead of *S. frugiperda* cells.

Polyacrylamide gel electrophoresis (PAGE) of several NPVs polyhedral proteins revealed the presence of multiple protein bands (Fig. 2). However pretreatment with SDS, DOC, heat and cold as described under materials and methods produced only a single protein band. Pretreatment of PIBs with 2.5% SDS from *A. californica* NPV produced *in vivo* and *in vitro* and *P. dispar* NPV produced *in vivo* all resulted in one protein band having a molecular weight of 29,000. A similar finding was demonstrated for *T. ni* SEV. It was also observed that there was a reduction in the number of protein bands from viral preparations when SDS was present in the alkali solution during dissolution of PIBs.

IV. Discussion and conclusions

An *in vitro* neutralization test employing the *S. frugiperda* cell line has been successfully employed to differentiate between several NPVs. Since the neutralization test is one of the most specific antibody-antigen reactions, it could prove useful for the serological characterization of other insect viruses. This test demonstrated a serological relationship between *A. californica* NPV and *T. ni* NPV employing antiserum prepared against the latter. However when the reciprocal test was run, no neutralization was observed. This lack of reciprocal cross reaction has also been

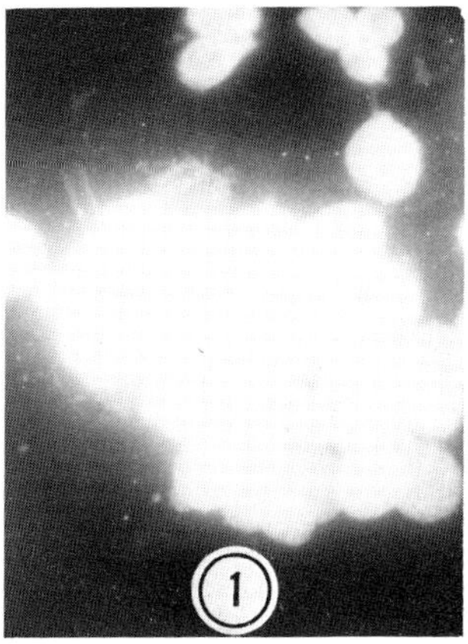

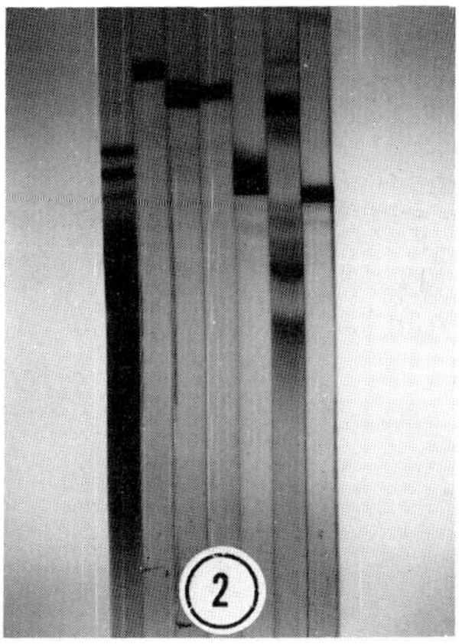

Fig. 1. Immunofluorescence of *S. frugiperda* cells 48 h postinoculation with *A. californica NPV. x128.*

Fig. 2. PAGE of the polyhedral proteins of two NPVs from left to right:
1. *A. californica* PIBs produced *in vivo:* without SDS pretreatment.
2. *A. californica* PIBs produced *in vivo:* with 1% DOC pretreatment.
3. *A. californica* PIBs produced *in vivo:* pretreatment -90°C for 10 min.
4. *A. californica* PIBs produced *in vivo:* dissolved at 4°C.
5. *A. californica* PIBs produced *in vitro (S. frugiperda):* with SDS pretreatment.
6. *P. dispar* PIBs produced *in vivo:* without SDS pretreatment.
7. *P. dispar* PIBs produced *in vivo:* with SDS pretreatment.

reported by Krywienczyk *et al.*, 1960, who used a complement fixation test to compare other viruses infecting Lepidopterous insects. Possible explanations for this phenomenon include mode of preparation of antiserum, low titer antiserum or actual differences in the antigenic composition of the viruses. The other NPVs employed in this study, namely *T. ni* SEV, *H. zea* SEV and *P. dispar* MEV could readily be distinguished from *A. californica* NPV by failure of their antisera to neutralize this virus. *S. frugiperda* was employed as the cell line of choice because like TN-368 it supports the replication of *A. californica* NPV and *T. ni* NPV. However, unlike TN-368, it is not as sensitive to the rabbit serum employed in the neutralization test.

The fluorescent antibody tests confirmed the findings of the neutralization tests. Cross reactions were minimized by employing the heterologous *S. frugiperda* cell line, since both viruses were produced in the cabbage looper and the TN-368 cell line. Furthermore, in our laboratory, it has been shown that there is very little cross reaction between *T. ni* and *S. Frugiperda* cells employing antiserum prepared against *T. ni* cells. An interesting observation from the immunofluorescent studies was the finding that *A. californica* PIBs did not stain with conjugated antiserum prepared against virus particles. Krywienczyk (1963), reported the staining of *Bombyx mori* PIBs with conjugated antiserum. However her tests showed that antiserum prepared against alkali released virions cross-reacted with inclusion body protein in a ring test. In our studies occasional peripheral stainings of PIBs were observed, although the entire inclusion body did not stain. This reaction may be due to partially exposed virions on the surface of the PIB which did not become entirely occluded by the protein matrix. Such virions can be readily discerned by electron microscopy. Because of the numerous NPVs which can be propagated *in vitro*, it will be possible to harvest free virions from tissue culture fluids, purify them and therefore greatly reduce the contamination with other proteins. In this way it will be possible to resolve the question as to whether or not PIBs and free virions share antigens in common.

The results of PAGE indicate that the inclusion body proteins of both MEVs and SEVs have approximately the same molecular weight subunit (29,000). Recently, Kozlov *et al.*, 1975, using SDS electrophoresis, reported the molecular weight of the polyhedral proteins of two other NPVs *(B. mori* and *Galleria mellonella*) to be ± 28,000, when inclusion bodies were dissolved in 67% acetic acid. They also found that dissolution in Na_2CO_3 at pH 11.00 produced several protein components and suggested that this cleavage of the polyhedral protein is a result of alkaline proteinase(s) activity. The first report of an alkaline protease in NPV inclusion bodies was by Yamafuji *et al.*, 1958. The source of this enzyme is not certain as they also isolated an alkaline protease with similar characteristics from healthy host tissue. A pH of 10.6 was found to be the optimum for the enzyme in both host and polyhedron. The alkaline solution normally used to dissolve polyhedral protein has a pH of 10.5 thus further suggesting the presence of an alkaline protease which degrades the polyhedral protein. It was also reported by Yamafuji *et al.*, 1960, that there is an alkaline protease in the NPV viral particle. This is consistent with our findings that pretreatment with SDS reduced the number of viral protein bands.

If there is an alkaline protease present which apparently degrades the inclusion body protein, it is important to establish conditions under which the enzyme will be inactivated. Treatment with SDS has such an effect. Also, besides causing apparent enzyme inactivation, SDS is useful in purification of inclusion bodies obtained *in vitro.* Harvesting NPVs from tissue culture is an excellent way of obtaining highly purified PIBs having no microbial contamination, but it was found quite difficult to lyse the *in vitro* grown cells to release the PIBs. Freezing and thawing several times in water suspension had little effect on cell lysis while 2.5% SDS was found to be quite effective.

Determination of the molecular weight of inclusion body protein may prove useful in differentiation between CPV, GV, and NPV as SDS electrophoresis of the inclusion body protein of a CPV from *Nymphalis io* produced a molecular weight of 37,000 (Payne and Tinsley, 1974) in contrast to the 28,000-29,000 found for multiply embedded NPVs. Lewandowski and Traynor (1972) reported that the polyhedral protein of *B. mori* CPV had two major protein components of 20,000 and 30,000 and several minor proteins.

The particular method of electrophoresis employed in this study is quite a sensitive procedure for examination of proteins. As little as 0.1 μg of protein will produce a visible band. Molecular weight determination by electrophoresis is a comparatively simple and satisfactory way of protein subunit identification and differentiation. Sodium dodecyl sulfate as a method of determining molecular weights was first reported by Shapiro *et al.*, (1967). Weber and Osborn then established its reliability for molecular weight determination of many different proteins. For these reasons, SDS electrophoresis appears to be a useful tool for NPV characterization and identification. Perhaps utilizing these same methods for the virus particles will indicate the presence of different molecular weight protein bands, enabling the differentiation of NPVs. In this regard Padhi *et al.*, 1975, reported the finding of 14 different protein subunits in virus particles of *P. dispar* NPV ranging in molecular weight from 12,000 to 140,000. On the other hand Young and Lovel (1973) using a different SDS electrophoresis procedure reported 12 protein subunits for virus particles of *T. ni* NPV ranging in molecular weight from 61,500 to 81,600. These two reports suggest a considerable difference in protein subunits between *P. dispar* and *T. ni* NPV particles. Perhaps PAGE of virus particles themselves will become a useful method by which NPVs can be identified.

In conclusion, the neutralization, immunofluorescent and electrophoresis methods have proven to be useful in the identification and characterization of a number of NPVs. It is expected that such methods will also prove useful for other insect viruses.

Acknowlegments

This work was supported, in part, by NSF Grant BMS 74-13608 and by a Charles and Johanna Busch Postdoctoral Fellowhip.

TABLE 1.

*A comparison of several nuclear polyhedrosis viruses (NPVs) by the neutralization test**

Virus	Antiserum Versus	Neutralizing Titer
A. californica MEV	*A. californica* MEV	1:64
	T. ni MEV	1:128
	T. ni SEV	0 (1:8)**
	H. zea SEV	0 (1:8)
	P. dispar MEV	0 (1:4)
	Chilo	0 (1:8)

* Titrated in the PRL-21 *Spodoptera frugiperda* cell line of Vaughn.

** Figures in brackets represent dilutions of antisera tested.

V. References

Benton, C.V., Reichelderfer, C.F. and Hetrick, F.M. (1973). *J. Invertebr. Pathol. 22,* 42.

Coons, A.H. Fluorescent antibody methods. (1958). *Gen. Cytochem. Methods 1,* 399-422. (J.F. Danielli ed) Academic Press, New York.

Cunningham, J.C., Tinsley, T.W., and Walker, J.M. (1966). *J. Gen. Microbiol. 42,* 397.

Couch, J.A. (1974). *J. Invertebr. Pathol. 24(3),* 311.

Faulkner, P., and Henderson, J.F. (1972). *Virology 50,* 920.

Goodwin, R.H., Vaughn, J.L., Adams, J.R., and Louloudes, S.J. (1970). *J. Invertebr. Pathol. 16,* 284.

Grace, T.D.C. (1967). *In Vitro 3,* 104.

Hink, W.F. (1970). *Nature (London) 226,* 466.

Kozlov, E.A., Sidorova, N.M., and Serebryani, S.B. (1975). *J. Invertebr. Pathol. 25,* 97.

Krywienczyk, J., and Bergold, G.H. (1960a). *Virology 10,* 308.

Krywienczyk, J. and Bergold, G.H. (1960b). *J. Immunol. 84,* 404.

Krywienczyk, J. (1962). *J. Insect Pathol. 4,* 185.

Krywienczyk, J. (1963). *J. Insect Pathol. 5,* 309.
Laemmli, U.K. (1970). *Nature 227,* 680.
Lewandowski, L.J. and Traynor, B.L. (1972). *J. of Virology 10,* 1053.
McIntosh, A.H., Maramorosch, K., and Rechtoris, C. (1973). *In Vitro 8,* 375.
McIntosh, A.H., and Maramorosh, K. (1973). *J.N.Y. Entomolog. Soc. 81,* 175.
Padhi, S.B., Eikenberry, E.F., and Chase, Jr., T. (1975). *Intervirology* (in press).
Payne, C.C. and Tinsley, T.W. (1974). *J. Gen. Virol. 25,* 291.
Shapiro, A.L., Vinuela, E., and Maizel, J.V. (1967). *Biochem. Biophys. Res. Commun. 28,* 815.
Shapiro, M. and Ignoffo, C.M. (1970). *Proc. IV Int. Colloq. Insect Pathol., College Park, Md.*
Sohi, S.S., and Bird, F.T. (1971). *IV Ann. Meet. Soc. Invertebr. Pathol. Montpellier,* France.
Sohi, S.S. and Cunningham, J.C. (1972). *J. Invertebr. Pathol. 19,* 51.
Vail, P.V., Jay, D.L., and Hunter, D.K. (1970). *Proc. IV Int. Colloq. Insect. Pathol., College Park, Md.*
Vail, P.V., Sutter, G., Jay, D.L., and Gough, D. (1971). *J. Invertebr. Pathol. 17,* 383.
Vail, P.V., Jay, D.L., and Hink, W.F. (1973). *J. Invertebr. Pathol. 22,* 231.
Weber, K., and Osborn, M. (1969). *J. Biol. Chem. 244,* 4406.
Wildy, P. (1971). In: *Monogr. Virol. 5,* 81. (S. Karger, Basel).
Yamafuji, J., Yoshihara, F., and Hirayama, K. (1958). *Enzymologia, 19,* 53.
Yamafuji, J., Mukai, J., and Yoshihara, F. (1960). *Enzymologia 22,* 1.
Young, S.Y., and Lovell, J.S. (1973). *J. Invertebr. Pathol. 22,* 471.

Chapter 33

CHARACTERIZATION OF INFECTIOUS COMPONENTS OF *AUTOGRAPHA CALIFORNICA* NUCLEAR POLYHEDROSIS VIRUS PRODUCED *IN VITRO*

W.A. Ramoska

I. Introduction

The nuclear polyhedrosis virus (NPV) of the alfalfa looper *Autographa californica* multiplies in the TN-368 *Trichoplusia ni* cell line (Vail *et al.*, 1973). The agents responsible for initiation of *in vitro* infection have, until this time, been undetermined. Investigators working with different NPVs have reported that the non-enveloped nucleocapsid, the enveloped nucleocapsid, infectious nucleic acid or a combination of the above can be capable of initiating infection in various host cells. The enveloped nucleocapsid was shown to be the infectious entity of *Trichoplusia ni (T. ni)* NPV replicating *in vitro* on the TN-368 cell line (Henderson *et al.*, 1974). Infectious DNA was reported to be infective to the *Heliothis zea* cell line (Ignoffo *et al.*, 1971). Few other investigators have dealt with the agent responsible for *in vitro* virus infection.

Studies on infectious NPV viral entities *in vivo* have employed either released virions from dissolved inclusion bodies or free unoccluded virions from the diseased insect. Utilizing membrane filters, Stairs and Ellis (1970) reported that the smallest infecting entity of *Galleria mellonella* NPV was the non-enveloped nucleocapsid. Scott *et al.*, (1971) reported that the enveloped nucleocapsids released from polyhedra of *T. ni* NPV were infectious to *T. ni* larvae via injection. Other investigators (Knosaka & Himeno, 1972), Kawarabata, 1973) using either free non-occluded or released virus material, demonstrated that the non-enveloped nucleocapsids of *Bombyx mori* NPV were infectious to the host insect via intercoelomic injection.

In this study we report a procedure for separation of the *in vitro* produced *Autographa californica* virus entities. They are identified and assayed to determine their infectivity *in vitro*.

II. Methods and methods

The virus was originally isolated from an *Autographa californica* larva (Vail *et al.*, 1971). The virus used in this study was produced in the *T. ni* (TN-368) cell line. The inoculum produced approximately an 80:20 ratio of FP to MP plaque variants described by Hink and Vail (1971) and Ramoska and Hink (1973).

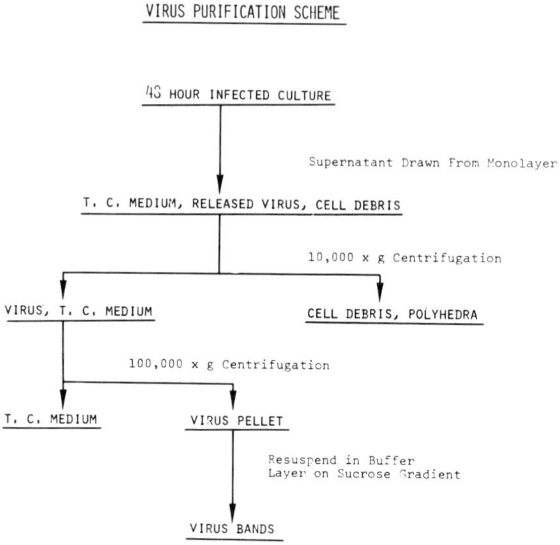

Two-hour old TN-368 monolayers, in 75 cm^2 plastic tissue culture flasks, were infected with an inoculum containing a multiplicity of infection of five plaque forming units (pfu) per cell. After a one hour adsorption period, the virus was poured off and 12 ml of TNM-FH (Hink, 1970) was poured over the cell monolayer.

The cells were incubated at 28°C for 48 hours, at which time they showed typical NPV cytopathology and were still intact. Tissue culture medium was poured off the monolayer and the released virus in the tissue culture medium was purified as outlined in Fig. 1. The virus suspension was centrifuged at 10,000 g for 30 minutes in a Sorvall PC2-B centrifuge with an ss-34 anglehead rotor. This was done to remove any loose cells that might lyse and release their contents at a later time. The supernatent was centrifuged at 100,000 g for 30 minutes in a Beckman ultracentrifuge using either an angle 50 or sw-27 rotor. The resultant translucent-white pellet was resuspended in phosphate buffered saline (PBS, Henderson *et al.*, 1974) pH 7.2 at approximately .005% of the original volume.

Density Gradient

Density gradient sucrose was dissolved in PBS and used in preparing 10-40% w/v continuous gradients in 5 ml sw-39 nitrocellulose centrifuge tubes. 0.2 ml aliquots of the virus preparation were layered on top of the gradients which were then centrifuged for 45 minutes at 87,000 g in a Beckman model L ultracentrifuge. The gradients were immediately fractionated using an Isco (Instruments Specialties Co., Lincoln, Nebraska) fraction recovery system. During fractionation the gradients were continuously monitored for light absorbance at 260 nm. 0.5 ml fractions were collected from the gradients.

CsCl Density Gradient

The virus preparation was subjected to equilibrium isodensity centrifugation (Bowen, 1970). 0.5 ml of the virus preparation was added to 4.5 ml of CsCl dissolved in PBS and the density adjusted to 1.35 gm/cc^3. Using an sw-39 rotor this preparation was centrifuged for 48 hours at 100,000 g at 4.0^{o}C. Banding could be seen in the centrifuge tube and fractions were taken as described for sucrose gradients.

Column Chromatography

The virus preparation was subjected to agarose gel column chromatography using Sepharose 2B which contains bead sizes of 60-250 in diameter (Pharmacia Fine Chemicals, Upsala, Sweden). A 2.0 ml sample of the virus preparation was layered onto the 12 x 30 cm column and eluted with PBS. The flow rate was adjusted to 4 ml/hr. The eluate was continuously monitored at 260 nm for ultraviolet light absorbance.

Electron Microscopy

Fractions in which absorption peaks occurred were pooled and dialysed against PBS (pH 7.2) overnight. The fractions were then either fixed in 25% gluteraldehyde or dialysed against .005 M Na_2CO_3 for 12 hours. Following dialysis the fractions were mixed with an equal volume of 4% uranyl acetate and sprayed onto 200 mesh formvar coated grids. The grids were examined using an RCA EMU 3-G electron microscope operated at 50kv acceleration.

IN VITRO Assay of Infectivity

All fractions from the sucrose gradients were assayed on TN-368 cell monolayers utilizing the plaque assay technique for determining virus titers (Hink and Vail, 1973). Four replicates were used for each titration.

Spectrophotometry

The ultraviolet light absorbance spectrum of the virus was determined using a Beckman DU spectrophotometer with a H_2 lamp. Light scattering was corrected for by the method of Bonhoeffer and Schachman (1960).

Density Measurements

The density of the particles in each band produced by the CsCl centrifugation was measured by drawing a portion of each band into a 200 ul pipette of known weight, and weighing it.

III. Results

All three separation methods, rate zonal centrifugation, isodensity centrifugation and gel chromatography resulted in the separation of 3 ultraviolet light absorbing regions (peaks A, B, and C). Sucrose gradients produced one adsorbance peak at the top of the 5 cm tube (A). A second peak (B) is located approximately 1 cm from the top of the tube and a third peak (C) is located approximately 4 cm from the top of the tube (Fig. 2).

Ten 0.5 ml fractions from the gradients were assayed on cell monolayers and as figure 2 indicates two peaks of infectivity occur which correspond to peaks B and C of the ultraviolet light absorbance patterns.

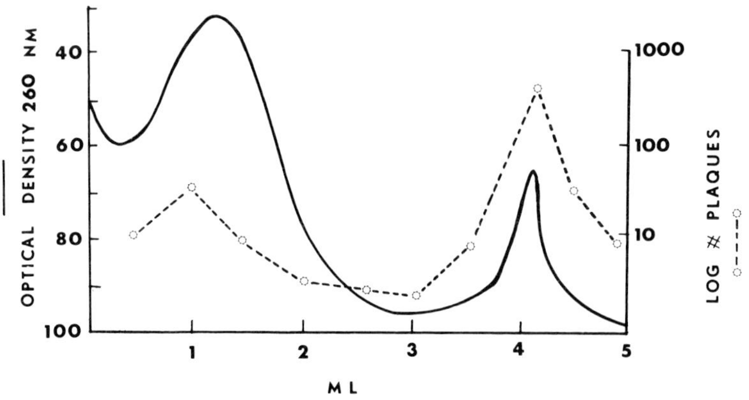

Fig. 2. Ultraviolet light elution profile (solid line) and mean plaque counts (dotted line) for fractions of virus preparation centrifuged at 100,000 g for 30 minutes on a continuous 10-40% w/v sucrose gradient.

Electron micrographs of the peak fractions show that peak A contains heterogenous material and debris. Peak B contains typical non-enveloped nucleocapsids approximately 280 nm in length and 32 nm in diameter. Enveloped nucleocapsids are found in peak C. They are 330 nm in lenght and 36 nm in diameter.

The ultraviolet light absorbance spectrum for fractions from peaks A, B, and C (figs. 3, 4 and 5) shows that the respective 260:280 absorbance ratios are 101, 176, and 119. These figures suggest the presence of nucleoprotein in peaks B and C.

Three peaks also appear in CsCl isodensity centrifugation. The ultraviolet light absorbance spectra and electron microscope studies indicate that enveloped nucleocapsids are found in the bottom band in CsCl. The middle band contains unenveloped nucleocapsids and the top band consists of undefinable debris. Figure 6 shows the particle density of the enveloped nucleocapsid is 1.28 gm/cc3 and that of the non-enveloped nucleocapsid 1.32 gm/cc3.

The elution profile for the preparation separated by Sepharose again demonstrates 3 peaks of UV absorbance (Fig. 7). As with the sucrose gradient separations, the first peak was found, under electron microscopic examination, to consist of enveloped nucleocapsids, the second peak contains non-enveloped nucleocapsids, and debris is located in the third peak.

IV. Discussion

Our results show that the virus released before lysis of the host cell consists of both enveloped and non-enveloped virus particles. All three methods of separation produced similar results in that the enveloped and non-enveloped nucleocapsids were separated. A third band (A) always appeared at the top of the sucrose gradient. Other investigators have reported such a fraction in their purifications (Summers and Pashke, 1970; Matta, 1969; Miyajima *et al.*, 1969). These investigators generally conclude that the top fraction is composed of decomposition units of the virus

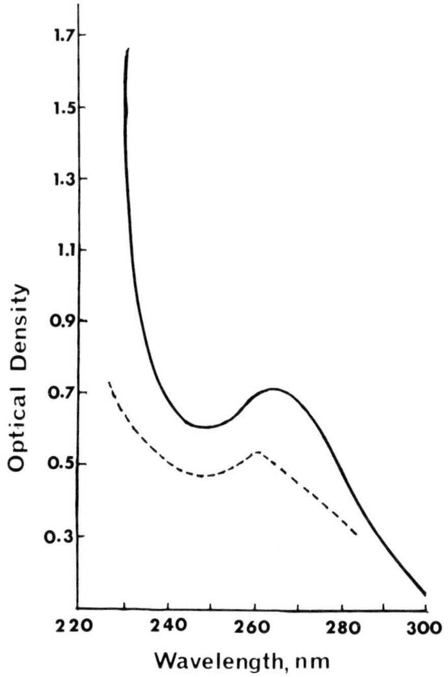

Fig. 3. Absorbance spectrum of peak A. 260:280 = 101.

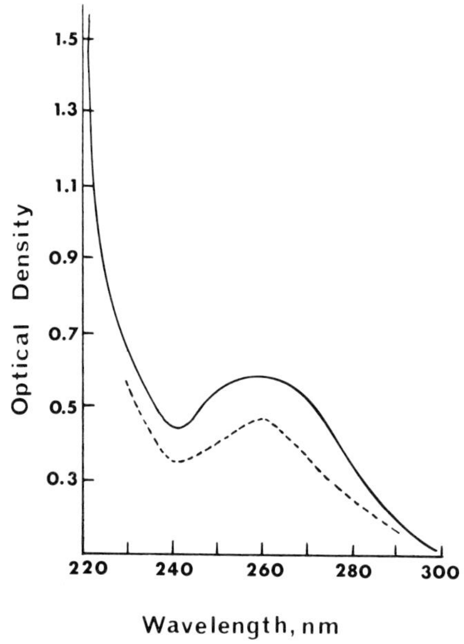

Fig. 4. Absorbance spectrum of peak B. 260:280 = 176.

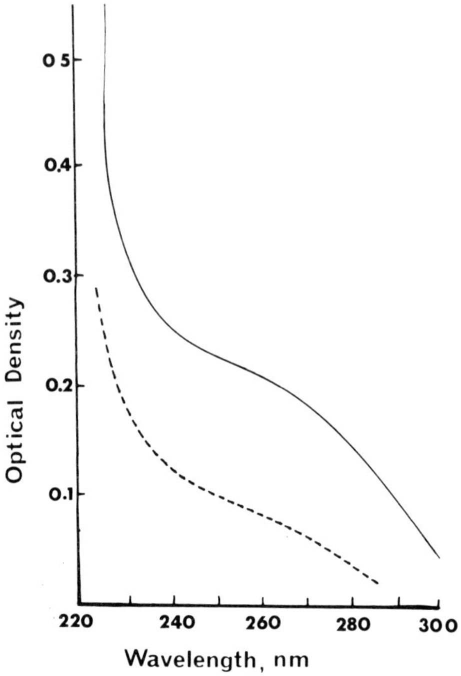

Fig. 5. Absorbance spectrum of peak C. 260:280 = 119.

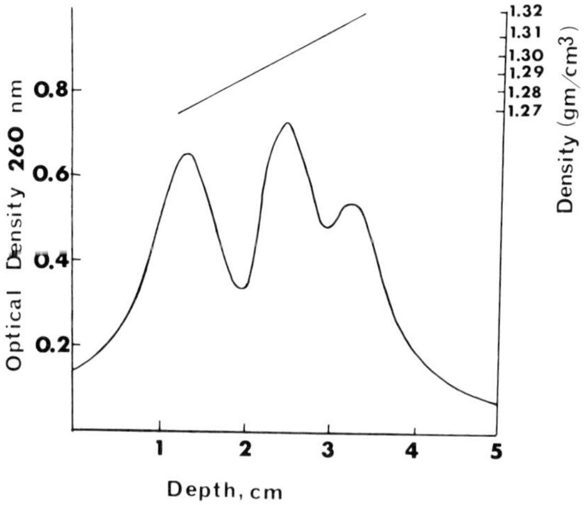

Fig. 6. Ultraviolet light elution profile from virus preparation centrifuged 48 hours in a 1.35 gm/cc^3 CsCl isodensity equilibrium gradient. The particle density of the first peak is 1.32 gm/cc^3, the second is 1.28 gm/cc^3 and the density of the third peak is 1.26 gm/cc^3.

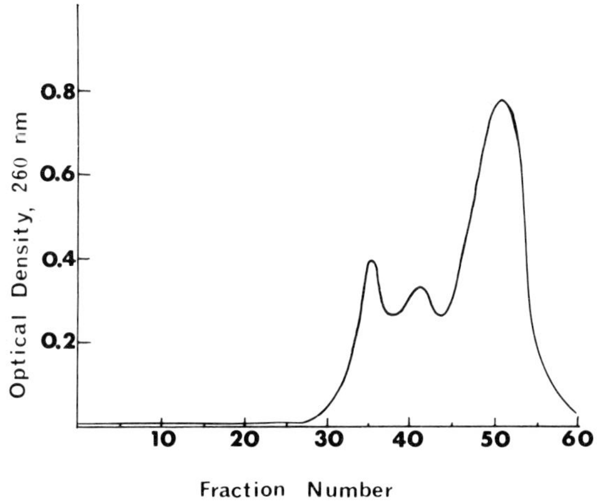

Fig. 7. Sepharose 2B elution profile. Bed dimensions: 12 x 30 cm. Eluant: PBS pH 7.2. Flow rate: 4 ml/hr. Sample volume: 2 ml.

and viral membranes. Our data tend to agree with their conclusions. We were unable to detect any amount of free infectious DNA in our experiments.

Peaks B and C had a low level (5%) of cross contamination at the .005% original volume suspension as shown by electron microscope counts. Dilution of the preparation did not lower this cross contamination but more concentrated suspensions produced aggregation of the virus particles and increased the cross contamination.

Infectivity tests show that the contents of peak A as the data suggest are not highly infectious. The infectivity at peaks B and C demonstrates that both contain infectious material at a higher titer than anywhere else in the gradient. The high titer of infectivity at peak C over that of peak B indicates that the enveloped nucleocapsid is more infectious than the unenveloped nucleocapsid. Because of the low level of cross contamination we do not feel that enveloped nucleocapsids are responsible for the infectivity in peak B.

This study was supported in part by Grant R-802516 from the Environmental Protection Agency.

V. References

Bonhoeffer, F., and Schachman, H.K., (1960). *Biophys. Res. Commun. 2,* 336.

Bowen, T.J., (1970). *An Introduction to Ultracentrifugation,* pp. 1-171, J. Wiley & Sons, London.

Henderson, J. F., Faulkner, P., and MacKinnon, E. A., (1974). *J. Gen. Virol. 22,* 143.

Hink, W.F., (1970). *Nature (London) 226,* 467.

Hink, W.F., and Vail, P.V., (1973). *J. Invertebr. Pathol. 22,* 168.

Ignoffo, C.M., Shapiro, M., and Hink, W.F., (1971). *J. Invertebr. Pathol. 18,* 131.

Kawarabata, T., (1974). *J. Invertebr. Pathol. 24,* 196.

Khosaka, T., and Himeno, M., (1972). *J. Invertebr. Pathol. 19,* 62.

Matta, James F., (1970). *J. Invertebr. Pathol. 16,* 157.

Miyajima, S., Kimura, I., and Kawase, S., (1969). *J. Invertebr. Pathol. 13,* 296.

Ramoska, W.A., and Hink, W.F., (1974). *J. Invertebr. Pathol. 23,* 197.

Scott, H., Young, S.Y., and McMasters, J., (1971). *J. Invertebr. Pathol. 18,* 179.

Stairs, G.R., and Ellis, B.J., (1970). *J. Invertebr. Pathol. 17,* 350.

Summers, M.D., and Paschke, J. D., (1970). *J. Invertebr. Pathol. 16,* 227.

Vail, P.V., Jay, D.L., and Hink, W.F., (1973). *J. Invertebr. Pathol. 22,* 231.

Vail, P.V., Sutter, G., Jay, D.L., and Gough, D., (1971). *J. Invertebr. Pathol. 17,* 383.

Chapter 34

UTILIZATION OF TISSUE CULTURE TECHNIQUES TO CLONE AN INSECT CELL LINE AND TO CHARACTERIZE STRAINS OF BACULOVIRUS

P. Faulkner, M. Brown, and K.N. Potter

I. Introduction

Advances in the past few years have made it possible to grow several baculoviruses which are highly virulent against some agricultural and forest pests in invertebrate tissue cultures. At least three continuous insect cell lines (from *Bombyx mori*, Grace, 1974; from *Trichoplusia ni*, Hink, 1970; and from *Spodoptera frugiperda*, Vaughn, 1970) support the replication of a baculovirus causing nuclear polyhedrosis disease in the host insect from which the line was derived. In addition, baculoviruses isolated from other Lepidopteran species have also been shown to replicate in these cell lines (Table 1). Laboratory bioassays (Faulkner & Henderson, 1972) and field trials (Ignoffo *et al.*, 1974) have demonstrated that polyhedra produced in cell culture are as virulent as those propagated *in vivo.*

Those who work with baculovirus infected cell cultures commonly observe a wide variation in the number of polyhedra in individual cells in a culture. In the case of the NPV of *T. ni* the range is from 5-200 inclusion bodies per nucleus (Faulkner & Henderson, 1972). We have examined some parameters that may influence viral yield in individual cells in culture. *Trichoplusia ni* cell line, TN 368, (Hink, 1970) and a multiply enveloped* NPV (MEV) of *T. ni,* (Heimpel & Adams, 1966) were used for most of our studies. The cells were infected by passing non-occluded virus (NOV) from an infected culture to uninfected cells.

---

* The following abbreviations are used: NPV, nuclear polyhedrosis virus; MEV, multiply enveloped virus; SEV, singly enveloped virus; NOV, non-occluded virus; MOI, multiplicity of infection.

TABLE 1.

Invertebrate cell lines that permit serial transfer of baculoviruses and yield polyhedra in cells

Cell Line Derived From	Baculovirus
1. *Bombyx mori* (Grace, 1967)	(a) *Bombyx mori* NPV (Raghow & Grace, 1974)
2. *Trichoplusia ni* (Hink, 1970)	(a) *Trichoplusia ni* NPV (Faulkner & Henderson, 1972)
	(b) *Autographa californica* NPV (Vail *et al*, 1973)
3. *Spodoptera frugiperda* (Goodwin *et al*, 1970)	(a) *Spodoptera frugiperda* NPV (Goodwin *et al*, 1970) (Knudson & Tinsley, 1974)
	(b) *Trichoplusia ni* NPV (Goodwin *et al*, 1973)
	(c) *Autographa californica* NPV (Goodwin *et al*, 1973) (Dougherty *et al*, 1975)

Infected cell cultures release NPV in two forms: (1) Inclusion bodies (the polyhedra) which develop in nuclei. These are the principal vehicles for transmission of the baculovirus under natural conditions. They are released during decomposition of killed insects and stick to foliage. After ingestion by a caterpillar the inclusion bodies pass through the highly alkaline gut region of an insect where it is believed they dissolve and release infectious agents. Approximately 5% of the mass of inclusion bodies consists of enveloped nucleocapsids (virions). (2) Tissue cultures also release NOV through the cell membrane. These consist of single enveloped nucleocapsids which can subsequently infect other cells in culture (Fig. 1). They do not appear to be infectious *per os* (Dougherty *et al.*, 1975) but are virulent when injected into the hemocele of larvae. NOV have also been observed in infected larvae and may be responsible for systemic spread of the virus in an individual insect.

II. Effects of serial passage of a virus on yield of polyhedra and NOV

MacKinnon *et al.*, (1974) found that the yield of polyhedra in infected cells fell dramatically when the NOV was passed serially *in vitro* (Fig. 2). The MOI in these experiments was 0.1-1.0 and after 15 passages the average number of polyhedra per cell in culture fell from about 28 to 5 or less. The passaging was continued for more than 50 transfers, but the yield of polyhedra was not increased. MacKinnon *et al.* (1974) compared the *in vivo* infectivity of inclusion bodies from the 6th and 43rd passage and found that whereas a normal dose/response relationship existed for early passage inclusion bodies, the late passage inclusion bodies were not infectious (Table 2). In addition, it was apparent that a large number of the inclusion bodies found at the late passage contained abberant forms of virus. Electron micrographs of sections of polyhedra that have been partially digested with 0.05M sodium carbonate are shown in Fig. 3. Whereas typical bundles of nucleocapsid enclosed by a common membrane are present in early passaged infective polyhedra, late passage inclusion bodies contain embedded nucleocapsids which are on average shorter than those present in infectious material.

In these experiments it was considered necessary to show by the use of appropriate controls that observed results were not due to changes in the cell line as opposed to changes in the virus. This was done in two ways. In the first instance,

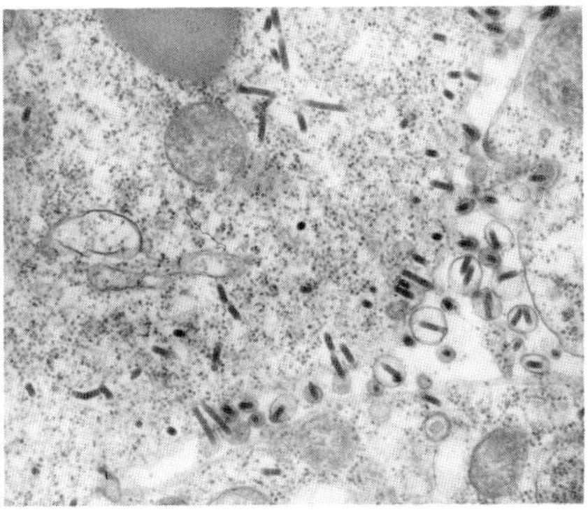

Fig. 1. Non-occluded virus of *T. ni* NPV released at cell membrane (x20,000) (Courtesy of Dr. E.A. MacKinnon).

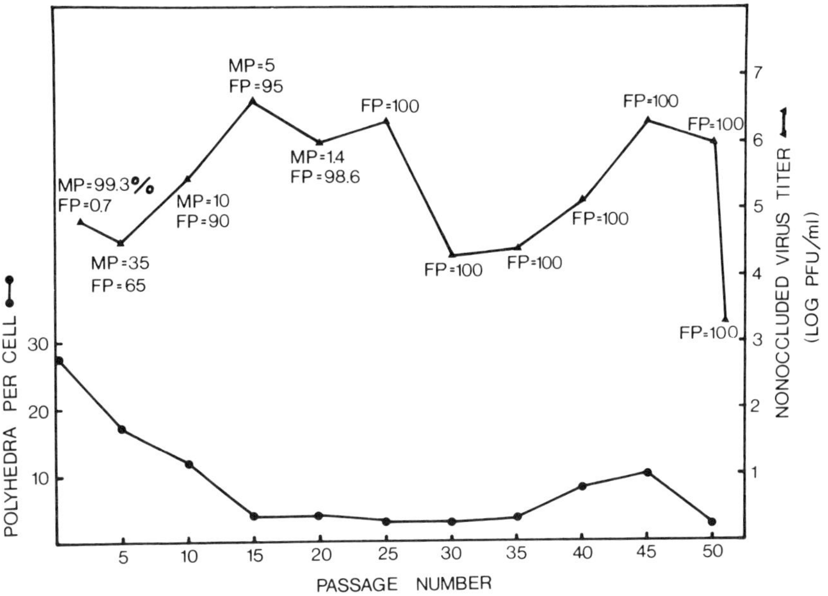

Fig. 2. Effect of serial undiluted passage of *T. ni* NPV on yield of polyhedra and phenotype of virus.

cell cultures were monitored at all times for susceptibility to primary NPV infection using infectious hemolymph. Over the course of the experiments during which approximately 100 subcultivations of the cells occurred, there was no detectable decrease in susceptibility to primary infection. Secondly, infected tissue culture supernatants from all passages were tested *in vitro* at the same time. In other words, the relevant data which are given in Fig. 2 and Table 2 were not performed until samples from all 50 passages had been obtained. These were then tested for *in vitro* induction of inclusion body formation using aliquots of the same cell population (MacKinnon *et al.*, 1974).

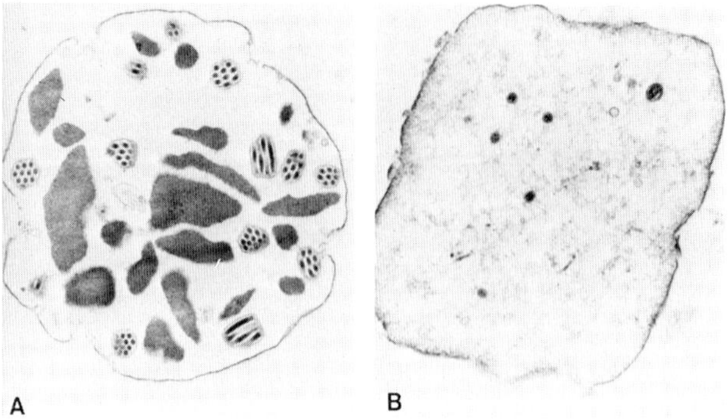

Fig. 3. Thin sections of partially digested polyhedra. A. Polyhedra harvested at passage level 5 (x17,000). B. Polyhedra harvested at passage level 51 (x16,000) (Courtesy Dr. E.A. MacKinnon).

TABLE 2.

Infectivity (per os) *of early and late passaged polyhedra produced in tissue culture*

Dose/Larvae (No. of polyhedra)	Polyhedra from 6th passage (% Mortality)	Polyhedra from 43rd passage (% Mortality)
2500	100	0
500	94.1	0
250	65.2	0
125	71.4	0
50	31.6	0
0	0	0

Infectivity of early and late passaged polyhedra produced in tissue culture. Purified polyhedra were applied to discs cut from cabbage leaves and fed to *T. ni* larvae. After 24 hours insects that had consumed > 80% of the disc were transferred to specimen vials and reared for 12 days on an artificial diet (Jaques, 1967). Dead larvae were examined by phase contrast microscopy for the presence of polyhedra.

III. Yield of polyhedra in cloned cell sub-lines.

The yield of NPV in fresh clones isolated from the TN 368 cell line was investigated. The original cell line had undergone several hundred splits in our hands since we first obtained it from Dr. Hink and it was possible that sublines with varying capacity to support the virus had arisen within it. The cells were cloned in wells of Microtest plates (Falcon Plastics #3034). Each well was seeded with 10 μl cell suspension containing 1-10 cells. Plates were placed in a humid box at 27°C for 2 hours, then scored for wells that contained a single cell. Wells containing single cells were observed over the next few days and at the 8-10 cell stage developing clones were transferred to 24 cm^2 flasks in 5 ml medium (Brown &

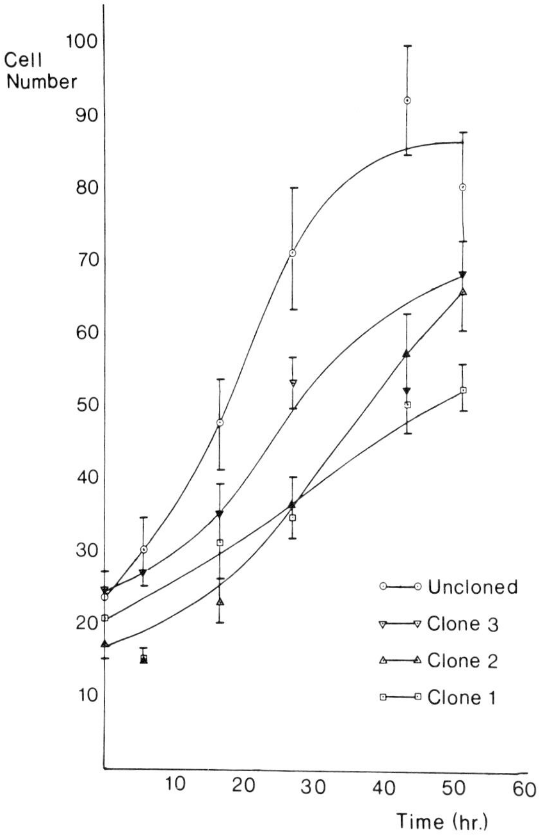

Fig. 4. Growth curves of uncloned and cloned sublines of TN 368 cells.

Faulkner, 1975). After reaching confluency the cells were passaged 4 times before they were stored in liquid nitrogen. Cloned cells used in experiments were between 5 and 30 passages after selection. In 2 plates seeded with 1-2 cells per well, 60-65% of cells gave rise to clones which were still viable 96 hours after seeding. From the growth curves for the uncloned cell line and for three clones at passage level 5 (Fig. 4) it may be calculated that the cell doubling times were 15.8 ± 1.5 hours for the TN 368 cell lines and 27.6 ± 3.4 hours, 21.9 ± 1.7 hours, 27.4 ± 5.9 hours for the three clones. *T. ni* cells are normally subdivided twice a week to preserve viability. It is evident that the cloned cell lines divide at a rate slower than the uncloned line. Similar observations have been made with other insect cells. Clones derived from mosquito cells (Suitor *et al.,* 1966) and also from *Trichoplusia ni* (McIntosh & Rechtoris, 1974) had larger doubling times than the uncloned culture. After 30 passages in culture growth rates were again measured (Fig. 5). The doubling time of the clones had decreased to that of the uncloned line. Thus during passage there was selection for faster growing cells.

The yield of polyhedra in uncloned cells infected at MOI = 4 seems to be of a similar order to that of the 3 cloned lines when they were examined at the 7th and

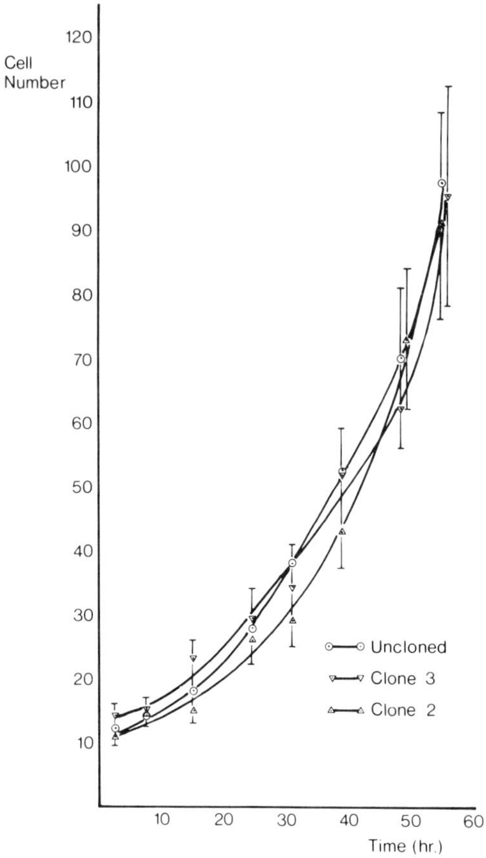

Fig. 5. Growth curves of uncloned and cloned sublines after 30 subcultivation.

19th passages (Table 3). Individual cells observed in the microscope contained variable numbers of inclusion bodies per cell. These data indicate that it is most unlikely that the extreme variations seen in yield of polyhedra in cells can be ascribed to different sublines of cells present in the population.

The data in Table 4 show that cell multiplication is arrested following infection with passage level 5 virus or with passage level 51 virus. Approximately 5 x 10^5 cells were seeded into a 25 cm^2 flask and were infected. Cell numbers were counted *in situ* using a graticule in the eyepiece of a microscope. At 24 hour intervals 10 random fields were counted in each flask. An arrest of cell multiplication has also been reported when *T. ni* cells are infected with *A. californica* NPV (Vail *et al.*, 1973) and when *S. frugiperda* cells are infected with their NPV (Knudson and Tinsley, 1974).

IV. Relationship between virus MOI and yield of polyhedra

We considered that the number of inclusion bodies found in cells in culture may be related to the amount of virus infecting individual cells. We found that the

TALBE 3.

Polyhedra yield of cloned cells

Cells	*Polyhedra/cell*
Uncloned	20 ± 3
	19 ± 3
Clone 1/19	17 ± 2
1/7	19 ± 3
Clone 2/19	25 ± 3
2/7	21 ± 5
Clone 3/19	15 ± 2

Polyhedra yield of cloned cells. Cells were infected at MOI = 4. At 65 hours post-infection cells and debris were harvested by centrifugation and polyhedra were released (Faulkner & Henderson, 1972) and counted in a hemocytometer.

TABLE 4.

Inhibition of cell multiplication following virus infection

	Time after infection			
Cells	0 hr	21 hr	45 hr	69 hr
Uncloned:				
Control	3 ± 1	7 ± 5	16 ± 8	26 ± 11
NPV_5 – MOI = 1	5 ± 2	4 ± 3	7 ± 4	10 ± 4
NPV_5 – MOI = 4	4 ± 1	5 ± 3	3 ± 2	5 ± 2
NPV_{51} – MOI = 1	4 ± 2	4 ± 2	5 ± 2	5 ± 3
Clone 2:				
Control	6 ± 3	9 ± 4	15 ± 8	21 ± 9
NPV_5 – MOI = 1	6 ± 2	8 ± 4	9 ± 4	8 ± 3
NPV_5 – MOI = 4	6 ± 3	8 ± 3	8 ±,3	7 ± 6
NPV_{51} – MOI = 1	5 ± 2	7 ± 1	7 ± 4	6 ± 3

Inhibition of cell growth following virus infection. Experimental details in text. The data reported are the mean cell numbers counted at each time interval ± standard error.

average yield of polyhedra per cells is affected by the MOI of input virus (Fig. 6). At multiplicities from 0.01 to 4 the average yield of polyhedra was about the same, but average yields tripled at MOI 20 to 30. Still higher multiplicities up to MOI = 500 led to progressive reduction of average polyhedra yield. The existence of an optimum MOI probably indicates that a large number of extracellular and intracellular sites need to be saturated to maximize polyhedron production. The particle/IU ratio has not been determined for the NPV of *T. ni* but is in the range of 266 ± 177 particles per PFU for another baculovirus, the NPV of *S. frugiperda* (Knudson & Tinsley, 1974). Thus at very high multiplicities of infection the cells would be

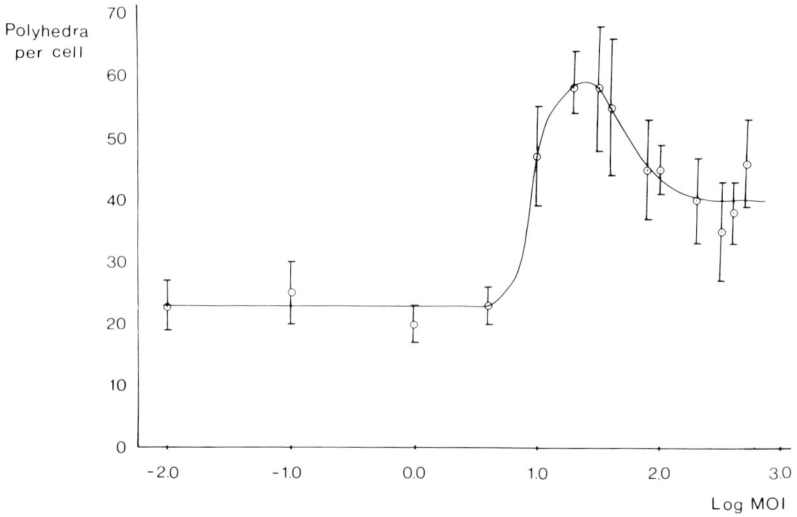

Fig. 6. Relationship between yield of polyhedra/cell and MOI of *T. ni* NPV in culture.

expected to interact with very large numbers of particles and some competition could be expected which may reduct the number of infectious particles entering cells which give rise to polyhedra. In addition, large numbers of particles may damage cell membranes and result in suboptimal conditions for polyhedra production.

V. Attempts to synchronize *T. ni* cell cultures

A parameter that could influence the production of baculovirus in culture is currently being studied in our laboratory. We are investigating the consequence of infecting cells at different stages in the cell cycle. Threre is no report in the literature of the description of the cell cycle of a continuous lipidopteran line. We have attempted to synchronize cloned *T. ni* cells for several years with little success. The cells do not attach firmly to glass or plastic, hence techniques which involve concentration of rounded cells in mitosis cannot be used; thus mitotic cells blocked with colcemid cannot be readily separated from other cells in culture. Our best success has been in using the double thymidine block technique to arrest cells at the G_1-S boundary of the cycle. An effective concentration of 5 mM has been established and the cells unblock rapidly when rinsed 3 times in medium. Cells are incubated in the presence of 5mM thymidine for 18 hours and are unblocked for 9 hours after washing 3 times with medium. They are then blocked for a further 18 hour period with thymidine and subsequently washed 3 times with fresh medium. DNA synthesis starts immediatly and is almost maximum during the first hour after unblocking. We have yet to perform an experiment in which we have been able to follow the complete cell cycle with regard to DNA synthesis and mitotic index of cells, and the only reproducible information that we have is that the S phase is 3-5 hours long.

VI. Characterization of NPV strains.

The isolation of strains of baculovirus from a given stock is a prerequisite for characterization and genetic studies. A most satisfactory method of cloning viruses utilizes a plaquing technique. Individual plaques represent a focus of infected cells that have arisen from a single infectious unit of virus. A major advance in insect virology occurred when Hink and Vail (1973) described a plaque assay for the titration of *A. californica* NPV in *T. ni* cells. Monolayers of these cells were infected with appropriate dilutions of virus and the resulting infectious centers were overlayed with 0.6% methyl cellulose. A linear dose/response relationship is observed when *T. ni* NPV is plaque assayed using this method (Fig. 7). We have applied the plaquing technique for two purposes. (1) To plaque purify strains of virus originally present in hemolymph of insects that have been fed polyhedra collected from the field or reared in the laboratory. (2) To examine the phenotype of the foci formed in monolayers by isolated virus strains. Hink and Vail (1973) found two types of plaques to be present in monolayers infected with *A. californica* NPV.

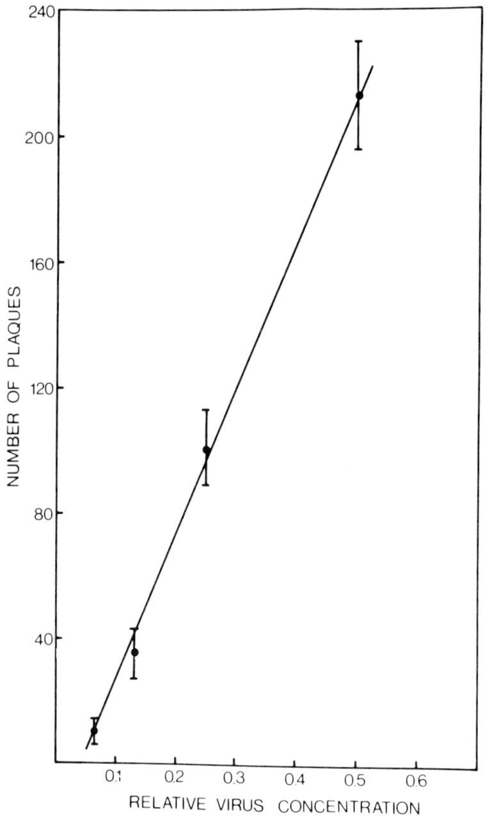

Fig. 7. Plaque assay of MP strain of *T. ni* NPV using method of Hink and Vail (1973). Three replicate plates were set up at each virus dilution. Plaques were counted (unstained) three days after infection of monolayer.

The first contained many polyhedra per nucleus and the second, few polyhedra per nucleus. Subsequent EM examination by Ramoska and Hink (1974) indicated that the MP foci contained polyhedra of the MEV morphology whereas the FP strains contained singly enveloped virus in the inclusion bodies. Both strains gave rise to inclusion bodies in tissue culture which were infectious *in vivo,* and both strains of virus reappeared in larvae which had been fed either the MP or the FP strain of polyhedra.

The virus that we work with, the NPV of *Trichoplusia ni,* also gives rise to two distinct types of plaques in *T. ni* cells (Fig. 8). Cells in the focus either contain more than 30 polyhedra per nucleus (MP type) or less than 10 polyhedra per nucleus (FP type). These two variants were 3 times plaque purified and are referred to as MP and FP strains of the virus. Stocks of both variants were grown up using a low MOI (0.01) to infect cultures.

VII. Virulence *In Vivo* of MP and FP strains of virus.

Inclusion bodies arising from the two strains of virus were compared for infectivity in feeding experiments. We are grateful to Dr. R.P. Jaques at the Canada Department of Agriculture, Research Station, Harrow, Ontario, for cooperating in these experiments. The feeding assay consisted in applying known numbers of polyhedra or known doses of NOV to collard (cabbage) discs. Each disc is placed with a single cabbage looper larva in a specimen vial. Larvae which consumed the disc within 24 hours were used for a dose/response LD_{50} assay and were subsequently fed on an artificial diet. The data in Table 5 show that polyhedra from the MP strain of virus is virulent whereas we did not observe deaths with larvae fed the FP polyhedra. Assay of the hemolymph of larvae fed MP and FP polyhedra clearly indicated that NOV of the MP strain was present in larvae fed MP polyhedra, but we were unable to detect any cytopathogenic effect (CPE) in tissue culture inoculated with hemolymph from larvae fed the FP strain. This result is quite different from that reported for *A. californica* strains of virus by Hink and Vail (1973). In their case, both FP and MP plaque suspensions were virulent and each gave rise to both FP and MP plaque morphology when hemolymph from infected larvae was assayed for virus.

Both strains yield non-occluded virus in tissue culture, and when injected into the hemocele these NOV proved pathogenic for larvae. Examination of the virus present in the hemolymph of diseased insects by plaque inspection showed that insects injected with FP virus had only FP strain in their hemolymph, whereas insects injected with MP strain contained some FP virus (Table 5).

VIII. Morphology of MP and FP virus.

Preliminary EM examination of thin sections of MP and FP polyhedra shows considerable differences in the two strains. The MP polyhedra appear to have the usual structure of NPV polyhedra and contain bundles of nucleocapsids which are surrounded by a common developmental membrane and occluded in the polyhedra (Fig. 9). On the other hand aberrant particles are present in the FP polyhedra (Fig. 10). Bundles of nucleocapsids were rarely found. The thin sections of FP polyhedra are not the same as the published micrographs of thin sections of the SEV *T. ni* polyhedra of Jaques (1967) and Goodwin *et al.,* (1973), and resemble the polyhedra that we observed when we serially passed the *T. ni* NPV for 50 passages *in vitro* (MacKinnon *et al.,* 1974).

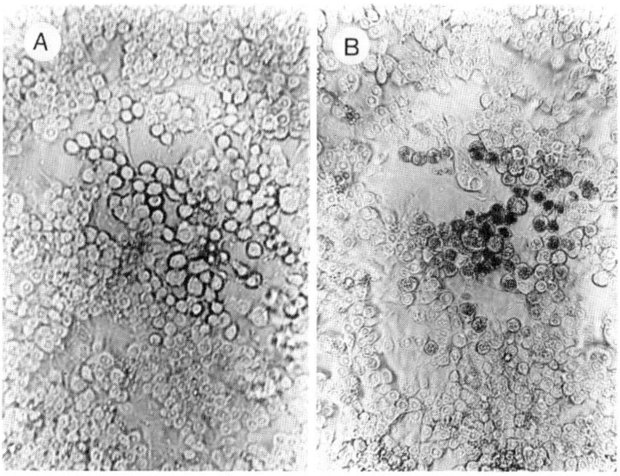

Fig. 8. Appearance of FP and MP plaques. *A.* FP plaque contains cells with swollen nuclei and few, if any polyhedra. *B.* MP plaque contains cells with nuclei packed with polyhedra.

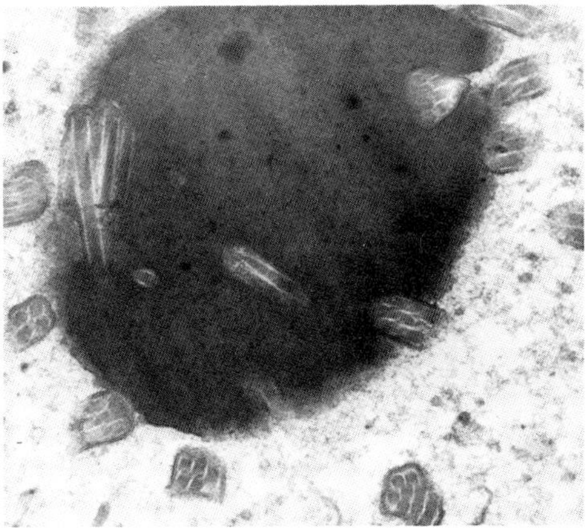

Fig. 9. Thin section of polyhedron in a culture infected with MP strain of virus (x51,000). (Courtesy Dr. E.A. MacKinnon).

Thin sections of cells infected by MP and FP strains show that single enveloped viruses are released at the cell membrane. A subjective assessment is that more MP viruses are liberated at the cell membrane.

A possible explanation for the two forms of inclusion bodies is that nucleocapsids are compressed, distorted or cleaved when the inclusion body protein of the FP strain crystallizes. This leads to the stunted appearance of the occluded nucleocapsids. This distortion could arise by a mutation which results in tighter packing of the polyhedron protein lattice or to a change in the composition of the envelope that surrounds nucleocapsids.

TABLE 5.

Bioassays of viral cell culture products

Form of Virus	Strain	LD_{50} Fed	LD_{50} Injected ($TCID_{50}$/insect)
NOV	MP	0	16 (97% MP, 3% FP)
	FP	0	29 (100% FP)
Polyhedra	MP	56 (100% MP)	
	FP	0 (No CPE in cell culture)	

Bioassays of viral cell culture products.

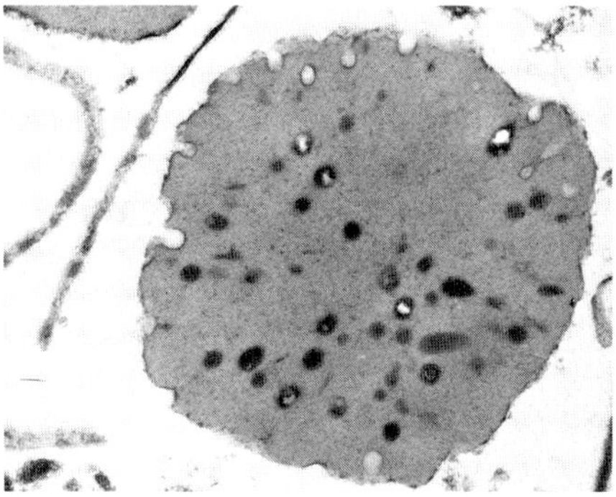

Fig. 10. Thin section of polyhedron in a culture infected with FP strain of virus. (x40,000). (Courtesy Dr. E.A. MacKinnon).

IX. *In Vivo* and *In Vitro* selection of viral strains.

The data given in Fig. 2 show that when NOV is passed undiluted *in vitro* that there is a reduction of the average number of polyhedra per cell. We assayed the distribution of MP and FP strains in the NOV of these serially passed cultures. The initial virus was derived from the hemolymph of an insect that had been fed polyhedra which had been collected from laboratory infected larvae and is believed to be in a similar condition to that found in the field. The data in Fig. 11 show that at the second passage level from the original insect hemolymph, the strain is predominantly the MP type. By passage level 10, the FP strain was the predominant type and at passage level 25, had become the sole species. Thus a 'conversion' in phenotype had taken place. When plaque purified MP was used to initiate a cycle, conversion to FP phenotype also occurred after the 8th passage in culture. These observations are best explained by suggesting that the NP strain undergoes mutation and selection during passage and the selective pressure *in vitro* is towards the FP phenotype. It should be noted that the vehicle for transmission of infectivity *in vitro*

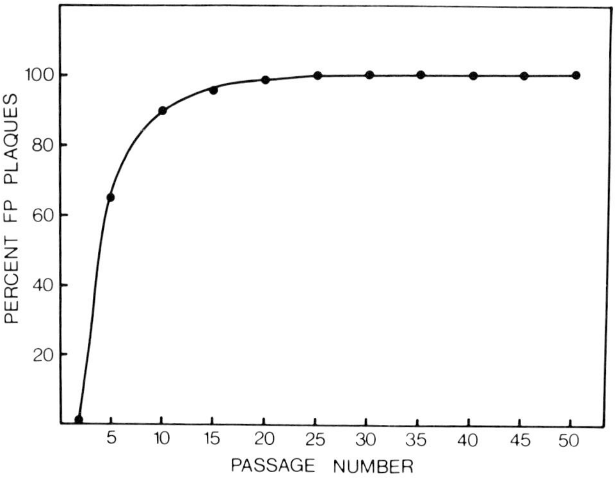

Fig. 11. Progressive appearance of FP strain of virus in tissue culture. The first culture at passage level 0 was infected with hemolymph from insects fed a field collected strain of *T. ni* NPV (MEV type). In subsequent passages NOV was transferred at MOI ∽ 1.0. Proportions of FP and MP strains were determined by plaque assay.

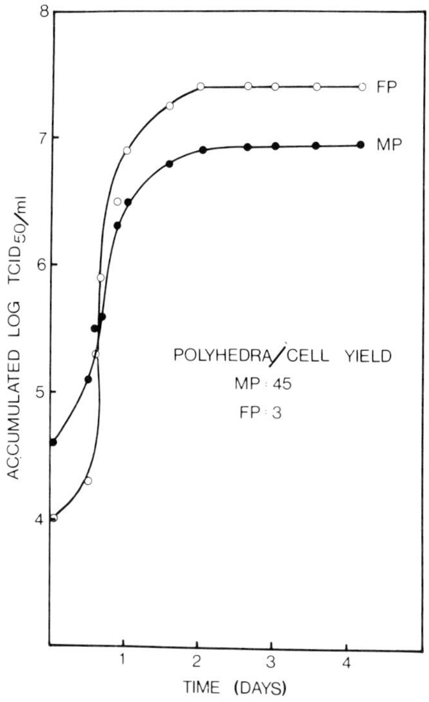

Fig. 12. Growth curves of MP and FP strains of virus in tissue culture.

is via non-occluded virus (NOV). Hence, there is no selective pressure to produce polyhedra for purposes of transmitting the infection. This explanation is borne out by examination of the growth curves of plaque purified FP and MP NOV as given in Fig. 12. FP has a steeper logarithmic rise period which indicates faster release of the virus and subsequent predominant accumulation of FP strain.

We suggest that selection *in vivo* is different than that seen *in vitro* since the vehicle for transmission of the virus is the inclusion body. We showed that the FP type inclusion bodies are not infectious for insects (Table 5). However, hemolymph from insects that have been fed polyhedra obtained from the field contain some FP strain virus (Fig. 2). Thus we suggest that when an insect ingests MP virus polyhedra, mutation and selection occur in most insects, and any FP virus that does arise retains its virulence and can potentiate systemic disease in an individual insect. However, only those MP nucleocapsids which become occluded into inclusion bodies are concerned in the transmission of the virus in nature. We consider that systemic infection with baculoviruses may progress by replication of the NOV of both MP and FP strains, but that the virus which is transmitted and infectious for other insects, is of the MP strain.

X. References

Brown, M. and Faulkner, P., (1975). *J. Invert. Path. 26,* 251.

Dougherty, E.M., Vaughn, J.L., and Reichelderfer, C.F., (1975). *Intervirology. 5,* 109.

Faulkner, P. and Henderson, J.F., (1972). *Virology 50,* 920.

Goodwin, R.H., Vaughn, J.L., Adams, J.R. and Louloudes, S.J., (1970). *J. Invert. Path., 16,* 284.

Goodwin, R.H., Vaughn, J.L., Adams, J.R., and Louloudes, S.J., (1973). *Misc. Public Entomolog. Soc. Amer. 9*, 66.

Grace, T.D.C., (1967). *Nature (London), 216,* 613.

Heimpel, A.M. and Adams, J.R., (1966). *J. Invert. Path. 8,* 340.

Hink, W.F., (1970). *Nature (London), 226,* 466.

Hink, W.F. and Vail, P.V., (1973). *J. Invert. Path. 22,* 168.

Ignoffo, C.M., Hostetter, D.L. and Shapiro, M., (1974). *J. Invert. Path. 24,* 184.

Jaques, R.P., (1967). *Can. Entomol., 99,* 785.

Knudson, D. and Tinsley, T.W., (1974). *J. Virol. 14,* 934.

MacKinnon, E.A., Henderson, J.F., Stoltz, D.B. and Faulkner, P., (1974). *J. Ultrastruct. Res., 49,* 419.

McIntosh, A.H. and Rechtoris, C., (1974). *In Vitro, 10,* 1.

Ramoska, W.A. and Hink, W.F., (1974). *J. Invert. Path. 23,* 197.

Sultor, E.C., Jr., Chang, L.L. and Liu, H.H., (1966). *Exptl. Cell Res. 44,* 572.

Chapter 35

REPLICATION OF A NUCLEAR POLYHEDROSIS VIRUS OF *CHORISTONEURA FUMIFERANA* (LEPIDOPTERA: TORTRICIDAE) IN *MALACOSOMA DISSTRIA* (LEPIDOPTERA: LASIOCAMPIDAE) HEMOCYTE CULTURES

S.S. Sohi and F.T. Bird

I. Introduction

The replication of a nuclear polyhedrosis virus (NPV) of *Lambdina fiscellaria somniaria* in the cultures of *M. disstria* hemocytes has been described earlier (Sohi and Cunningham, 1972). It was of further interest to see if the NPV of *Choristoneura fumiferana* would also grow in these cells. The results of this work are presented here.

II. Materials and methods

A. Cell Cultures

Two cell lines of *M. disstria* hemocytes (IPRI 66 and IPRI 108) were used in this work. Detailed reports on these cell lines have been published elsewhere (Sohi, 1971, 1973). IPRI 66 cells were in the 25th-27th subculture when the work was carried out. IPRI 108 cells were tested for susceptibility to the NPV at the 2nd, 3rd and 5th-7th subcultures in September-November 1970, and at the 31st, 37st, 43rd, 49th, 50th, 51st, 66th, 67th, 73rd and 74th subcultures in December 1972 to February 1973. In the 1972-73 series of tests, cells of different subcultures were obtained in 2 ways. Firstly, there were 3 sub-lines of these cells which grew at slightly different rates. Consequently we had these cells at different levels of subculturing at any given point of time. Secondly, some cells of earlier subcultures had been stored in liquid nitrogen using the procedure published earlier (Sohi, *et al.*, 1971). These cells of earlier subcultures were also revived and used in these tests.

All cultures were grown in Grace-Wyatt insect tissue culture medium (Grace, 1962). The medium was supplemented with 5% fetal bovine serum (FBS) and 3% *B. mori* hemolymph (BMH) for IPRI 66 cells, and with 20% FBS for IPRI 108 cells. Cultures were grown in 30-ml polystyrene disposable flasks (Becton, Dickinson and Co., Canada Ltd., Clarkson, Ontario).

B. Virus and Inoculation of Cell Cultures

An NPV originally isolated from *C. fumiferana* (Bergold, 1951) was used in these experiments. Fifth instar larvae of *C. fumiferana* were infected with the virus by feeding them on diet sprayed with suspension of polyhedra of the NPV. Diseased larvae were bled 10 days later by cutting a pair of prolegs, and the hemolymph was collected in a test tube held in crushed ice. The diseased hemolymph was then diluted 25-to 30-fold with the Grace-Wyatt culture medium containing no FBS or BMH. It was filtered through a 0.45 µm Millipore filter to remove bacteria and polyhedra, and added to cell cultures (1-2 ml/culture) from which spent medium had been removed. After 24 hours the appropriate growth medium was added to the cultures.

After infection the culture medium was used to inoculate further healthy cells. The medium was centrifuged at 500 g for 15 min to remove cells and debris. The supernatant was filtered through a 0.45 µm Millipore filter to remove polyhedra and was used to inoculate other cultures as described above.

All cultures were kept at 28°C before inoculation with the virus and at 25°C after inoculation. Cultures were examined periodically under phase contrast in a Unitron inverted research microscope.

C. Electron Microscopy

For electron microscopy, cells were removed from flasks 6 to 9 days after inoculation with the virus and centrifuged at 800 g for 15 min. The pellet of cells was fixed for one hour in 8% glutaraldehyde in cacodylate buffer (pH 7.0), and rinsed in buffer. It was then fixed for one hour in 2% osmium tetroxide in veronal acetate buffer, and dehydrated in ethanol. After dehydration the pellets were embedded in Epon, sectioned, stained with uranyl acetate and lead citrate, and examined in a Philips EM200 electron microscope.

D. Pathogenicity in Vivo

Semi-quantitative tests were made to determine if the *C. fumiferana* NPV propagated in tissue cultures was still pathogenic to the host insect. In one experiment, the NPV grown in IPRI 108 cells was tested by feeding to 4th instar larvae of *C. fumiferana*. Cells were removed from flasks by mechanical scraping 17 days after inoculation with NPV in the first passage in tissue culture, and centrifuged at 500 g for 5 min. The supernatant was used for inoculating further healthy cell cultures as described above. We waited for 17 days before processing the infected cultures in order to get maximum yield of polyhedra. The pellet (cells and cell debris) was resuspended in a small volume of the supernatant and fed to larvae (Table I, Test 1). A small drop of the suspension was placed on a clean glass slide, and 4th instar larvae placed near the drop. Each larva was watched till it consumed the drop of suspension. The larvae were then transferred to 1-oz plastic cream cups (5 larvae/

cup) containing insect diet (McMorran, 1965). One hundred larvae were inoculated with NPV in this manner. Also, another 100 larvae were inoculated by feeding them on diet on the surface of which the above suspension had been applied. An additional 100 larvae were kept as untreated controls.

TABLE 1

Infectivity of C. fumiferana *NPV propagated in* M. disstria *hemocyte cultures to* C. fumiferana *larvae*

Test	Treatment[a]	Total tested	Mortality NPV	Mortality Non-NPV	Healthy
1	Pellet 108 fed (on slide)	100	0	0	100
	Pellet 108 fed (on diet)	100	0	0	100
	Control	100	0	0	100
2	Pellet 108 fed	10	0	2	8
	Pellet 108 injected	10	4	4	2
	Supernatant 108 fed	5	0	2	3
	Supernatant 108 injected	10	3	7	0
	Pellet 66 fed	10	1	0	9
	Pellet 66 injected	10	0	6	4
	Supernatant 66 fed	5	0	1	4
	Supernatant 66 injected	10	3	4	3
	Control	15	0	0	15

[a] Supernatant refers to the medium from NPV-infected cultures of IPRI 108 or IPRI 66 hemocyte cultures, and pellet to the sedimented cells.

In the second experiment (Table I, Test 2) the NPV was tested after its second passage in tissue culture. Fifth instar larvae were used in this test. The infected cells of both the lines were removed from culture flasks 15 days after inoculation, and centrifuged at 1,500 g for 15 min. Both the sediment and the supernatant were tested for infectivity.

The pellet, which contained cells, cell debris and polyhedra, was resuspended in a small volume of distilled water. A quantity of this suspension was applied to the surface of artificial insect diet in one part of the experiment, and in the other part of the experiment the suspension was injected into the larvae intrahemocoelically (5 µl/larva).

The supernatant was recentrifuged (26,000 g, 60 min) to recover nonoccluded virions, and the pellet was resuspended in a small volume of distilled water. It was tested for infectivity by applying it to the surface of the insect diet and by intrahemocoelic injection into larvae.

All larvae were reared on the artificial insect diet (5 larvae/cup) before and after inoculation with the NPV, and were kept at 22°C and 60-70% relative humidity. Mortality counts were made for 21 days after inoculation, and the dead larvae were examined for NPV infection under a phase contrast microscope. The larvae that survived for 21 days were dissected and examined for infection as above.

III. Results and discussion

A. Light Microscope Observations

In the 1970 trials, polyhedra were seen in the nuclei 5 days after inoculation when the hemocytes were in early subcultures. Healthy cells of a control culture of IPRI 66 are shown in Fig. 1, and NPV-infected cells of this cell line in Fig. 2. Healthy cells of IPRI 108 are illustrated in Fig. 3, and cells of this line heavily infected with NPV in Fig. 4. Multiplication of the virus in IPRI 108 cells was more than in IPRI 66 cells (Figs. 2 and 4).

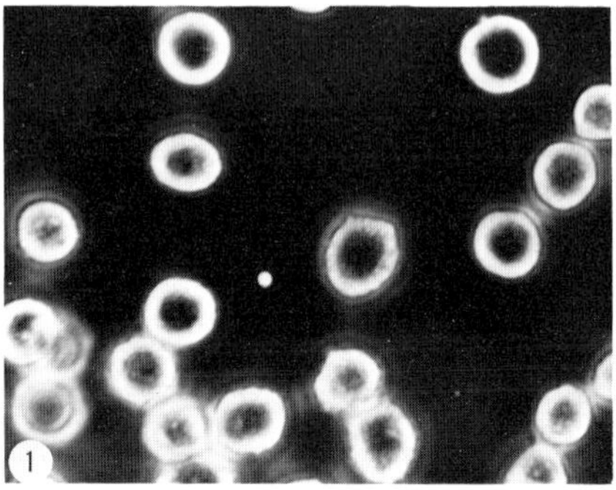

Fig. 1. Control IPRI 66 cells 6 days after 26th subculture, total time *in vitro* 393 days. Brightphase contrast. X710.

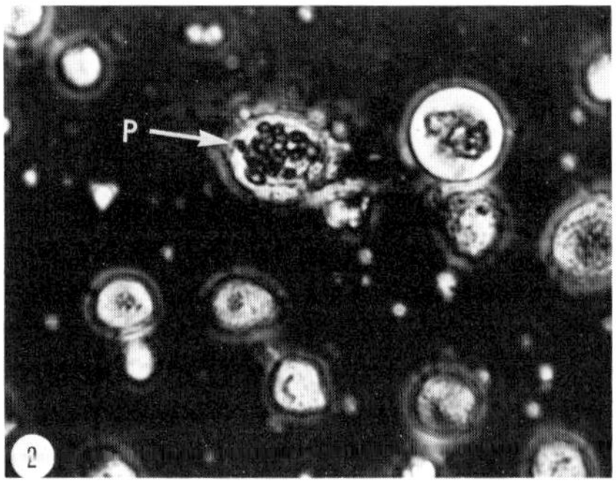

Fig. 2. IPRI 66 cells 15 days after inoculation with NPV. The cells had been subcultured 25 times over a period of 385 days prior to inoculation. Last subculture was made 7 days before cells were inoculated. Note cell with polyhedra (P). Bright phase contrast. X710.

In the 1970 experiments, the cells were inoculated with diseased blood in the first passage of the NPV *in vitro*. For the subsequent passages, medium from the infected cultures was used as the inoculum. In the IPRI 66 cells, virus multiplication was observed in the 1st and 2nd passage of the virus *in vitro*, but no multiplication was seen in the 3rd passage. In the IPRI 108 cells multiplication of virus was good through 4 passages of virus *in vitro*, but at this time these cells were growing very slowly because they were in the early subcultures of 2nd to 7th. We could not produce enough cells to continue passaging the virus in cultures beyond its 4th passage. The experiment was therefore terminated.

The IPRI 108 cells were growing well by December 1972. It was now possible to produce them in sufficient quantity for studying the long term multiplication of the NPV. Cells at different levels of subculturing from 31st through 74th were inoculated with diseased blood during December 1972 to February 1973. No multiplication of virus was seen in any of the cultures as judged by the criterion of polyhedron formation. Also, the inoculated cells did not show any cytopathic effect, and appeared to be similar to the uninoculated controls. Thus, the cells appear to have lost susceptibility to the virus.

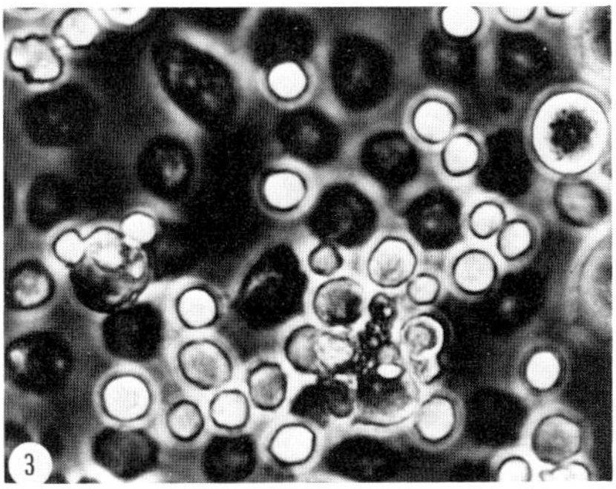

Fig. 3. Control IPRI 108 cells 41 days after 5th subculture, total time *in vitro* 217 days. Bright phase contrast. X710.

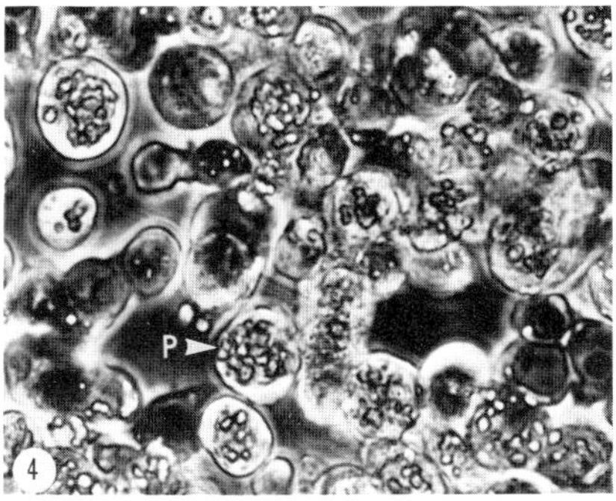

Fig. 4. IPRI 108 cells 6 days after inoculation with NPV. These cells had been subcultured once after being in tissue culture for 107 days, and were inoculated 85 days after subculturing. Thus total time *in vitro* was 198 days. Note many cells with polyhedra (P). Bright phase contrast. X710.

B. Electron Microscope Observations

Multiplication and morphogenesis of the virus are illustrated in Figs. 5 and 6. The electron microscopy of infected cultures revealed that the gross histopathology of the NPV in tissue cultures was in general similar to that observed *in vivo* (Bird,

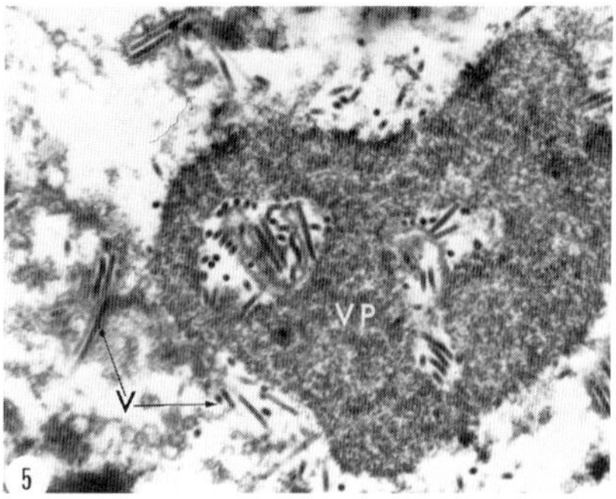

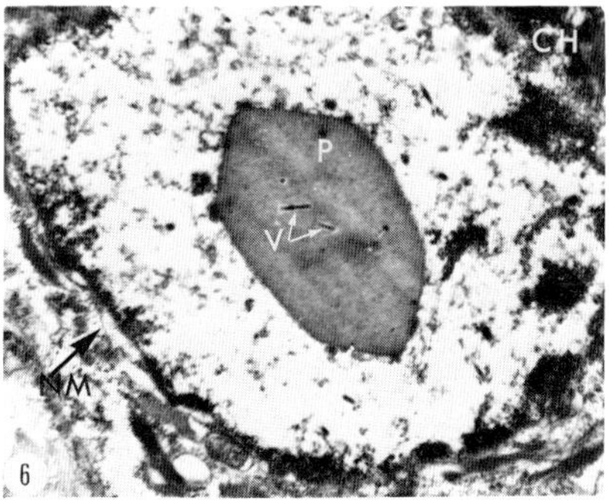

Fig. 5 and 6. Nuclei of NPV-infected IPRI 108 cells 6 days after inoculation. These cells had been subcultured 2 times over a period of 153 days, and the last subculture was made 35 days prior to inoculation, total time *in vitro* 198 days. CH = chromatic material, NM = nuclear membrane, P = polyhedron, V = virion, and VP = viroplasm. Fig. 5 X28,000. Fig. 6. X20,000.

1964). An extensive viroplasm, nonoccluded virions, and polyhedra with occluded virions were present in the nuclei (Figs. 5 and 6). Also, crystalline inclusions (not shown in the illustrations) were seen in the cytoplasm of some infected cells. No virions were detected in these cytoplasmic crystalline inclusions, and their exact nature is not yet fully understood. They are, however, associated with NPV infection, and have not been observed in healthy cells. The cytoplasmic inclusions probably represent the excess polyhedral protein that has not moved into the nucleus to form polyhedra.

Although the gross histopathology *in vitro* was fairly similar to that *in vivo*, there were some aberrations in the *in vitro* morphogenesis of the virus. Firstly, the size of the virions was quite variable (Fig. 5). Although it is difficult to judge the length of virions from sectioned material as it depends upon the plane and location of sectioning, some of the virions illustrated in Fig. 5 appear to be much longer than others. Electron microscope studies of the *in vivo* infected tissues did not show this much variation in the size of virions (Bird, 1964). Secondly, there were fewer polyhedra per cell in the cultures than in infected larvae (Figs. 5 and 6). Thirdly, the virions were occluded in the polyhedra singly (Fig. 6) rather than in multiple bundles (Bird, 1964). And fourthly, the number of virions occluded in the polyhedra *in vitro* seemed to be much lower (Fig. 6) than *in vivo* (Bird, 1964).

C. Pathogenicity to the Host

The results of pathogenicity tests are summarized in Table 1. No virus infection or mortality of *C. fumiferana* larvae was seen in test 1, but some infection and mortality were observed in the 2nd test in the larvae that were injected intrahemocoelically with the virus. It would appear that the titer of virus in the inoculum was quite low as no infection was obtained in test 1 in which the larvae were inoculated by *per os* feeding, and in test 2 also infection by *per os* inoculation was negligible (1 out of 30) whereas 10 out of 40 larvae were infected with the virus when they were inoculated by intrahemocoelic injection. It seems that the small amount of virus administered *per os* was not sufficient to cause infection, but the same amount was adquate to initiate infection when injected intrahemocoelically.

IV. Conclusions

These tests show that it is possible to propagate the NPV of *C. fumiferana* in insect tissue culture, and that the virus so produced retains its infectivity to the original host insect. However, the tissue culture system needs further improvements to make it more useful for *in vitro* investigations of insect viruses. For example, the culture medium needs to be improved so that it provides the cells with optimal nutritional requirements and a suitable physical environment.

Acknowledgment

We wish to thank Mrs. Anne Kronberger and Mrs. Dorcas Higginson for technical assistance.

V. References

Bergold, G.H. (1951). *Can. J. Zool. 29,* 17.

Bird, F.T. (1964). *Can. J. Microbiol. 10,* 49.

Grace, T.D.C. (1962). *Nature 195,* 788.

McMorran, A. (1965). *Can. Entomol. 97,* 58.

Sohi, S.S. (1971). *Can. J. Zool. 49,* 1355.

Sohi, S.S. (1973). *Proc. 3rd Intern. Colloq. Invertebr. Tissue Culture,* Bratislava, Czechoslovakia (1971): 27-39.

Sohi, S.S. and Cunningham, J.C., (1972). *J. Invertebr. Pathol. 19,* 51.

Sohi, S.S., Sullivan, C.R. and Bodley, C.L., (1971). *Lab. Practice 20,* 127.

Chapter 36

REPLICATION OF ALFALFA LOOPER NUCLEAR POLYHEDROSIS VIRUS IN THE *TRICHOPLUSIA NI* (TN-368) CELL LINE

W.F. Hink and E. Strauss

I. Introduction

We are evaluating the feasibility of using the *Trichoplusia ni* (TN-368) cell line for production of alfalfa looper, *Autographa californica*, nuclear polyhedrosis virus (NPV). This virus is pathogenic to several species of lepidopterous insect pests and is an important candidate for use as a viral insecticide (Vail *et al.*, 1970, 1973). This paper reports some of the results in manipulation of cells and virus to make the *in vitro* system as productive as possible.

II. Methods and materials

The virus in this study was propagated in the TN-368 cell line (Hink, 1970). The cells were grown in modified TNM-FH medium. The new formula was: 90.0 ml Grace (1962) medium, 8.0 ml fetal bovine serum, 0.3 gm lactalbumin hydrolysate, and 0.3 gm TC yeastolate. At the time of this study the cells had been in continuous culture for 5 3/4 yrs and subcultured 750 times. Cell cultures were incubated for 7 days after exposure to virus, the cell-virus suspension centrifuged at 1500xg for 15 min, and the supernatant filtered through a 0.45µm Millipore filter. These viral preparations were stored at 4°C and used to infect cultures.

The procedures for infecting different cell densities with varying titers of virus were as follows. Aliquots of cells were taken from 3 day-old cell cultures. Different volumes were used to obtain the various cell numbers. These cell suspensions were centrifuged at 1500 xg for 10 min. The cell-free supernatant removed and cells resuspended in 0.5 ml virus for 1 hr. The cell-virus suspension was gently agitated every 10-15 min during the 1 hr adsorption period. After adsorption, the cells were centrifuged at 1500 xg for 10 min, the virus pipetted off, 2.0 ml medium added, cells resuspended, and transferred to 60 mm polystyrene Petri dishes.

To determine the percentage of infected cells, the cultures were examined at 24 and 48 hrs post infection. An inverted microscope, with a grid superimposed on the field of view, was used to examine 200 to 300 cells per dish. Cells were considered to be infected if they exhibited cytopathology characteristic of NPV infection (Bird and Whalen, 1954).

The numbers of polyhedra produced per cell were determined by using 2.0 ml cultures 72 hrs after infection. The cells and polyhedra were scraped off dishes with rubber policemen. The cells were lysed and clumps of polyhedra broken up by sonication for 1 min at 70 watts. The polyhedra were quantitated with a hemacytometer and 3 to 4 counts were made on each culture. The number of cells per culture, the percentage of infected cells, and the polyhedral counts were used to calculate the yields of polyhedra per cell.

Cells used for the growth curve were infected with an input multiplicity of 20 plaque forming units (pfu)/cell to ensure a single cycle of virus replication. Cells from 2 day-old parent cultures (0.7 ml, 1×10^6 cells) were transferred to 2.0 ml medium in 60 mm polystyrene petri dishes. These were rotated to obtain a homogenous cell suspension and incubated at 28°C for 2 hrs during which time the cells settled and attached. The medium was poured off the cell monolayer and 0.5 ml virus inoculum added. The virus was adsorbed at 28°C for 60 min. After adsorption, the virus was removed and the cell monolayer washed twice with 2.0 ml Hanks' PBS (Hanks and Wallace, 1949) to which additional glucose (14.0g/1) was added to make it isotonic to the tissue culture medium. Tissue culture medium (2.0 ml) was put on the cells and the cultures returned to the 28°C incubator. During these manipulations some cells were washed off the petri dishes and the final cell counts were 8.7×10^5 cells/dish. Since cells do not multiply after exposure to virus, this figure was used to determine the approximate yield of pfu/cell.

Petri dishes were removed randomly at specified intervals (1 hr, 2 hr, 24 hr, etc.). The media and any free cells or inclusion bodies were poured into centrifuge tubes and centrifuged at 1500xg for 15 min. The supernatants were pipetted off and these aliquots contained the released infectious material. To recover the free cells, the pellets were resuspended in 1.0 ml TNM-FH and returned to the original dishes containing the rest of the cells. To prevent desiccation of the cells during centrifugation, 1.0 ml fresh media were put on these cell sheets in the petri dishes. All cells were freed with rubber scrapers and these preparations contained the cell associated virus in volumes of 2.0 ml. The infected cells were lysed by 3 cycles of freezing in a dry ice-acetone mixture and thawing in a 30°C water bath. The released virus was treated the same way prior to plaque assay.

The virus was titrated by using a plaque assay technique (Hink and Vail, 1973). Virus was serially diluted ten-fold in TNM-FH and 0.25 ml put on cells, 4 petri dishes of cells per dilution.

The procedures used to evaluate the effects of temperature on virus replication were as follows. Cells were grown in 100 ml spin flasks for 24 hrs. Aliquots (5.0 ml) were transferred from the spin flasks to 25 cm^2 Falcon polystyrene flasks and 0.5 ml virus inocula added. These were gently mixed and placed at various temperatures. At 48 hrs after infection, the percentages of infected cells were determined according to procedures previously outlined in paragraph three of this section. At 4 to 5 days after infection procedures given in paragraph four of this section were used to count polyhedra.

III. Results and discussion

The cell densities and multiplicities of infection (pfu/cell) were varied to find the minimal concentration of virus that will initiate a high percentage of infection. We also wanted to determine if there was a maximum cell density at which percent infection and yield of polyhedra were optimal (Tables 1 and 2).

At 24 hrs after infection, there were no significant differences in percentages of infection between cells exposed to 5 or 20 pfu/cell at all cell densities. Therefore, we may use 5 pfu/cell to initiate synchronous infections. However, we observed that cells treated with 20 pfu/cell developed polyhedra more rapidly than those exposed to 5 pfu/cell.

At 24 hrs post infection and either 0.5 or 1.0 pfu/cell, the greater the cell density the higher the percentage of infection. At low cell densities, the relationship between pfu/cell and percent infection is most apparent. At higher cell densities, the

TABLE 1.

Percentage of Infected Cells at Different Cell Densities and Titers of Plaque Forming Units

Cells/2 ml Culture	pfu/cell			
	0.5	1.0	5.0	20.0
$2.5x10^5$	33% [a]	48% [a]	95% [a]	98% [a]
	90% [b]	91% [b]		
$5.0x10^5$	34% [a]	57% [a]	98% [a]	100% [a]
	86% [b]	87% [b]		
$1.0x10^6$	60% [a]	78% [a]	98% [a]	99% [a]
	84% [b]	87% [b]		
$2.0x10^6$	68% [a]	92% [a]	99% [a]	99% [a]
	89% [b]			
$4.0x10^6$	92% [a]	99% [a]	99% [a]	99% [a]

[a]percent infected cells at 24 hrs post inoculation

[b]percent infected cells at 48 hrs post inoculation

TABLE 2.

Yield of Polyhedra Per Cell at Different Cell Densities and Titers of Plaque Forming Units

Cells/2 ml Culture	pfu/cell			
	0.25	0.5	1.0	5.0
$2.4x10^5$	131	100	170	140
$7.2x10^5$	122	170	137	179
$2.7x10^6$	-	148	172	174

At 3 days post infection the cultures had 94 to 99% infected cells percent infection was not effected as greatly by the multiplicity of infection (MOI). This suggested that for mass production of virus, a dense cell culture would be more efficient because fewer pfu/cell would be required to initiate high rates of infection.

Data from cultures at 24 and 48 hrs after infection showed that numbers of infected cells increased as cultures were incubated. This is because the infection probably cycled within the cell population.

Table 2 gives the yields of polyhedra per cell. At a given cell density, there appeared to be no correlation between MOI and polyhedra per cell. When infecting with low MOI many cells were apparently infected via lateral transmission and these cells probably produced as many polyhedra as those infected with the original inocula.

The virus growth curves revealed that lateral transmission could begin at about 14 hrs after infection because production of cell associated virus and released virus began at this time (fig. 1). The first observable cytopathological effects (hypertrophy of nuclei and rounded cells) were observed 12 hrs after infection and by 14 hrs 95% of the cells exhibited these conditions. These cytopathologies corresponded, in time, with the initial manufacture and release of virus. As virus is produced, some remained in the cells and other progeny left the cells and were immediately detectable in the culture medium. Elution of virus from the cell surface probably occurred and this was most evident in the released virus growth curve at 6 to 14 hrs post inoculation. If virus was being eluted during the period from 14 to 24 hrs post inoculation, it was unlikely that it contributed significantly to the 1.5 log increase in pfu during this period. Since viral inclusion bodies were first observed 20 hrs after exposure of cells to virus, the infectious material was released before polyhedra were formed.

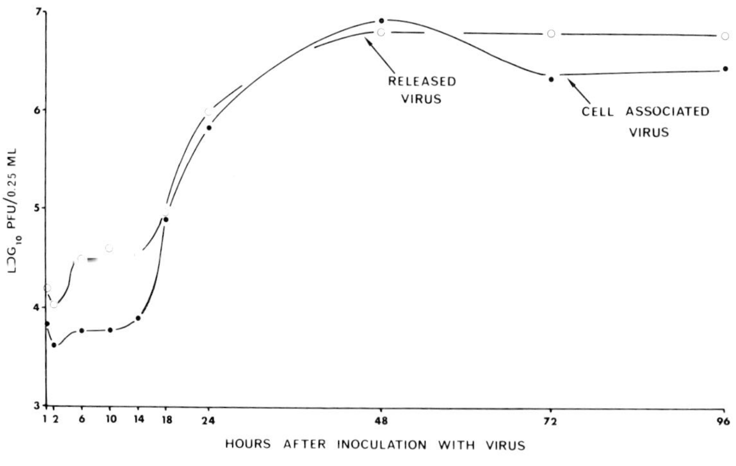

Fig. 1. Growth of alfalfa looper nuclear polyhedrosis virus in TN-368 cells. Cultures were titrated at specified times after inoculation with virus and each dot represents the average of 4 plaque assays on each culture.

After the initiation of virus production at 14 hrs post infection, it continued for 34 hrs and reached the maximum level at 48 hrs. During this time, cell associated virus and released virus titers increased at equal rates. The drop in cell associated virus after virus replication was completed at 48 hrs was probably due to observed cell lysis which began at this time and may have released virus into the culture medium.

The cell associated virus increased from 7×10^3 pfu/0.25 ml to 1×10^7 pfu/0.25 ml in 48 hrs, an approximate 1,000 fold increase. There was a similar increase in released virus in the cell-free culture medium, from 1×10^4 pfu/0.25 ml to 7×10^6 pfu /0.25 ml in 48 hrs.

The approximate yield of pfu per cell was calculated in the following manner. The total volume of the infected cultures was 2.0 ml and 0.25 ml diluted aliquots were plaque assayed so the total yield of both cell associated virus and released virus was 8×10^7 pfu/culture. Each culture contained 8.7×10^5 cells with 95% of the cell population becoming infected. This means that 8.3×10^5 cells were probably manufacturing viral progeny. Therefore, each cell yielded about 100 pfu of cell associated virus and 100 pfu of released virus. These figures do not represent all virus manufactured by the cells because nucleocapsids occluded in polyhedra were not quantitated and the plaque assay technique is not sensitive enough to detect all pfu in the viral inocula (Hink and Vail, 1973). Also, it is probable that this assay does not quantitate all possible forms (naked DNA, DNA-protein complexes, viral subunits, enveloped and non-enveloped nucleocapsids) of infectious viral entities.

Virus replication probably occurred more rapidly in these cultured cabbage looper cells than in most other previously investigated *in vivo* infections. In infections *in vivo*, viral progeny were observed in cells of larvae 22 to 24 hrs after per os or intrahemocoelic exposure (Younghusband and Lee, 1969; Harrap, 1970; Summers, 1971).

In other insect virus-cell culture systems, the first viral progeny have been detected at various periods. In *Sericesthis* iridescent virus infected *Antheraea eucalypti* cells, the virus particles were observed 2 to 3 days after inoculation (Bellett and Mercer, 1964). In *Aedes aegypti* cells infected with mosquito iridescent virus, mature and incomplete virions were seen after 2 1/2 days (Webb, *et al.*, 1973). Nuclei of NPV infected *Spodoptera frugiperda* cell line contained nucleocapsids after 20 hrs (Vaughn *et al.*, 1972) and infectious material was released from cells 12 hrs post inoculation (Knudson and Tinsley, 1974). Crystalline-array virus replication occurred in cultures of grasshopper dorsal vessels within 24 hrs (Henry, 1972).

The temperature of incubation has a dramatic effect on infection (Table 3). At 18°C and 37°C the numbers of infected cells were reduced. There were no differences in percentages of infection when cells were cultured at 25°C, 28°C, or 32°C. However, in all three replicates, incubation at either 25°C or 32°C produced more polyhedra per cell than cells cultured at 28°C. This suggests that the optimum temperature for cell growth (28°C to 30°C) may not necessarily be optimum for viral inclusion body formation.

Acknowledgements

This research was supported, in part, by Grant R-802516 from the Environmental Protection Agency and Grant GB-38154 from the National Science Foundation.

TABLE 3.

Infection of Cells at Various Temperatures

Incubation Temperature	Percent Infection	Polyhedra/ ml (x 10^7)	Polyhedra/ Cell
18°C	22%	–	–
25°C	76%, 77%, 92%	5.4, 7.15, 8.0	126, 130, 163
28°C	77%, 82%, 87%	5.05, 5.6, 5.7	89, 94, 121
32°C	84%, 84%, 85%	6.0, 6.15, 7.5	104, 124, 142
37°C	6%	–	–

Each vertical column contains data from one experiment.

IV. References

Bellett, A.J.D. and Mercer, E.H. (1964). *Virology, 24,* 645.

Bird, F.T. and Whalen, M.M. (1954). *Can. J. Microbiol., 1,* 170.

Grace, T.D.C. (1962). *Nature, 195,* 788.

Hanks, J.H. and Wallace, R.E. (1949). *Proc. Soc. Exp. Biol. Med. 71,* 196.

Harrap, K.A. (1970). *Virology, 42,* 311.

Henry, J.E. (1972). *J. Invertebr. Pathol., 19,* 325.

Hink, W.F. (1970). *Nature, 226,* 466.

Hink, W.F. and Vail, P.V. (1973). *J. Invertebr. Pathol., 22,* 168.

Knudson, D.L. and Tinsley, T.W. (1974). *J. Virol., 14,* 934.

Summers, M.D. (1971). *J. Ultrastruct. Res., 35,* 606.

Vail, P.V., Jay, D.L. and Hunter, D.K. (1970). *Proc. IV Int. Colloq. Insect Pathol., College Park, Maryland.* p. 297.

Vail, P.V., Jay, D.L. and Hink, W.F. (1973). *J. Invertebr. Pathol., 22,* 231.

Vaughn, J.L., Adams, J.R. and Wilcox, T. (1972). *Monogr. Virol., 6,* 27.

Webb, S.R., Paschke, J.D., Wagner, G.W. and Campbell, W.R. (1973). *Proc. V Int. Colloq. Insect Pathol. Microbial Control,* Oxford, England, p. 32.

Younghusband, H.B. and Lee, P.E. (1969). *Virology, 38,* 247.

Chapter 37

AN ELECTRON MICROSCOPE STUDY OF THE SEQUENCE OF EVENTS IN A NUCLEAR POLYHEDROSIS VIRUS INFECTION IN CELL CULTURE

D.L. Knudson and K.A. Harrap

I. Introduction

The processes of nuclear polyhedrosis virus (NPV) uptake, morphogenesis and release are fairly well described at the ultrastructural level from studies of infected tissues of the host insect (Harrap and Robertson, 1968; Summers, 1969; Summers and Arnott, 1969; Harrap, 1970; Summers, 1971; Harrap, 1972; Kawanishi *et al.*, 1972; Harrap, 1973; Stoltz *et al.*, 1973; Robertson *et al.*, 1974). Recently the increasing ease with which certain NPV-susceptible lepidopteran cell lines can be handled has resulted in similar ultrastructural studies, in particular with the NPV-cell culture systems of *Trichoplusia ni* and *Bombyx mori* (MacKinnon *et al.*, 1974; Raghow and Grace, 1974). We have studied the processes of uptake, morphogenesis and release with another NPV in cell culture, that of the fall armyworm *Spodoptera frugiperda*, an insect also of some economic importance. One of us has already reported the purification, assay of infectivity and growth characteristics of the virus in this cell culture system (Knudson and Tinsley, 1974).

II. Material and results

The conditions employed for the culture of the cells and the growth of the virus for this work were fully described by Knudson and Tinsley (1974). Briefly *S. frugiperda* cells in BML-10 medium + 10% calf serum were synchronously infected at a MOI of 100 $TCID_{50}$ per cell and after an adsorption period of one hour the cultures were harvested and the cells pelleted at various intervals post-infection (1,3,4,6,8, 9,12,15,18,21,24,36, and 48 hours). The pelleted cells were processed for electron microscopy by fixation for one hour in 5% (v/v) glutaraldehyde followed by two washes and a second fixation in 2% (w/v) osmium tetroxide again for one hour. The fixed pellets were dehydrated through a graded ethanol series and embedded in Epikote. Sections were stained in uranyl acetate followed by lead citrate.

III. Results

Mock-infected cells showed a typical ultrastructure with well-preserved cell organelles. Some of the cells were extensively vacuolated.

The events observed in the NPV-infected cells will be described in the chronological order in which they appear to occur.

Both enveloped virus particles, usually as 'bundles', and naked virus particles could be found adjacent to the plasma membrane and in vacuoles in the cell cytoplasm by one hour post infection. Enveloped virus particles were seen in close contact with the plasma membrane and virus uptake seems to occur by invagination as enveloped virus particles could be found in vacuoles in the cell cytoplasm. Such viropexis could perhaps have occurred with both nakes and enveloped virus particles though it is possible that the naked virus particles seen in vacuoles might have lost their envelope after uptake either through some degradative process in the vacuole or by fusing with the vacuole membrane. Knudson and Tinsley (1974) demonstrated that infective virus derived from this cell culture system is likely to be naked virus particles as on sucrose gradients virus-containing fractions fell into a homogeneous single peak rather than a series of peaks as was found when a 'bundled' enveloped population of *S. frugiperda* NPV particles was centrifuged in an identical way. Also the banding density of the homogeneous peak seemed to be consistent with that of naked virus particles. In another cell system Henderson *et al.* (1974) demonstrated properties for the infectious entity of *T. ni* NPV in cell culture consistent with it being a fragile enveloped particle. In a morphological study Raghow and Grace (1974) observed only enveloped virus particles in contact with *Bombyx mori* cells early in the infection sequence. The work we report here still does not resolve this question as either naked or enveloped virus particles could have been taken up into vacuoles by viropexis. In the insect host it is generally accepted that infection of the gut columnar cells occurs as a result of fusion of the virus envelope and the cell plasma membrane so here the infectious entity has to be the enveloped virus particles. However in the haemocoele of NPV-infected silkworms Kawarabata (1974) found that the infectious form of the virus is probably a naked rod. Similarly with an insect poxvirus Granados (1973) has reported fusion of the virus envelope with the microvillus plasma membrane of the gut cells yet pinocytotic uptake of the same virus by haemocytes. Such findings are in agreement with Dales (1973) who in an excellent review of this topic considered that cells possessing deformable surfaces could acquire virus by engulfment whereas the mechanism associated with differentiated non-deformable surfaces could be fusion. We attempted to study this problem further by examining the entry of purified, alkali-released, insect-grown virus into cultured cells. Although the system was perhaps somewhat artificial it had the apparent advantage of introducing large quantities of virus particles into the cultures. In fact 2 ml of 10^7 cells were mixed with 0.25 ml of virus freshly prepared using essentially the methods of Harrap and Longworth (1974). The virus preparation contained 25 µg of virus protein. The suspension was incubated at 27°C for two hours with gentle shaking. Aliquots were removed at 30 min intervals and prepared for thin sectioning. Various stages of association of virus and cell were observed. Naked and enveloped virus particles could be seen outside the cells. In several instances the envelope of enveloped virus particles was seen to be closely adsorbed to the plasma membrane of the cell. At this point of adsorption the plasma membrane was often deformed or indented. Internalization of the virus particles appeared to occur by viropexis and both naked and enveloped virus particles could again be seen in vacuoles in the cytoplasm. In some instances the vacuole membrane appeared to have disintegrated leaving virus particles free in the cytoplasm. Here again then it is conceivable that both forms of virus particle could be taken up by the cell though it is true that we observed what appeared to be stages of viropexis only with enveloped virus so it is more likely that it represents the infectious entity.

Vacuoles containing virus particles apparently ruptured as naked virus particles could be found free in the cytoplasm. Only rarely was a naked virus particle seen aligned perpendicularly with the nuclear membrane possibly at a nuclear pore. Naked virus particles were seen on two occasions within the nucleus at one hour and three hours post infection. The presence of such virus particles in the nucleoplasm would seem to preclude the hypothesis of injection of the virus deoxyribonucleoprotein into the nucleus through a nuclear pore, a process which makes uncoating unnecessary (Summers, 1969; 1971; Raghow and Grace, 1974). Examination of cultures three hours post infection indicated that fewer intact virus particles were present and by four hours virus could not be detected. This suggests that uncoating does occur.

The first signs of virus replication were detected at eight hour post infection when enlargement of the cell nucleus and peripheral displacement of the nuclear chromatin along the nuclear membrane were seen. At nine hours post infection naked virus particles associated with a disperse virogenic stroma were apparent in many cells. By 12 hours the virogenic stroma was condensed and envelopment of the virus particles could be observed between the condensed stroma and the nuclear membrane and less frequently in spaces in the stroma itself. Between 12 and 18 hours virus could be seen leaving the cell in various ways. Naked virus particles were found 'budding' from the inner nuclear membrane into the perinuclear space, an area contiguous with the endoplasmic reticulum cisternae, even though the same nucleus contained virus particles which had acquired envelopes within the nucleoplasm. The fate of the enveloped virus particles in the perinuclear space is not clear. They could remain in the endoplasmic reticulum cisternae or perhaps eventually leave the cell by way of the cisternal spaces. Alternatively they may fuse with, or be ingested pinocytotically by, the endoplasmic reticulum membrane allowing either entry of a naked virus particle into the cytoplasm or transport of an invacuolated particle, still bearing an envelope derived from the nuclear membrane, across the cytoplasm to the cell surface to be released by reverse pinocytosis. Indeed both mechanisms might occur. Another possibility is that the naked virus particles seen in the cytoplasm might be there as a result of breaks occurring in the nuclear membrane though we observed no such breaks at this infection time. The important point though is that regardless of the way in which such naked virus particles might have reached the cytoplasm they could be seen leaving it by 'budding' from the plasma membrane. MacKinnon *et al.* (1974) have reported similar types of release mechanism in the *Trichoplusia ni* NPV-cell culture system though they were seen somewhat later in the infection sequence. Thus it appears that envelope acquisition can occur in three different ways during infection. In this connection it is worth noting that Robertson *et al.* (1974) have reported granulosis virus particles 'budding' from the plasma membrane of gut cells in infected larvae.

Polyhedron formation first was detected at 18 hours post infection. Only those virus particles acquiring their envelopes, either singly or in 'bundles', within the nucleus were occluded by the polyhedron matrix protein. By 21 hours post infection nuclei contained large numbers of developing polyhedra as well as naked and enveloped virus particles. The occlusion of the virus particles into polyhedra was very similar to descriptions of this process in *in vivo* NPV infections. By 36 hours electron-dense fragments could be seen on the periphery of the polyhedra and fibrillar areas containing similar fragments, but of around twice the width of the peripheral polyhedron fragments, were present in the fibrillar mass. By 48 hours the dense peripheral layer (the socalled polyhedron membrane) was almost continuous around the polyhedron. Its synthesis seems to represent the terminal event of morphogenesis.

The original stock of *Spodoptera frugiperda* NPV used in this work was received in our laboratory as infected cultures with no information as to the number of passages of the virus in cell culture. Electron microscopy of such infected cultures showed abnormal morphogenetic characteristics. These included electron-lucent tubule-like structures of variable length which could appear to be 'partially filled' with structures any enveloped virus particles had amorphous material adhering to them and polyhedron formation was rare. Polyhedra containing either few or no virus particles, or virus particles of a shorter length than normal or of indistinct structure were observed. Accumulation of structures resembling polyhedron peripheral layers back to back were also seen. Similar morphogenetic aberrations have also been observed on prolonged passage from *Trichoplusia ni* NPV-cell culture systems (MacKinnon *et al.*, 1974).

The morphogenetic events of *S. frugiperda* NPV in cell culture are broadly similar to what is observed in the infection of host tissues. However there are some significant differences. Firstly, fusion of the virus envelope and the plasma membrane is not a prerequisite for infection and in consequence both enveloped and naked virus particles may well initiate infection. Secondly, virus release from the cell can occur in different ways before polyhedron formation commences. Thirdly, gross morphogenetic aberration can occur with highly passaged virus. The entry of the virus and its subsequent penetration in cell culture is not unlike the events that have been observed for other animal viruses (Dales, 1973). The general pattern for other DNA animal viruses such as adenoviruses, herpesviruses and papovaviruses which replicate in the nucleus is viropexis into vacuoles, lysis of the vacuole membrane and partial uncoating and alignment of the virus with the nuclear membrane or nuclear pores with complete uncoating. The first two events were definitely observed in this system but the latter was seen only rarely.

IV. References

Dales, S. (1973). *Bacteriol. Rev. 37,* 103.

Granados, R.R. (1973). *Virology 52,* 305.

Harrap, K.A. (1970). *Virology 42,* 311.

Harrap, K.A. (1972). *Virology 50,* 133.

Harrap, K.A. (1973). In: Viruses and Invertebrates (A.J. Gibbs ed.) North Holland, Amsterterdam, pp. 271-299.

Harrap, K.A., and Longworth, J.F. (1974). *J. Invertebr. Pathol. 24,* 55.

Harrap, K.A., and Robertson, J.S. (1968). J. Gen. Virol. 3, *221.*

Henderson, J.F., Faulkner, P., and MacKinnon, E.A. (1974). *J. Gen. Virol. 22,* 143.

Kawahishi, C.Y., Summers, M.D., Stoltz, D.B., and Arnott, H.J. (1972). *J. Invertebr. Pathol. 20,* 104.

Kawarabata, T. (1974). *J. Invertebr. Pathol. 24,* 196.

Knudson, D.L., and Tinsley, T.W. (1974). *J. Virology 14,* 934.

MacKinnon, E.A., Henderson, J.F., Stoltz, D.B., and Faulkner, P. (1974). *J. Ultrastruct. Res. 49,* 419.

Raghow, R., and Grace, T.D.C. (1974). *J. Ultrastruct. Res. 47,* 384.

Robertson, J.S., Harrap, K.A., and Longworth, J.F. (1974). *J. Invertebr. Pathol. 23,* 248.

Stoltz, D.B., Pavan, C., and daCunha, A.B. (1973). *J. Gen. Virol. 19,* 145.

Summers, M.D. (1969). *J. Virology, 4,* 188.

Summers, M.D. (1971). *J. Ultrastruct. Res. 35,* 606.

Summers, M.D., and Arnott, H.J. (1969). *J. Ultrastruct. Res. 28,* 462.

Chapter 38

REPLICATION OF *AMSACTA MOOREI* ENTOMOPOXVIRUS AND *AUTOGRAPHA CALIFORNICA* NUCLEAR POLYHEDROSIS VIRUS IN HEMOCYTE CELL LINES FROM *ESTIGMENE ACREA*

R.R. Granados and M. Naughton

I. Introduction

The preparation of primary cell cultures from *Estigmene acrea* larvae and their infection with *Amsacta moorei* entomopoxvirus (EPV) was recently described (Granados and Naughton, in press). This paper reports on the establishment of two continuous hemocyte cell lines from *E. acrea* larvae and their infection with *A. moorei* poxvirus, *Autographa californica* nuclear polyhedrosis virus (NPV), *Trichoplusia ni* cytoplasmic polyhedrosis virus (CPV), and iridescent virus type 6 (Chilo iridescent virus). A preliminary summary of these findings has been reported (Granados, in press).

II. Materials and methods

1. Hemocyte cultures

Grace's tissue culture medium (Grace, 1962) supplemented with three components was used for culture of the cell lines. The ingredients, pH, and osmolality of the medium are listed in table 1. The 0.001 M cysteine was added to the non-heat inactivated cell-free hemolymph (CFH) component in order to retard melanization. Primary hemocyte cultures were prepared with the same medium described in table 1 except that CFH was omitted (Granados and Naughton, in press). The preparation of primary hemocyte cell cultures was reported previously (Granados and Naughton) and will only be described briefly here. Hemolymph was obtained by cutting a proleg of a surface sterilized *E. acrea* larva and collecting 1 drop in a 30-ml tissue culture flask (Falcon Plastics). To increase the cell concentration, additional drops of hemolymph and fresh 0.001 M cysteine were added at 2-day intervals until there was a total of 3 or 4 drops of hemolymph in each culture.

TABLE 1.

Ingredients of medium used to culture ESTIGMENE ACREA hemocytes

Component	Amount
Grace's tissue culture medium	90 ml
Fetal calf serum	10 ml
Heat inactivated *Estigmene* hemolymph	5 ml
Non-heat inactivated *Estigmene* hemolymph in 0.001 M cysteine	2 ml
pH = 6.4	
Osmolality = 330-340 milliosmols.	

2. *Virus preparation and inoculation*

Initially, for *Amsacta* EPV and *Autographa* NPV, hemolymph derived from virus-infected *Estigmene* larvae were used to inoculate the cell line cultures. After the first replication in culture, the standard virus source was the tissue culture medium from infected cell cultures. Cells and medium from cultures infected for 7 to 10 days were centrifuged for 10 min at 10,000 *g* and 15 min at 1,400 *g* for *Autographa* NPV *Amsacta* EPV, respectively. The virus-containing supernatant fraction was used as the inoculum. Except when indicated in the text, only early passage virus inocula (<6 passages) were used.

Trichoplusia CPV was prepared from intestines disected from diseased larvae as reported earlier (Granados *et al.*, 1974). Iridescent virus type 6 was propagated in *T. ni* (TN 368) cell cultures and 7 to 10 days postinoculation (p.i.) the infected cell cultures were centrifuged for 10 min at 1,000 *g*. The virus-containing supernatant fraction was used as inoculum.

Healthy cultures were inoculated by replacing the culture medium with 0.5 ml of virus inoculum and 3.5 ml of fresh tissue culture medium. After a 2 to 3 hr adsorption period the inoculum and medium were replaced with 4.0 ml of fresh medium. Inoculated cultures were placed in an incubator (26° C) and periodically examined with a Zeiss phase-contrast inverted microscope.

3. *Electron microscopy*

Cultures to be examined by electron microscopy were mechanically agitated to release any attached cells and centrifuged at 200 *g* for 5 min. The medium was discarded and replaced with cold 3% glutaraldehyde in 0.1 M cacodylate buffer, pH 7.4. After 2 hr at 4°C the cells were centrifuged, washed with fresh buffer and postfixed in 1% osmium in 0.1 M cacodylate buffer, pH 6.5 - 6.8. Following post-fixation for 1 hr at 4°C the cells were centrifuged, washed in fresh buffer and the cell pellet, embedded in 2% agar by the addition of warm agar into the centrifuged tube. The agar-embedded cell pellet was then dehydrated in a graded series of ethanol solutions and embedded in Epon. Ultrathin sections were cut with diamond knives and collected on bare 300 or 400 mesh copper grids. A Zeiss EM9-IIS electron microscope was used for the examination of ultrathin sections.

III. Results

1. Establishment of cell lines

Ten primary hemocyte cultures were prepared on November 13, 1974. Two days later these cell cultures received an additional 1 ml of fresh medium and 1 drop of hemolymph from surface sterilized *Estigmene* larvae. Approximately 10 days later the suspended cells began to divide and the primary cultures were subcultured for the first time on November 27, 1974. For each primary culture, the spent medium and the suspended cells were split 1:2. The remaining attached cells were removed by the addition of 0.05% trypsin (increased to 0.125% on 2nd subculture) to each flask. After 5 minutes, 3 ml of fresh medium were added to each flask to stop the trypsinization. The cell suspension was centrifuged for 2 min at 200 *g* and the cell pellet resuspended in 3 ml of fresh medium. This cell suspension was divided into the 2 flasks containing the original suspended cells and spent medium. Eight days following the initial subculture the procedure was repeated, using for the first time, the initial subculture the procedure was repeated, using for the first time, medium containing the CFH component (table 1). The cells in two culture flasks showed improved cell growth and only these cell lines were kept and maintained in culture. They were usually split 1:2 at 7 to 10 day intervals. These cell lines were designated EA 1174 A and EA 1174 H. On the 7th subculture the cell lines were growing well and two sublines, EA 1174 AS1 and EA 1174 HS1, were adapted to a medium without CFH. Through early passages, all the cell lines were infrequently split 1:2 with fresh medium. As cell growth improved, later passages were split weekly at lower cell concentrations (1:3 and 1:4). A summary of the growth characteristics of the cell lines and sublines is presented in table 2.

As of June 1, 1975 the cell lines and two sublines had been subcultivated over 40 times. The lines are now routinely subcultured at 3 or 4 day intervals at an initial cell seeding density of 2 to 3 x 10^5 cells/ml in 4 ml of fresh medium. These cultures usually attain a density of 2 to 3 x 10^6 cells/ml, 72 to 96 hr after subculture. Both EA 1174 H grow at approximately equal rates and have a doubling time of 20 to 21 hr. On the basis of light and electron microscopic observations the cells in culture appear to be plasmatocyte-type blood cells (Figs. 1A and 1B). A third subline, EA 1174 AS2, was established from EA 1174 A at its 32nd passage by selection of only those cells which remained attached to the bottom of the flask (Fig. 1C). This subline is routinely subcultured once a week by using 0.25% trypsin and splitting 1:2.

2. Infection with Amsacta EPV

The first signs of infection by *Amsacta* EPV were observed by phase-contrast microscopy 36 to 40 hr postinoculation. There was an increased accumulation of refractive inclusions in the cell cytoplasm accompanied by a hypertrophy of the cell. By 48 hr p.i. typical virus-containing inclusions (VCI) could be discerned. The rate of virus multiplication varied between the two cell lines. In general, 72 hr p.i., 90 to 95% of the cells in line EA 1174 A were infected compared to only 20 to 25% of the cells in line EA 1174 H (Figs. 2A and 2B). By 96 hr p.i. 100% of the cells in line EA 1174 A were usually infected, whereas this level of infection in line EA

1174 H was not achieved until 120 hr or more p.i.

Electron microscopic examination of infected cells showed that *Amsacta* EPV replication *in vitro* was similar to that observed *in vivo* (Granados, 1973). Immature virus particles were formed at the periphery of viroplasms, and these particles underwent a differentiation from the spherical immature form into the mature particles. Mature virions moved from the viroplasmic areas and were either occluded in VCIs or released into the medium by exocytosis (Fig. 3). Lysis of infected cells did not occur. All of the sublines (table 2) were susceptible to *Amsacta* EPV infection.

TABLE 2.

Summary of growth characteristics of ESTIGMENE ACREA cell line and sublines

Cell line or subline	Medium	Doubling time (hrs)	Passage *	Growth characteristics
EA 1174 A	Grace's + CFH **	—	40	Suspended and attached cells
EA 1174 AS1	Grace's — CFH	20.4	41	Suspended cells
EA 1174 AS2	Grace's — CFH	—	8 ***	Attached monolayer
EA 1174 H	Grace's + CFH	—	45	Suspended cells
EA 1174 HS1	Grace's - CFH	20.8	44	Suspended cells

* Passage as of June 1 1975

** CFH = non-heat inactivated *E. acrea* cell-free hemolymph

*** Established from EA 1174 A at the 32nd passage

The susceptibility of *Estigmene* larvae to VCI propagated in cell cultures was determined. VCI produced in all cultures, following infection with 8th passage virus inoculum, were harvested 7 to 10 days p.i. Aliquots (20 µl) of varying dilutions of the VCI suspension were layered on a cube (2 cm^2) of artificial medium in 10-60 x 10 mm plastic (Falcon) Petri dishes. One 8-day-old *Estigmene* larva was placed in each Petri dish. After the contaminated artificial medium was consumed by the test larvae, fresh medium was added and the dishes were held at 26°C until all test insects had died or pupated. Table 3 shows that *Estigmene* larva are susceptible to the VCI propagated in cell cultures.

3. *Infection with Autographa NPV*

Hemocytes in cell lines EA 1174 A and H were susceptible to *Autographa* NPV and within 20 hr p.i., cells were observed with hypertrophied nuclei containing virogenic stroma. By 24 hr p.i. polyhedra were observed in the nuclei of a few cells and by 48 to 72 hr p.i. 100% of the cells were infected. As was the case with *Amsacta* EPV, the rate of NPV development in both cells varied. In general, by 72 hr p.i. 100% and 75% of the cells in lines EA 1174 A and EA 1174 H were infected, respectively (Figs. 4A and 4B). Eventually all of the cells in line H became infected.

Electron microscopic examination of infected cells showed the normal viral developmental stages reported in NPV-infected cell cultures by MacKinnon *et al.* (1974) and Raghow and Grace (Fig. 5).

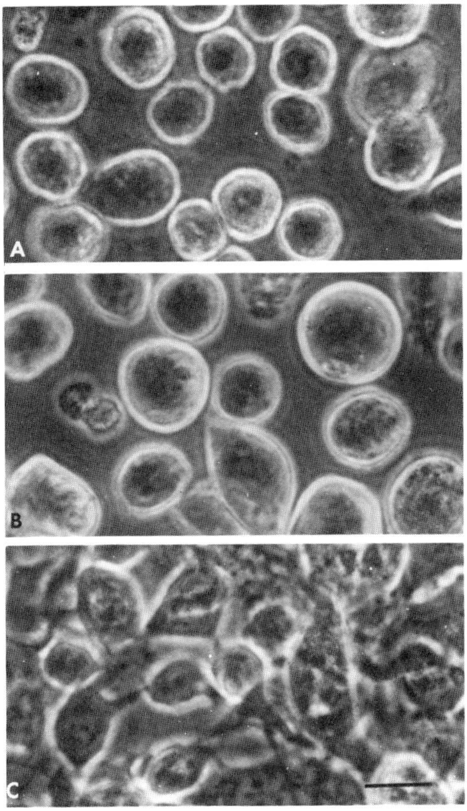

Fig. 1. (A-C). Continuous hemocyte cell lines established from *Estigmene acrea* larvae. Bar = 15.0 µm. (A) Cell line EA 1174 A. (B) Cell line EA 1174 H. (C) Subline EA 1174 AS2.

TABLE 3.

IN VIVO pathogenicity of AMSACTA MOOREI entomopoxvirus-containing inclusions produced IN VITRO to ESTIGMENE larvae

Dose/larva (No. of VCI) *	VCI from 8th virus passage (% mortality)
10^6	100
10^5	100
10^4	90
10^3	40
10^2	10
0 (control)	0

*VCI - virus-containing inclusions

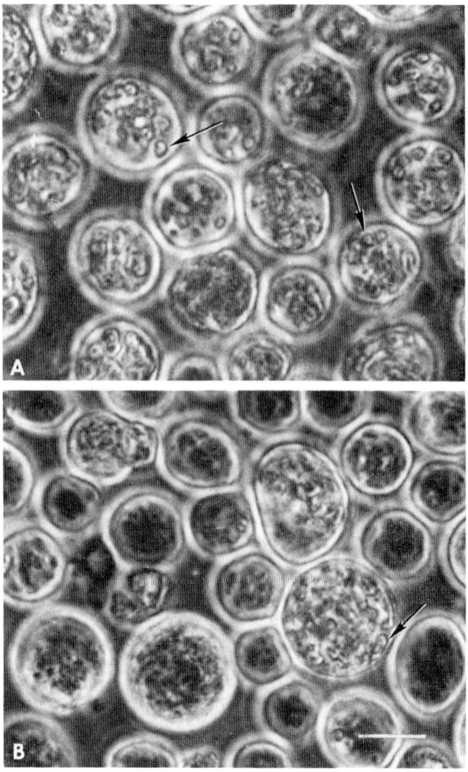

Fig. 2.(A-B). Estigmene hemocyte cell lines infected with *Amsacta moorei* entomopoxvirus 96 hr postinoculation. Virus-containing inclusions can be seen in infected cells (arrows). The percentage of infected cells is greater in cell line EA 1174 A (A) then in line EA 1174 H (B). Bar = 15.0 μm.

4. *Infection with Trichoplusia CPV*

Both cell lines EA 1174 A and H were susceptible to *Trichoplusia* CPV but the percentage of infected cells was low. By 72 hr p.i., 20% and < 1% of the cells in lines EA 1174 A and EA 1174 H were infected, respectively (Fig. 6A). The ultrastructural sequence of events in CPV replication (Fig. 6B) were similar to those reported for the same virus in *T. ni* cell cultures (Granados *et al.*, 1974).

5. *Infection with iridescent virus type 6*

Only cell line EA 1174 A was inoculated with iridescent virus type 6 and this line proved to be susceptible. By 72 hr p.i. some hypertrophied and vacuolated cells (Fig. 7A) could be observed. Ultrastructural examination of these cell cultures 96 hr p.i. revealed that the majority of the cells were infected (Fig. 7B).

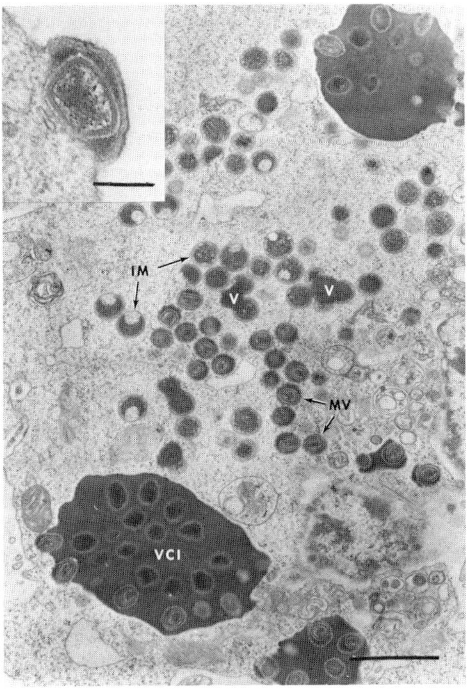

Fig. 3. Electron micrograph of cell line EA 1174 A infected with *Amsacta* poxvirus 96 hr postinoculation. Note viroplasm (V), immature virus particles (IM), mature virons (MV), and virus-containing inclusions (VCL). Bar = 0.5 um. Insert. Virus particle budding through the plasma membrane of an infected cultured cell. Bar = 0.2 μm.

IV. Conclusions

Two continuous hemocyte cell lines were established from *Estigmene* larvae and designated EA 1174 A and EA 1174 H. The cells in culture were identified as plasmatocyte-type blood cells. The cultured cells in both lines grow at equal rates and have a doubling time of approximately 20 hr. The hemocyte cell lines are unique in that they are susceptible to an EPV, NPV, CPV, and an iridescent virus. The ability of these continuous hemocyte cell cultures to support the replication of several different types of insect viruses may make them a very useful tool in insect virology.

Acknowledgments

This study was supported in part by USPHS Grant No. AI-08836. We are grateful to Ms. Beatrice Cato for technical assistance.

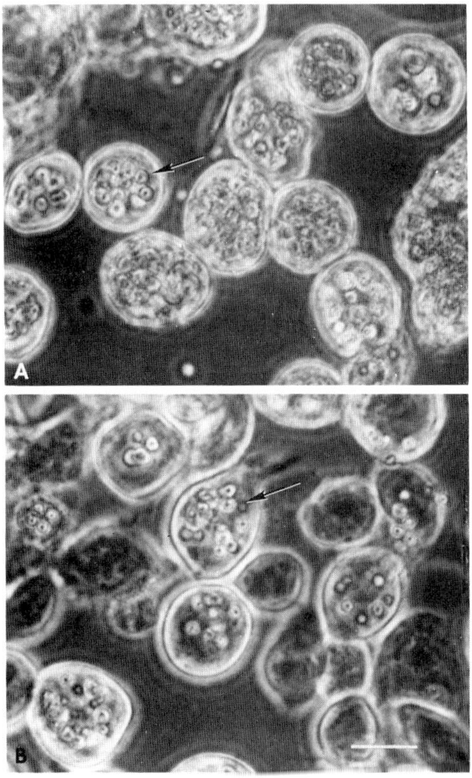

Fig. 4. (A-B). Estigmene hemocyte cell lines infected with *Autographa californica* nuclear polyhedrosis virus 72 hr postinoculation. Refractive polyhedra (arrows) can be seen in infected cells. The percentage of infected cells is greater in cell line EA 1174 A (A) than in line EA 1174 H (B). Bar = 15.0 μm.

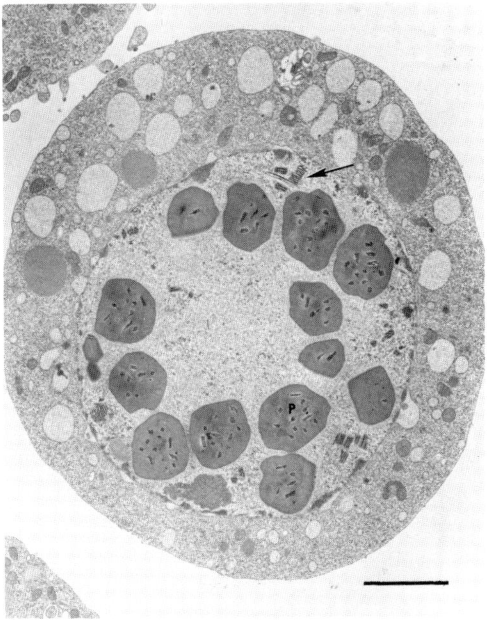

Fig. 5. Electron micrograph of cell line EA 1174 A infected with *Autographa* nuclear polyhedrosis virus 48 hr postinoculation. Note hypertrophied nucleus containing polyhedra (P) and nonoccluded, enveloped nucleocapsids (arrows), Bar = 2.0 μm.

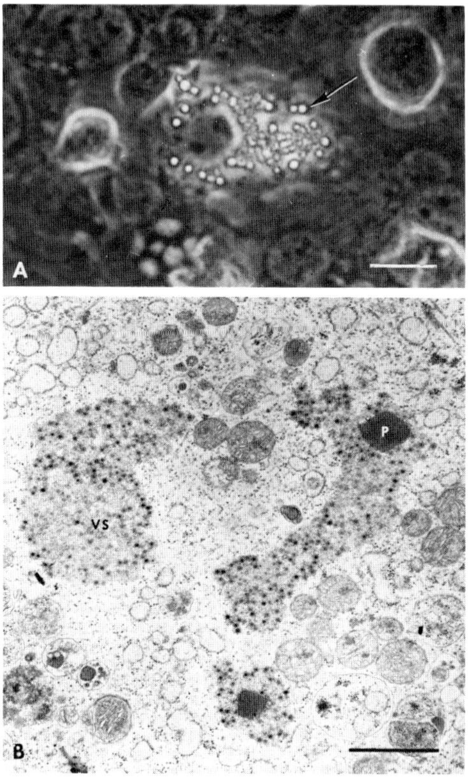

Fig. 6. (A-B). Cell line EA 1174 A infected with *Trichoplusia ni* cytoplasmic polyhedrosis virus 96 hr postinoculation. (A) Phase-contrast photograph of infected cell culture. Note refractive polyhedra (arrows) in the cell cytoplasm. Bar = 15.0 μm. (B) Electron micrograph of an infected cultured cell. Note the virogenic stroma (VS) and the developing polyhedron (P) in the cytoplasm. Bar = 1.0 μm.

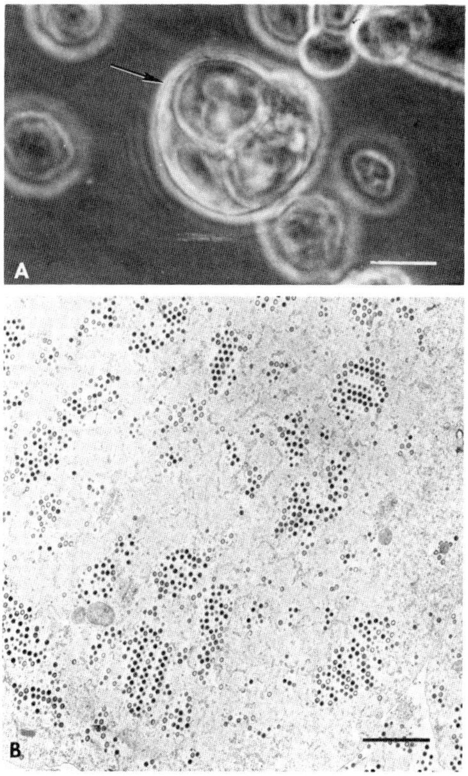

Fig. 7. (A-B). Cell line EA 1174 A infected with iridescent virus type 6. (A) Infected cell culture 72 hr postinoculation. Note swollen and vacuolated cell (arrow). Bar = 15.0 μm. (B) Electron micrograph of an infected cultured cell showing numerous virions in the cytoplasm. Bar = 2.0 μm.

V. References

Grace, T.D.C. (1962). *Nature 195,* 788.

Granados, R.R. (1973). *Misc. Publ. Entomol. Soc. Amer. 9,* 73.

Granados, R.R. (1975). *Advan. Virus Res. 20,* (in press).

Granados, R.R., and Naughton, M. (1975). *Intervirol. 5,* 62.

Granados, R.R., McCarthy, W.J., and Naughton, M. (1974). *Virology 59,* 584.

MacKinnon, E.A., Henderson, J.F., Stoltz, D.B., and Faulkner, P. (1974). *J. Ultrastruct. Res. 49,* 419.

Raghow, R., and Grace, T.D.C. (1974). *J. Ultrastruct. Res. 47,* 384.

Chapter 39

DUAL INFECTION OF THE *TRICHOPLUSIA NI* CELL LINE WITH THE *CHILO* IRIDESCENT VIRUS (CIV) AND *AUTOGRAPHA CALIFORNICA* NUCLEAR POLYHEDROSIS VIRUS

M. Kimura and A.H. McIntosh

I. Introduction

Insect viruses which produce or cause the production of inclusion bodies in infected cells, offer a unique system for the study of dual infections. The reason for this is that the production of inclusion bodies as distinct form virus production can be studied by light microscopy since the former are readily discernible in infected cells. Although dual infections have been extensively reported for vertebrate cell cultures, the same is not true for insect cell cultures. Most of the studies concerning dual infections with insect viruses have been conducted *in vivo* (Smith and Xeros 1953; Aizawa 1963; Smith 1967; Amargier *et al.*, 1968; Garzon and Kurstak 1969; Kurstak and Garzon 1971, 1975; Kurstak *et al.*, 1972). However Garzon and Kurstak (1972) successfully demonstrated dual infection of *Galleria melonella* cells *in vitro* with a nuclear polyhedrosis virus (NPV) and the *Tipula* iridescent virus (TIV). In the present report the dual infection of Hink's (1970) established cell line (TN-368) from the cabbage looper *Trichoplusia ni* has been studied. Two DNA viruses were employed, namely the NPV of *Autographa californica* and the cytoplasmic *Chilo* iridescent virus (CIV). Cells were infected with one virus and later challenged with the other to determine the effect on viral replication and polyhedral inclusion body (PIB) production.

II. Cell line.

The established TN-368 (Hink 1970) cell line from *T. ni* as adapted to growth in TC 199-MK (McIntosh *et al.*, 1973) was employed in all studies.

Viruses.

The NPV of *A. californica* (Vail *et al.*, 1973) and CIV were propagated in the TN-368 cell line at 28°C. The former was used at the third passage level and the latter at the sixty-second passage level in TN-368.

III. Infection of cultures.

Falcon T-flasks (25 cm^2) containing 2 x 10^5 cells/ml were challenged with 1 ml of a 10^6 $TCID_{50}$ and placed on a rocking platform. After various periods of time, the infected culture was washed twice with Hanks' balanced salt solution (HBSS) and then challenged with the second virus at the same dosage. Cultures were observed daily for signs of cytopathology, production of PIBs and prepared for electron microscopy by a previously described method (McIntosh and Kimura, 1974).

IV. Results and discussion

TN-368 cultures inoculated with *A. californica* NPV at 0,2,4,6 and 18 h and challenged with CIV produced PIBs 48h postinoculation (Fig. 1.). However, when the infection process was reversed, namely inoculation first with CIV, followed 24h later by super-infection with *A. californica* NPV, no PIBs were observed when cultures were held for as long as 96h postinoculation (Fig. 2.). On the other hand infection of cultures with CIV for 6h did not inhibit PIB production following inoculation with *A. californica* NPV. Infection of permissive insect cell lines with *A. californica* NPV leads to the production of free virions as well as virions which are occluded into the inclusion body protein matrix. Since the mechanism of inclusion body protein synthesis is not clearly understood, the possibility exists that virus synthesis may occur separately from PIB synthesis. To examine this possibility, electron microscopy was performed on dually infected TN-368 cells. Electron microscopy results confirmed the light microscopy findings, namely no PIBs were observed. However, both viruses were visualized within the same cell (Fig. 3). Kurstak *et al.*, (1974), similarly found a decrease in inclusion body proteins during multiple viral infections of *G. mellonella*. NPV particles appeared normal, with some displaying the characteristic envelope (Fig. 3) in cells containing no PIBs. No attempt was made to determine whether the titer of either virus was affected.

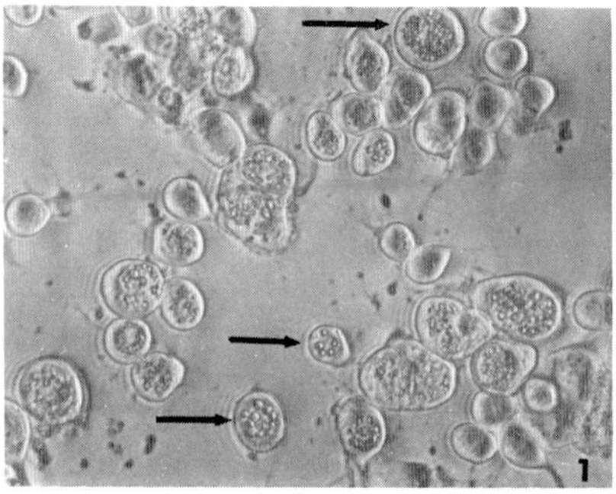

Fig. 1. PIBs in *T. ni* cells (arrows) 48h postinoculation. Cells were superinfected with CIV 18h postinoculation with *A. californica* NPV. x 393.

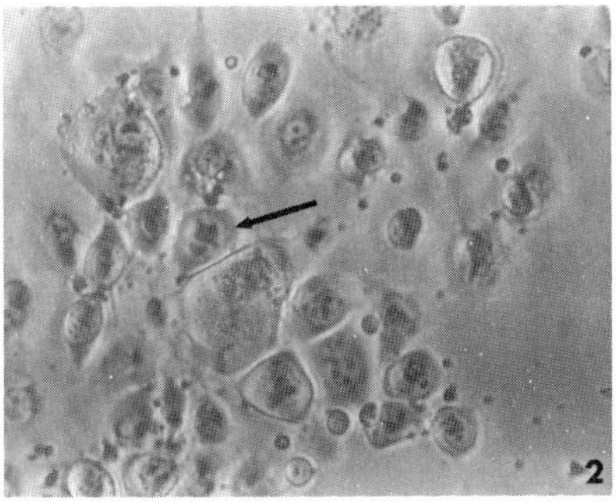

Fig. 2. T. ni cells display no PIBs 48h postinoculation. Cells were superinfected with *A. Californica* NPV 24h postinoculation with CIV. Arrow indicates cell in mitosis. x 393.

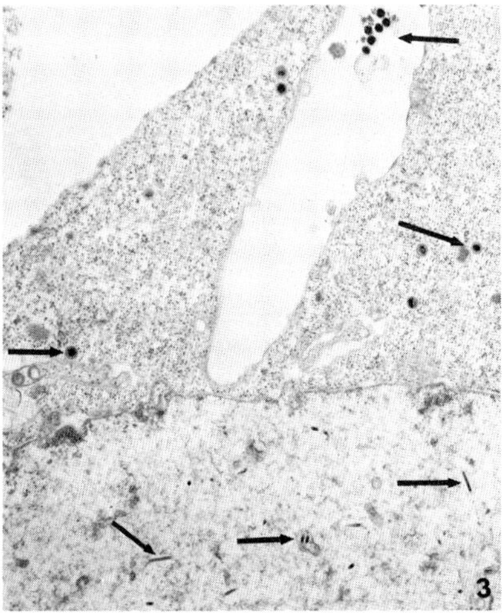

Fig. 3. Dual infection of a *T. ni* cell with CIV in the cytoplasm and extracellular space (top arrows) and *A. californica* NPV in the nucleus (bottom arrows). Note absence of PIBs. Experimental conditions as described in Fig. 2. x 19,200.

In the present report, it has been established that the TN-368 insect line can be dually infected with a nuclear polyhedrosis virus and a non-occluded cytoplasmic virus. Inhibition of PIB production indicates that CIV interfered with inclusion body protein synthesis. It is not known whether such interferrence occurs at the level of transcription. translation, or assembly. However, it does indicate that production of virus particles is independent of PIB production. Further studies are necessary to fully elucidate the mechanism of inhibition of inclusion body synthesis.

Acknowledgments

Supported in part, by NSF Grant BMS 74-13608.

V. References

Aizawa, K. (1963). In: Insect Pathology. (E.A. Steinhaus, ed.). Vol. 1, 382-412, Academic Press, N.Y.

Amargier, A., Meynadier, G. and Vago, C. (1968). *Mikroskopie 23,* 245.

Garzon, S., and Kurstak, E. (1969). *Rev. Canad. Biol. 28,* 89.

Garzon, S., and Kurstak, E. (1972). *C.R. Acad. Sci. Paris, 275,* 507.

Hink, W.F. (1970). *Nature (London) 226,* 466.

Kurstak, E. and Garzon, S. (1971). *Proc. Fed. Soc. Biol. Canad. 14,* 162.

Kurstak, E., Garzon, S., and Onji, P.A. (1972). *Archiv. Ges. Virusforsch. 36,* 324.

Kurstak, E., Garzon, S., and Onji, P.A. (1974). *Proc. Internat. Assoc. Microbiol. Soc. Tokyo, Japan.*

Kurstak, E., and Garzon, S. (1975). *Ann. New York Acad. Sci. 266,* 232.

McIntosh, A.H., Maramorosch, K., and Rechtoris, C. (1973). *In vitro, 8,* 375.

McIntosh, A.H., and Kimura, M. (1974). *Intervirology, 4,* 257.

Smith, K.M., and Xeros, N. (1953). *Parasitology, 43,* 178.

Smith, K.M., (1967). *Insect Virology,* 256 pp. Academic Press, N.Y.

Vail, P.V., Jay, D.L., and Hink, W.F. (1973). *J. Invertebr. Pathol. 22,* 231.

Chapter 40

PROPAGATION OF A MICROSPORIDAN IN A MOTH CELL LINE

T.J. Kurtti and M.A. Brooks

I. Introduction

The microsporida are spore-forming intracellular protozoans, in which the spore serves as an extracellular vehicle for transmission. Infection occurs when a spore discharges a coiled tube, inoculating a sporoplasm into the cytoplasm of the host cell. Intracellular inoculation is a prerequisite for the growth of the microsporida.

The spores of certain microsporida can be primed to infect cells *in vitro* by incubating them in a solution of alkaline pH (Ishihara and Sohi, 1966; Ishihara, 1968; Kurtti and Brooks, 1971; Weidner, 1972). This paper is a preliminary report of experimentally infecting a cell line of the corn earworm with a microsporidan of the forest tent caterpillar, and the subsequent growth of the parasite in the foreign host cells.

II. Materials and methods

Organ cultures of salivary glands from naturally infected *Malacosoma disstria* larvae were used to obtain suspensions of spores free of bacteria (Kurtti and Brooks, 1971). The microorganisms were cultured in a cell line, IPLB 1075, derived from ovaries of pupal *Heliothis zea* (Goodwin, in press). The spores were primed by incubating them in a solution of 0.1 N KOH for 40 minutes. The cells were infected by adding 0.1 to 0.3 ml of the concentration of primed spores to a concentration of cells (5 to 6 x 10^6 cells per ml). The cell and spore mixture was then diluted to 5 x 10^5 cells per ml and replicate monolayer cultures on coverslips were set up.

In a sequence of times, coverslip cultures were selected and prepared for study by rinsing them with saline solution, fixing with absolute methanol, and staining with Wright's stain followed by dilute Giemsa's stain.

In other cases, infected cells were maintained and transferred in 30 ml culture flasks. Spores were harvested from the infected cultures after the third to fifth subculture, and stored in triple distilled water at 4^oC. The spores were primed and tested for their infectivity for *H. zea* cells.

III. Results

All spores which appeared refractile with phase-contrast microscopy were considered to be mature and infectious. Spores isolated from the salivary gland cultures were generally 85 to 95% refractile, and when they were incubated in the KOH solution, they did not lose refractility. However, after the spores were mixed with the the cells in culture medium, there was a loss of refractility in most of the spores caused by ejection of the sporoplasms. Empty spore cases and an occasional sporoplasm were observed among the cells. The presence of empty spore cases immediately after mixing the primed spores and the cells indicates that the cells were infected within the first few minutes. The proportion of cells which germinated varied with the sample; germination as high as 80% was observed in some cases.

The initial level of infection was dependent on the proportion of mature spores to host cells, and the number of cells and spores in the mixing medium. A high incidence of infection was fostered by mixing a high density of spores and cells in a small volume. The initial level of infection represented in Fig. 1 was 1.2%, and the percentage of infected cells increased with the duration of culture, as shown in Fig. 1. The spread of infection occurred at the rate of 0.19% infected cells per hour for the first 120 hours of cultivation.

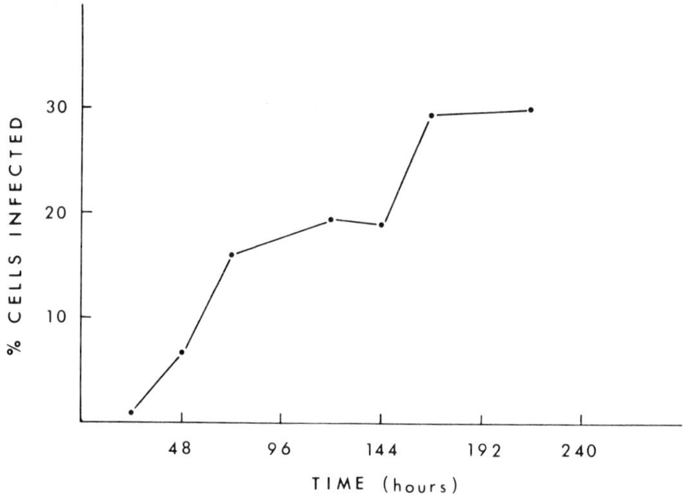

Fig. 1. Spread of microsporida infection in cultured *H. Zea* cells. Results shown are from a single experiment.

The growth and differentiation of the parasites within the host cells were analyzed by determining the average number of microsporida per infected cell and their stage of development (Fig. 2). A one-day lag phase occurred during which the parasites could not be demonstrated with stains. Cells examined 6 and 12 hours after mixing suggested that there was an eclipse period in which the parasite reorganized. Cells stained 24 hours after mixing contained a single microbe per infected cell. A two-day logarithmic growth phase followed in which the population doubled once every 11.5 hours. The trophozoites were elongate and increased by binary fission; schizogony was not observed. The parasites doubled approximately 5.5 times, so on day 5 there was, on the average, 48 to 50 parasites per infected cell. Sporogony was observed on the third day and on day 4 sporonts were common. In some cells trophozoites were intermixed with sporonts, and a proportion of these cells retained the trophozoites. Infected cells were capable of mitosis, which permitted the transmission of the infection through the subcultures. After several transfers, the spores were harvested. There were many immature (non-refractile) spores and trophozoites resulting from the asynchronous development of the parasite population after the completion of the first generation *in vitro*. Spores harvested from the cultures could be primed and used to infect other cultures of *H. zea* cells. The growth characteristics of the microsporida obtained from cell cultures were similar to those obtained from the salivary gland cultures.

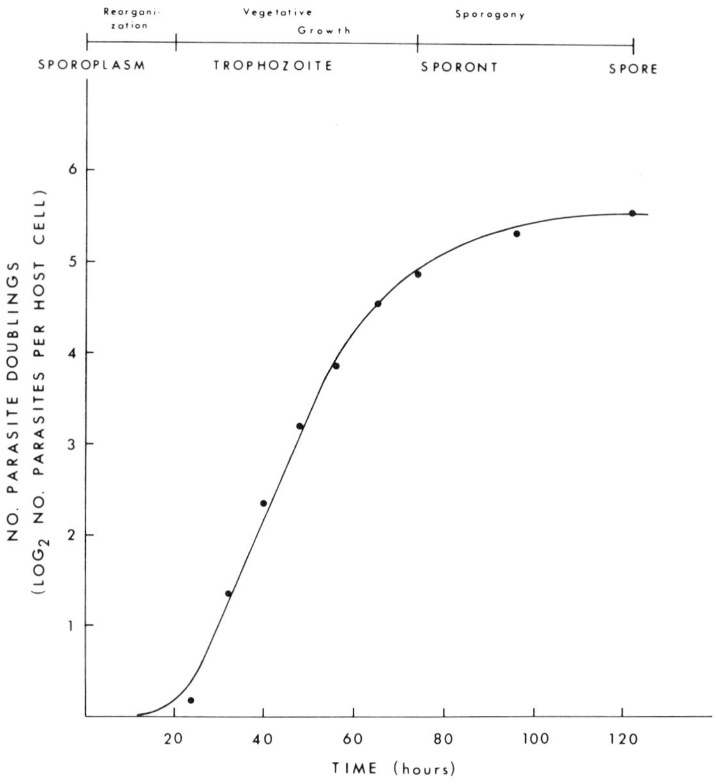

Fig. 2. First generation growth of microsporida in *H. zea* cell line. Each point represents logarithm to base 2 of the average number of microsporida per cell in 100 infected cells.

IV. Discussion

This work, as well as others cited in the introduction, documents conclusively that certain microsporidan spores can be induced to infect, and grow in, cells of organisms which are not their natural hosts. However, Weidner (1972) has demonstrated that either the sporoplasm or the host cell may react defensively in cross-infections. For example, *Nosema michaelis*, a microsporidan of the blue crab, *Callinectes sapidus*, retained two sporoplasmic envelopes when it was inoculated into the cytoplasm of human red blood cells or ascites leukemia EL4 cells. Normally, in crab cells, one of the envelopes is lost. The ascites leukemia cells or mouse macrophages responded with a "fibrous corona" around the sporoplasm; whereas the corona was absent from sporoplasms injected into the cytoplasm of crab cells. Cell reactions may be occurring in the lag phase which we observed.

The mechanism by which the infection spread in our cultures is unknown. Shadduck (1969) found that in cultures of rabbit choroid plexus cells infected with *Nosema cuniculi*, there was an increase in the incidence of infection with time. Invasive forms have been observed in cell cultures infected with other species of intracellular protozoans, e.g. the coccidians (Doran, 1973). Ishihara (1969) presented cytological evidence for a "secondary infectious form" of *Nosema bombycis* in cultures of *Bombyx mori* cells. We observed extracellular microsporida in our cultures which resemble the "secondary infectious forms." Studies are currently underway to determine the means by which the infection spreads in our cultures.

Acknowledgements

This research was supported in part by U.S. Public Health Service Research Grant No. AI 09914 from the National Institute of Allergy and Infectious Diseases. This is paper No. 9134, Scientific Journal Series, Minnesota Agricultural Experiment Station.

V. References

Doran, D.J. (1973). In: The Coccidia. *Eimeria, Isospora, Toxoplasma,* and related genera. (D.M. Hammond and P.L. Long, eds.). University Park Press, Baltimore, Md. Pages 183-252.

Goodwin, R.H. (1975). *In Vitro.* (In press).

Ishihara, R. (1968). *J. Invert. Pathol. 11,* 328.

Ishihara, R. (1969). *J. Invert. Pathol. 14,* 316.

Ishihara, R. and Sohi, S.S. (1966). *J. Invert. Pathol. 8,* 538.

Kurtti, T.J. and Brooks, M.A. (1971). In: Arthropod Cell Cultures and their Application to the Study of Viruses. (E. Weiss, ed.). *Current Topics Microbiol. Immunol. 55,* 204.

Shadduck, J.A. 1969. *Science 166,* 516.

Weidner, E. (1972). *Z. Parasitenk, 40,* 227.

A 6
B 7
C 8
D 9
E 0
F 1
G 2
H 3
I 4
J 5